Crop physiology

some case histories

W9-CIP-699

Plate 1.1   Some crop physiologists.

Above left: W. L. Balls (1882–1960)   *Above right:*   D. J. Watson

Below: M. Monsi (left) and T. Saeki

Crop physiology
some case histories

edited by
L. T. EVANS

Cambridge University Press

CAMBRIDGE

LONDON   NEW YORK   NEW ROCHELLE

MELBOURNE   SYDNEY

Published by the Press Syndicate of the University of Cambridge
The Pitt Building, Trumpington Street, Cambridge CB2 1RP
32 East 57th Street, New York, NY 10022, USA
296 Beaconsfield Parade, Middle Park, Melbourne 3206, Australia

Library of Congress catalogue card number: 73-91816

ISBN 0 521 20422 4 hard covers
ISBN 0 521 29390 1 paperback

First published 1975
Reprinted 1976
First paperback edition 1978
Reprinted 1980

Printed in Great Britain at the
University Press, Cambridge

Contents

List of contributors

I. C. ANDERSON: Department of Agronomy, Iowa State University, Ames, Iowa, USA.

D. N. BAKER: Cotton Production Research Institute, USDA, State College, Mississippi, USA.

T. A. BULL: David North Plant Research Centre, Toowong, Qld., Australia.

W. G. DUNCAN: Department of Agronomy, University of Kentucky, College of Agriculture, Lexington, Kentucky, USA.

L. T. EVANS: Division of Plant Industry, CSIRO, Canberra, ACT, Australia.

G. W. FICK: Department of Agronomy, Cornell University, Ithaca, N.Y., USA.

R. A. FISCHER: International Maize and Wheat Improvement Center, CIMMYT, Londres 40, Mexico.

A. H. GIBSON: Division of Plant Industry, CSIRO, Canberra, ACT, Australia.

K. T. GLASZIOU: David North Plant Research Centre, Toowong, Qld., Australia.

J. D. HESKETH: Dorman Hall, Department of Agronomy, State College, Mississippi, USA.

R. S. LOOMIS: Department of Agronomy, University of California, Davis, California, USA.

J. A. McARTHUR: Department of Biological Science, Florida International University, Tamiami Trail, Miami, Florida, USA.

S. MATSUSHIMA: Department of Physics and Statistics, National Institute of Agricultural Sciences, Tokyo, Japan.

F. L. MILTHORPE: School of Biological Sciences, Macquarie University, North Ryde, NSW, Australia.

J. MOORBY: School of Biological Sciences, Macquarie University, North Ryde, NSW, Australia.

Y. MURATA: Laboratory of Crop Science, Faculty of Agriculture, The University of Tokyo, Tokyo, Japan.

J. S. PATE: Botany Department, University of Western Australia, Perth, WA, Australia.

R. M. SHIBLES: Department of Agronomy, Iowa State University, Ames, Iowa, USA.

I. F. WARDLAW: Division of Plant Industry, CSIRO, Canberra, ACT, Australia.

W. A. WILLIAMS: Department of Agronomy, University of California, Davis, California, USA.

Foreword

Only nine crops are dealt with in this book, yet between them they contribute more than two-thirds of the dry weight, half of the protein and a quarter of the oil and fat in our diet, and half of the fibre we use. Our dependence on these major crop plants is by no means diminishing, as drought or disease in some part of the world reminds us every few years. The area of land under cultivation rises only slowly and the survival of a growing world population will depend on further increase in the yield of these crops, which hinges, in turn, on greater understanding of the physiological basis of yield in them. Hence this book, which is a physiological counterpart to *Essays on Crop Plant Evolution* edited by J. Hutchinson, and complements *An Introduction to Crop Physiology* by F. L. Milthorpe and J. Moorby, from Cambridge University Press.

The crop plants considered here were selected not only for their importance, but also for the variety of their characteristics. Two of them follow the C_4-dicarboxylic acid pathway of photosynthesis, while the others rely wholly on the Calvin cycle. Two are legumes, whereas the remainder cannot take advantage of symbiotic nitrogen fixation. The harvested storage organs are in many cases the seeds; but for sugar cane it is the stem, for sugar beet the root, for potato the tuber, and for cotton the lint hairs on the seed. Inflorescences may be terminal, as in wheat and rice, or axillary as in peas. The plants may be tall, like cane or maize, or rosettes like sugar beet. Clearly, there is considerable variety in the path to success as a crop plant.

A comparable variety of approach is also evident among crop physiologists and is exposed in the following chapters. Uniformity of treatment was not aimed at, but the authors were encouraged to attempt explanatory syntheses of what we now know about the functioning of these very important plants.

L. T. EVANS

18 *July* 1973

[vii]

1

Crops and world food supply, crop evolution, and the origins of crop physiology

L. T. EVANS

Man has made tools for more than a million years, but his crop plants have evolved under the influence of his powers of observation, selection and imagination for only about ten thousand years. Some of these crops, such as maize, would no longer survive without human intervention, but man's survival is equally dependent on his crops. Crops and man have evolved together in a kind of symbiosis. As Darlington (1969) has put it: 'In the silent millennia during the expansion of agriculture the men themselves were transformed by the new relations with the plant and animal which they themselves were in the process of establishing.' In the face of increasing dependence by an expanding world population on relatively few crop plants, the shape of our future will depend on our understanding of these plants and of the limits to their productivity.

In this introductory chapter we shall consider three aspects of the inter-relations between man and his crop plants: first, the role of crops in the world food supply, to emphasize the scale of our current and likely future depend-ence on them; secondly, some physiological aspects of crop plant evolution, to provide a background for the contemporary portraits of major crops given in the later chapters; and finally, a short history of crop physiology, to give some perspective to the current fashions and enthusiasms which shape the following case histories.

CROPS AND THE WORLD FOOD SUPPLY

Perhaps 3000 plants have been used as food by man in the course of his history, and about 200 of these have been domesticated. The Iron Age man from Denmark, whose stomach contents were examined by Helbaek (1950), included almost 60 species in his last meals. Today five cereals, three tuber crops, several legumes and sugar cane and beet provide most of the dietary dry weight and protein for man, as may be seen from Table 1.1.

The entries in the table involve many uncertainties, beginning with the estimates of world production (cf. Farmer, 1969), but their object is to indicate the relative importance of the various crops, animal products and fish as sources of food protein and dry weight. The first point to note is the scale of food production, which is of comparable magnitude, in terms of

Table 1.1. *World production of edible dry matter and protein*

Production figures were taken from the *1970 FAO Production Yearbook* and *United Nations Statistical Yearbook*. The supply of edible dry matter and protein was estimated from the production figures using data in The State of Food and Agriculture 1964 and other sources. Figures are bracketed when little or none of the product is used for human food.

	Dry matter (metric tons $\times 10^7$)		Protein (metric tons $\times 10^6$)	
Cereal grains				
Wheat	27.5		32.9	
Rice	26.7		23.2	
Maize	23.5		24.7	
Barley	11.4		11.6	
Sorghum/millet	8.2		7.4	
Others	7.6		1.1	
	——	104.9	——	100.9
Starchy roots				
Potato	6.6		6.0	
Sweet potato and yams	3.9		2.9	
Cassava	3.4		0.8	
	——	13.9	——	9.7
Sugar crops				
Cane	4.3 (sugar)		–	
Beet	3.0		–	
		7.3		
Legumes and oil seeds				
Soybean	4.2		16.7	
Peanuts	1.6		4.8	
Peas	1.3		3.5	
Beans	1.5		5.4	
Cotton – seed	(2.0)		(7.2)	
– fibre	(1.1)		–	
Others	(3.5)		(12.4)	
	——	10.2	——	35.6
Vegetables		2.8		8.0
Fruit		2.5		1.3
Animal Products				
Milk	5.2		14.5	
Meat	2.8		12.6	
Eggs	0.5		2.5	
Fish	1.7		8.5	
	——	10.2	——	38.1
		152.8		193.6

weight, to the annual production of coal or petroleum, and four times that of iron ore. Of the edible dry weight, more than two-thirds is contributed by the cereals and 80 % by only 11 plant species, compared with only 6 % from all animal sources. Of the protein supply, half is contributed by the cereals alone and less than a quarter from animal sources. Data presented by Autret (1970) suggest that animal proteins constitute a larger proportion of actual diet protein (32 %) than the FAO production figures indicate. The figures for legume and oil seed proteins in Table 1.1, and to a smaller extent those for cereal proteins, will be too large insofar as these are used for animal feed. The figures for vegetables and fruit, on the other hand, probably under-estimate their contribution to human diet. Nevertheless, crops are clearly the major source of both calories and protein in the human diet. They also provide more than two-thirds of the annual production of about 40 million metric tons of fats and oils.

To emphasize again how much we depend on so few crops, it may be noted that the crop plants examined in this book provide two-thirds of the dry weight and about half of the protein in the world's diet. In Mangelsdorf's (1966) words, 'Since these plants quite literally stand between mankind and starvation, . . . we should know as much about each of them as we know about some of the destructive agents of the world – as the medical profession, for example, knows about the world's principal human diseases . . .'. This book is meant to be a step in that direction.

Alternative sources of food

Before dealing with our major crop plants in more detail, we should first consider possible alternatives. In a book called *The Environment Game* Calder (1967) wrote that 'agriculture is simply failing us' and will need to be replaced by synthetic methods of food production, as McPherson (1965) has also argued. However, at present there is no known and tried way of synthe-sizing carbohydrates or similar compounds on a large scale by using solar or nuclear energy (Calder, 1967; Pyke, 1970), let alone one of an efficiency comparable with crop plants. Compounds such as 1,3-butanediol and 2,4-dimethyl-heptanoic acid, which can be produced cheaply from petrochemicals, have been found to be safe and acceptable energy sources for experimental animals (Scrimshaw, 1966), but long-term replacements for crops must be based on solar or nuclear energy conversion rather than on fossil fuels.

Table 1.1 shows that about 13 % of the edible dry matter is protein, the same as in the FAO production figures for 1934/38, when food production was only half the present level. On the basis that 10–12 % of diet calories should come from proteins, and that even 6 % may be sufficient in adult diets, the proportion of protein in the world diet is adequate, though it may not be so in areas where the diet consists mainly of low protein foods such as rice or the starchy roots. However, increase in the yield of grain crops is often associated with a decrease in protein percentage, for reasons discussed later, and alternative sources of protein are likely to play an increasing role in future.

According to the President's Science Advisory Committee (1967), estimates of the sustainable annual fish harvest vary from 5.5 to 200×10^7 metric tons liveweight. The lowest estimate, made in 1962, has been exceeded since 1964. Ryther (1969) has estimated separately the potential photosynthesis and length of the food chains in the open oceans, the coastal zones, and the upwelling areas on the western coasts of continents where potential productivity is highest. He concludes that the total annual fish production could be 24×10^7 metric tons, with the sustainable catch only 10×10^7 tons, i.e. less than twice the present harvest. Although some forms of aquaculture can have remarkably high yields (Bardach, 1968), and can undoubtedly make a major contribution to local supplies of edible protein, the potential total production is limited, and Calder's (1967) claim that 'it would be possible to replace land-based agriculture entirely by aquaculture of the oceans, and produce plenty for all' lacks substance.

Pilot plants for large-scale algal culture have been established in Japan, USA, Israel, Holland, Czechoslovakia and elsewhere to examine the potentialities for both food energy and protein production. Despite elaborate equipment for efficient circulation, needed both to give each algal cell its turn in the sun and to increase access to carbon dioxide, yields are commonly only $8-18$ g DW m^{-2} day^{-1} (Tamiya, 1957; Thomas, 1965; Prokes and Zahradnik, 1968), whereas crop growth rates are frequently much higher than this, once a fully intercepting light canopy has developed. The diffusion coefficient of carbon dioxide in water is only 1×10^{-4} of that in air, and whereas the liquid phase pathway for carbon dioxide is only a few microns in plants, it is considerably more in algal cultures. This may account for their relatively low growth rates per unit surface area, in spite of the high carbon dioxide concentration used for aeration. On the other hand, algal cultures can be harvested continuously throughout the year, they can be set up on rooftops, to reclaim the land lost to agriculture by urbanization, and they can be used to dispose of sewage. Their protein content is high both in absolute amount and in terms of essential amino acids (Tamiya, 1957). A further advantage is that their chemical composition can be controlled by varying the composition of the medium. However, the need to remove cell walls as well as to maintain continuous circulation of the cultures not only results in a fairly high cost of production (Enebo, 1970), but presumably also in a high ratio of input: output energy.

Recently there has been more emphasis on yeasts, bacteria and mould fungi as potential sources of single cell protein. Torula yeast, containing about

50 % of protein with a high content of essential amino acids, was used for food in Europe during both world wars, but its palatability is low and its cost high. It can be grown on the sulphite liquor from papermaking, or on cheap molasses. More recently strains have been selected to remove the higher *n*-alkanes, which are undesirable in lubricants, from petroleum. The current cost of such yeast protein is about twice that of soybean protein. If the process were applied to the total world petroleum production of 2.1×10^9 tons per year, of which these alkanes comprise 5.6×10^7 tons, about 32×10^6 tons of single cell protein could be produced, roughly the same amount as that contributed by the legume crops alone. To produce all the protein needed by man would require 15–20 % of present world petroleum production to be diverted to this end (Johnson, 1967). Bacteria such as *Micrococcus* can also be grown on petroleum fractions. Other systems under development involve the growth of fungi such as *Penicillium* and *Aspergillus* on corn or potato starch: these would not lessen our dependence on crops as sources of available energy, but they could usefully supplement future diets insofar as legume crops are often less productive than cereal and tuber crops. Where cereal and legume proteins are extracted to manufacture extenders and meat analogs, the use of mould fungi to generate protein from the residues – especially where they are indigestible, like those of soybean – has much to commend it. Similarly, systems involving the growth of bacteria on animal and industrial wastes, and on shrubby and other indigestible plant materials, will undoubtedly be developed as ways of recycling potential pollutants and of increasing the proportion of land surface involved in food production. Transport and equipment costs for the processing of materials harvested from a wide area may preclude their extensive development, and such systems have most promise in relation to concentrated wastes, such as animal manure and straw from feed lots, bagasse from sugar mills, and effluents from factories.

Problems of rapid transport and handling may also impose a limitation on the extent to which extracted leaf proteins can contribute to the supply of dietary protein. But, as Pirie (1970) has argued, small, simple extraction plants could certainly increase local protein supplies in many wet tropical areas where grain crops do not thrive, and where tuber and sugar crops and low protein fruits such as bananas predominate.

This brief review of alternative sources of food suggests that man's dependence on the major crop plants for his energy and protein is unlikely to decrease in the foreseeable future. He may depend less on them for vitamins and drugs (Pyke, 1970), but for the present even the precursors of birth control pills are derived from crop plants. Thus, it is essential that we understand better the development of, and limits to, yield in these plants.

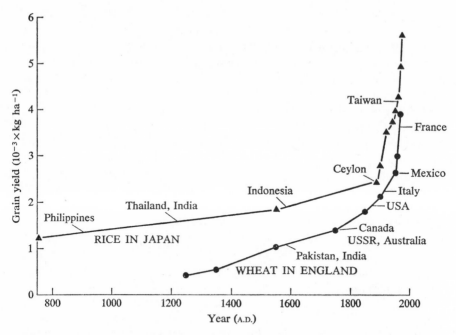

Figure 1.1. Historical trends in the grain yield of rice in Japan and of wheat in England compared with 1968 yields of wheat and rice in several countries. Early rice yields from Matsuo (1959), early wheat yields from Gavin (1951), and recent yields from FAO Production Yearbooks.

Crop yield

When Malthus concluded, in 1798, that agricultural production increased arithmetically, whereas population increased geometrically, he did not foresee the enormous increase in cultivated area that followed the colonization of the new world. Today, in both developed and developing countries, about 11 % of the total land area is cultivated. Estimates of the potentially cultivable land vary widely, but according to Kellogg and Orvedal (1968) the present arable area of 1.4×10^9 hectares could be increased to at least 3.2×10^9 hectares without assuming ready desalination of sea water. In recent years agricultural production has increased somewhat faster than has population in most countries. In the developed countries nearly all the increase in production has come from increase in yield per unit area, but in developing countries about half has come from increase in the area under cultivation for cereals (Revelle, 1966). In Africa and in Central and South America, the proportion of land under cultivation is very small, and considerable increase is possible. But in Asia, and in the developed countries, future increases in food production must come mainly from increased crop yields, and from multiple cropping where it is possible.

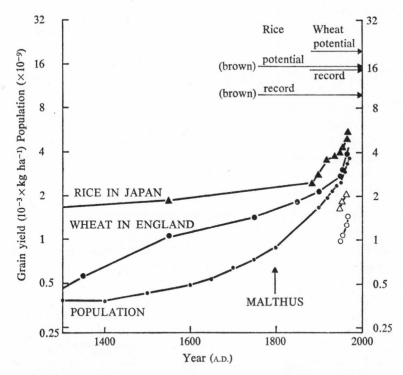

Figure 1.2. Historical trends in world population, rice yields in Japan (▲) and the world (△), and wheat yields in England (●) and the world (○). Sources as for Figure 1.1. World populations taken from Clark (1967). The record wheat yield is that for variety Gaines, grown in Washington, USA, and that of rice is for variety Ootori, grown in Japan (data of Y. Murata). The potential yield ceilings are those estimated for 400 cals cm^{-2} day^{-1} during grain filling (Evans, 1973). The vertical axis is on a logarithmic scale in all cases.

Historical trends in the yield of rice in Japan and of wheat in England are shown in Figure 1.1. The early yield data, derived from Matsuo (1959) and Gavin (1951), are, naturally, inadequate, but the arithmetic plot emphasizes the remarkable rates of increase sustained in these countries over recent years. Present yields in other countries are indicated, but should not be taken as reflecting their level of agricultural sophistication. For example, the low yields of wheat in Canada, USA and Australia reflect the more adverse environments in which the crop is grown in those countries. Clearly, the increase in the yield of these crops is due as much to improved agronomy as to increase in their yield potential.

Rice and wheat are the main food of more than half the world's population, and in Figure 1.2 changes in world population are compared with the changes in wheat and rice yields, the vertical scales being logarithmic. The world record yields for rice and wheat crops (variety 'Ootori' in Japan and 'Gaines' in USA respectively) are indicated, together with estimates of the maximum potential yields for a radiation level of 400 cal cm^{-2} day^{-1}. Both the record yields and the estimates of maximum yield will change, but they suggest that, at the most, only three doublings of average world yields, and not much more than one doubling of already high national yields, can be expected. To approach these potential yields on a world-wide scale would require an enormous increase in water use (already under pressure from urban and industrial demands), in the consumption of non-renewable fertilizers such as phosphate, in the release of nitrogenous and other fertilizers into rivers and lakes, in the use of herbicides, fungicides and pesticides, and in the consumption of fossil fuels for agricultural operations.

PHYSIOLOGICAL ASPECTS OF CROP PLANT EVOLUTION

Many of the changes wrought in the earliest stages of plant domestication were common to several crops, comprising what Vavilov (1951) referred to as homologous variation. Vavilov also concluded that nearly all our crop plants were derived from only eight centres of origin which were mostly in low latitudes, of hilly or mountainous terrain, and separated from one another by deserts or mountains. Later work suggests that Vavilov's centres of diversity were not always centres of origin, but rather secondary centres of hybridization, selection and cultivation, as was Ethiopia for wheat, barley and millet. There is now a great deal of evidence for a nuclear area of plant and animal domestication in the Near East at about 7000 B.C. and for other centres of origin of many crop plants in Mesoamerica and China. In Africa and South America, on the other hand, crop domestication appears to have been dispersed over a wide zone, or 'non-centre' in Harlan's (1971) terminology, as it may also have been in South East Asia, where evidence is now accumulating for the early domestication of a range of crops (Solheim, 1972). Harlan (1971) has suggested that there may have been considerable interaction between each centre and its adjacent non-centre in the development of their crops. Whatever the pattern, it is clear that the first steps in domestication were taken independently in several areas, and over long periods of time.

Several of the temperate cereals and pulses were probably domesticated in the 'Fertile Crescent' of the Middle East about 9000 years ago. Throughout that area extensive dense stands of the probable wild progenitors of wheat, barley and oats can still be found, and archaeological investigations at many sites within the Fertile Crescent have yielded clear evidence of sequential changes from gathered wild plant to cultivated cereal (Helbaek, 1966). In the early remains, from about 7000 B.C., wild forms and primitive domesticates of wheat were most abundant. By 5500 B.C. cultivated forms, including modern bread wheat, *Triticum aestivum*, were predominant.

The massive stands of the wild progenitors undoubtedly caught the eye of observant primitive man, and presented no problems in harvesting. Harlan (1967) has described how, either by hand or with a flint-bladed sickle, he harvested about two kilograms of grain per hour from stands of predominantly wild einkorn wheat in south eastern Turkey. The grain was found to have a much higher protein content than modern wheat, and would have provided a moderately well balanced, if monotonous, diet. Harlan concluded that 'A family group . . . working slowly upslope as the season progressed, could easily harvest wild cereals over a three-week span or more and, without even working very hard, could gather more grain than the family could possibly consume in a year.' From similar stands in East Galilee, Zohary (1969) has harvested up to 500–800 kg of grain per hectare, which compares with yields of wheat crops in England in the Middle Ages (cf. Figure 1.1). The gathering of such grain, once the means for its storage were developed, would have offered a more secure alternative to hunting for people living in the Fertile Crescent. Today gathered plants still play an important role, even in developed communities: Schwanitz (1966) lists many, including fruits, nuts, berries, herbs, sugar maple, wild rice and marine algae.

Reduced dissemination and increased seed size

The inflorescences of the wild wheat (*T. boeoticum* and *T. dicoccoides*) and barley (*Hordeum spontaneum*) have a brittle rachis which, when ripe, readily breaks up into dispersal units well adapted by virtue of their shape, awns and other characters for burying themselves where dropped. Gathering would have caused many of these to fall to the ground, and the brittle rachis character would have ensured the regeneration of the gathered communities. On the other hand, occasional plants with a tough rachis would comprise a higher proportion of the grain carried back to the settlement and dropped, or later sown, in the nitrogen-enriched soil near it. As cultivation replaced gathering there would be selection pressures for increased rachis toughness, making dissemination less effective. Survival of the crop plant would become dependent on man as he depended more on it. This change, one of the earliest and most marked differences between domesticated plants and their wild progenitors, occurred not only in einkorn and emmer wheats and barley, but also in many other crops such as maize, flax, peas, and peppers (Pickersgill, 1969) and beans (Kaplan, 1965), which were selected for indehiscent pods and fruit.

Along with reduced dissemination there was often a considerable increase in seed size. In the diploid wheats, for example, the grain of the cultivated *T. monococcum* is two to three times as large as that of the wild *T. boeoticum*. Likewise in the cultivated species of *Phaseolus* the seeds are five to eight times larger than those of the corresponding wild species (Kaplan, 1965; Smartt, 1969) and no transitional forms are known, either today or in the archaeological record. Plants with larger seeds can be harvested more quickly and may have been gathered preferentially by proto-agricultural man, and such selection could have been reinforced in the enriched and disturbed ground near settlements. In this environment competition with other plants is likely to have been intense, and those with larger seeds and embryos would have an advantage in the more rapid early growth of their seedlings, as found among the wheats (Evans and Dunstone, 1970). Thus, there could have been strong concurrent selection pressures for tough rachis or indehiscent fruit and increased seed size.

Crops, weeds and adaptability

Several authors (e.g. Hawkes, 1969; Zohary, 1969) have suggested that large seeds and 'weediness' preadapted certain plants for cultivation. The term weed, in the sense of an unwanted plant, would have little meaning under the conditions of early domestication, although the concept was well established by Biblical times. But in the sense of being plants adapted to ground that is disturbed and often enriched, by man and his domestic animals, crop plants and weeds have much in common (Harlan and de Wet, 1965). Many crops, such as maize, cucurbits, potatoes, hemp, amaranths, and castor beans, may first have caught man's attention by their abundance as pioneers in the disturbed ground near settlements. Other crops, such as rye and oats, began as weeds that became relatively more productive as the primary crop, wheat or barley, was extended into less favourable environments.

Anderson (1954) has emphasized that many successful crop plants share with weeds the property of adaptability to a wide range of environmental conditions. To some extent this is true in that the mechanisms controlling their germination and flowering have been relaxed under selection. In many modern varieties of soybean, potatoes, wheat and rice, for example, control of flowering by day-length is far less stringent than in related wild species. In fact, independence of day-length is specified as a desirable character in both the Mexican wheat and IRRI rice breeding programmes, to aid trans-world adaptation. However, wild progenitors and crop plants are not consistently different in this respect (cf. Smartt, 1969): modern wheat cultivars from high latitudes often have an absolute requirement for long days, possibly to avoid frost injury to the young inflorescence, while some tropical rices display an extreme sensitivity to day-length (Njoku, 1959; Dore, 1959) which may also be adaptive.

Change in form

According to Schwanitz (1966), one of the most consistent and striking differences between cultivated plants and their wild progenitors is the gigantism of the cultivated forms. The increase in size applies particularly to the organs harvested and selected by man, and therefore also to the proportion they comprise of the whole plant. Just as domestication permitted the natural dispersal mechanisms to become less effective, and in fact increased selection pressure in that direction, so presumably did cultivation free crop plants from other requirements for survival in the wild state. More of the accumulated plant substance could be stored in the organs man harvested for food or fibre, at the expense of other organs whose role was more important for survival in the wild. During grain filling in wheat, for example, more assimilate goes to the grain, and far less to the tillers, roots and stem in modern wheats than in their wild progenitors (Evans and Dunstone, 1970). Under natural selection such a change would decrease the advantage of further change in the same direction, but under cultivation and selection pressure from man the advantages of further change could be enhanced. Thus, there could be a rapid and progressive increase in the proportion of investment in organs harvested by man. The more control man had over cultural conditions and competition from weeds, the further this shift in proportions could be pushed. Our crops derive, therefore, from orthogenetic evolution under the influence of man, involving shifts in morphogenetic and integrating mechanisms within the plants. For some crops, especially those with multiple uses, the variety and extent of change in form is quite remarkable, as in the cabbage–kale–collard–kohlrabi–brussels sprout–cauliflower complex.

Changes in composition

Not only the form, but also the composition of the harvested organs has changed greatly. Toxic components, such as occurred in many of the wild yams, have been eliminated and bitter principles reduced, as in lupins. Other components of special interest to man, such as sugar in beet and cane, oil in maize, and oil or protein in peanuts and soybeans have been much increased. Beets, for example, were grown as salad plants along the Mediterranean coasts in the sixth century B.C. Selection for increased sugar content in the roots began early in the nineteenth century, and raised it from 6% to over 20% within one hundred years (Schwanitz, 1966). Selection for particular kinds of oil or protein components has also been highly effective, as in the search for maize with more lysine in its endosperm proteins, to improve its nutritional value. Zein, the prolamine which comprises about half of the total endosperm protein in wild type maize, has only about 0.1% lysine. Two mutants with far higher lysine contents were discovered (Mertz *et al.* 1964), and these have been used in breeding programmes for high lysine maize Neither of them affects the amino acid composition of the various proteins (Concon, 1966), but by suppressing zein production and permitting a compensatory increase in albumins, globulins and glutelins, they considerably increase both the percentage and the total content of lysine and other essential amino acids (Jimenez, 1966).

Drugs and their precursors are another class of compounds subject to selection. Primitive man had a remarkably comprehensive knowledge of the sources of drugs. However, for all his biochemical and botanical surveys, modern man has added no new source of caffeine (Anderson, 1954), in fact no new major food crop. Where industrialization has created a demand for a new product, new crops have been developed, as in the case of rubber. *Hevea brasiliensis*, the rubber tree, was originally grown for its nuts. The Russian dandelion, *Taraxacum kok-saghyz*, on the other hand, has only recently undergone primary domestication as a source of rubber.

Thus, most crop plants, including those examined in this book, have been subject to selection by man over long periods of time, and their physiological adaptations to agricultural conditions are likely to be complex, subtle and not rapidly extended to other plants. Their uses may change, together with the selection pressures on them, but their adaptation as crops constitutes one of man's crucial and remarkable achievements. It is difficult to understand how Darwin could have written, even in 1868, 'Although the principle of selection is so important, yet the little which man has effected, by incessant efforts during thousands of years, in rendering the plants more productive or the grains more nutritious than they were in the time of the old Egyptians, would seem to speak strongly against its efficacy.'

A SHORT HISTORY OF CROP PHYSIOLOGY

Scientific interpretation is a creature of fashion. Thus, the case histories which follow reflect not only the individuality of the crops and their evolutionary paths, but also the current perspectives and enthusiasms of crop physiology. A brief account of the origins and development of crop physiology is desirable, therefore, to provide a background for the following chapters.

Two hundred years ago, in 1771, Joseph Priestley discovered that plants could regenerate oxygen in an atmosphere rich in carbon dioxide. It was not until eight years later, however, that the essential role of light in this process was shown by Ingenhousz, and the study of photosynthesis, a process central to crop physiology, got under way.

The next major figure was de Saussure whose work, summarized in 1804, established the need by plants for the uptake of minerals and nitrate from soil. Experiments by Boussingault in 1837 led to some modification of this view, insofar as he found that whereas wheat plants could gain nitrogen only from the soil, clovers and peas could also gain significant amounts from the atmosphere. Fifty years later Hellriegel and Wilfarth established the nature of symbiotic nitrogen fixation by legumes and, thereby, one of the major differences among crop plants. In the intervening period, considerable confusion had been caused by Liebig's view that all plants obtained their nitrogen from ammonia in the air. He had, however, rejected the view that plants get their nourishment from humus and he carried out the first field experiments with synthetic fertilizers. These failed because of Liebig's omission of combined nitrogen and his use of insoluble mineral substances, but those begun at Rothamsted in 1843 by Lawes and Gilbert were more successful.

By 1865, when Sachs published his *Experimentelle Pflanzenphysiologie*, virtually all the major branches of that subject were established, but their development in the next fifty years showed little concern with crop productivity and yield, except in the area of mineral nutrition (cf. Reed, 1942). Not only was the essentiality and role of many mineral nutrients established, but the dynamics of their availability in soils and of crop response to them was the main preoccupation of plant physiologists, especially in the USA, through the first half of the twentieth century. Towards the end of that period herbicide physiology also became prominent: initially concentrated on herbicides of the auxin type, it now deals with growth regulators having a far wider spectrum of activity.

Crop physiology, with the aim of understanding the dynamics of yield development in crops, really began about sixty years ago with the work of W. L. Balls (see Plate 1.1) on the cotton crops of the Nile. Balls and Holton (1915*a*, *b*) analysed the effects of plant spacing and sowing date on the development and yield of Egyptian cotton plants within crop stands, not in isolated plants. In a third paper, Balls (1917) analysed the effects of various environmental factors on yield and found that many apparently contradictory phenomena could be explained by looking for steps in the life cycle of the crop that were important in determining subsequent yield and were sensitive to environmental stress. For example, he found 'that during July the flowering of a normal Egyptian cotton is unaffected by watering, and yet at the same moment the same watering is the only recognizable determinant of the length of the lint hair cells which are developing inside the bolls'. He concludes: 'Armed with the conception of Limiting Factors, and also the frequent Predetermination of their effects, there would seem to be notable advances in knowledge of what might be called "crop physiology" by applying our methods of continuous registration to the development of the plant. The subject is obviously a close relation of Ecology, but without the essential complications introduced in the latter subject by the competition of diverse species. At the same time it is very doubtful whether such investigations will do more – from the farmer's standpoint – than elucidate the reasons for conventional practice, though they might occasionally indicate the path by which advances in physiological knowledge could be directed to application in agriculture, and conversely.'

Note, first of all, Balls' reliance on the Law of the Limiting Factor, which F. F. Blackman had enunciated in 1905, as the key to the analysis of the intricate relations between a crop and its environment. This, combined with the biographical approach to the crop, led to his emphasis on the identification of critical stages in yield development and of the controlling factors at each stage. The concept of strong interactions between such factors developed much later. Note also Balls' prediction that crop physiology may often provide only a retrospective explanation of farming practice and plant breeding success. This doubt, which still haunts crop physiologists (cf. Bunting, 1971), will be considered below. Balls was of an original and inventive mind, and his prolific work during his first ten years in the Sudan advanced our understanding of the cotton crop in many ways. In spite of this, Kitchener decided not to renew Balls' contract and he was thereby forced to abandon his crop physiology programme (Harland, 1961).

In the ten years following Balls' work there was, in England, a rapid development of the methods of growth and yield analysis by a galaxy of investigators, as may be seen from the *Proceedings of the Imperial Botanical Congress* held in London in 1924, at which 'The physiology of crop yield: a survey of modern methods of attack' was reviewed by Balls, V. H. Blackman, F. G. Gregory, G. E. Briggs, R. A. Fisher, E. J. Maskell and F. L. Engledow among others.

Blackman (1919) provided the cornerstone for growth analysis with his recognition that plant weight could increase logarithmically, and that growth could be described by what he called an efficiency index, or relative growth rate. Gregory (1917) suggested that the rate of increase in dry weight per unit leaf area, the net assimilation rate, could be used as a measure of the net photosynthetic efficiency of the leaves, for which Briggs *et al.* (1920) preferred the term unit leaf rate. After some harrying by Fisher (1920), growth analysis was formulated in a way that allowed dry matter accumulation to be described in terms of net assimilation rate on the one hand and of growth in leaf area on the other.

Photosynthesis

Much attention was given, initially, to net assimilation rate. However, insofar as it was an average rate for all leaves on a plant and was usually determined for intervals of about a week, it was a rather insensitive measure of photosynthetic rate. This led Heath and Gregory (1938) to conclude that the net assimilation rate during the vegetative phase was approximately constant for a wide range of species and environments (although the rates actually ranged from 0.12 to 0.72 g dm^{-2} week^{-1}). They concluded that 'the main determining factor leading to very great differences in dry weight accumulation in different types of plants is the rate of extension of the leaf surface, in which is included both the size of individual leaves and the rate of production of new leaves'. In succeeding years much more attention was given to leaf growth of crops, but more recently two advances have restored the emphasis on photosynthetic rate as a major determinant of yield. The first was the development of infra-red gas analysis for the determination of carbon dioxide levels, permitting not only sensitive, short term measurements with single leaves but also, following the development of micrometeorology in the 1950s, the estimation of short term rates of photosynthesis and respiration by crops in the field. The other advance was the realization that there were major differences among crop plants in the rate and pathway of their photosynthesis. In 1963 Hesketh and Moss showed that photosynthesis by leaves of maize, sugar cane and related tropical grasses could reach much higher rates, with less marked light saturation, than leaves of other plants. At about the same time Tarchevskii and Karpilov (1963) and Kortschak *et al.* (1965) showed that the initial products of $^{14}CO_2$ fixation in these plants differed from those in other plants, being mostly in C_4-dicarboxylic acids. Subsequent work, particularly that by Hatch and Slack (cf. 1970), has clarified the differences between these C_4 pathway plants and those with only the C_3-Calvin cycle pathway. The differences in pathway are associated with differences in photosynthetic rate, in responses to light intensity, temperature and oxygen level, in photorespiration, in leaf anatomy and chloroplast morphology, in rate of translocation, and in the efficiency of water use, which can have profound effects on the physiology of yield determination. Plants with C_4 photosynthesis are represented by maize and sugar cane among the crops covered in this book.

Leaf growth and canopy development

Early growth analysis in England was much concerned with the mathematical formulation of leaf growth. Boysen-Jensen (1932) recognized that photosynthesis by a plant community was less subject to light saturation than that by individual leaves, because lower leaves were shaded, and that the inclination of leaves in the canopy would influence this difference. This led him to consider the accumulated foliage area in relation to ground area, an approach which was crystallized in 1947 by D. J. Watson (Plate 1.1) in his concept of leaf area index (LAI). The introduction of this index provided a more meaningful way of analysing growth in crops, and stimulated renewed interest in crop physiology. Then in 1953 Monsi and Saeki (Plate 1.1) in Japan combined the concept of leaf area index with that of the extinction of light penetrating a crop canopy in a manner analogous to that formulated in Beer's law, to make possible a quantitative description of the profiles of light and photosynthesis within a crop. Large differences were shown to exist between plant communities in their light extinction coefficient, as determined by the inclination, size and other attributes of leaves. Monsi and Saeki's paper had a profound influence on crop physiology, in two ways. In the first place it led to attempts by rice breeders in Japan to develop a more comprehensive definition of an ideal plant type. Secondly, it opened the way to the spate of work in the last decade on the modelling of crop photosynthesis. Early models, such as those of Davidson and Philip (1958) and Saeki (1960), assumed that crop respiration was directly proportional to leaf area index, and therefore concluded that there must be an optimum LAI and stand structure, and much effort went towards defining these. However, crop respiration was subsequently shown (e.g. by Ludwig *et al.* 1965) to increase asymptotically with increase in LAI, with the consequence that net photosynthesis showed a broad plateau relation with LAI, rather than a sharp optimum. A great deal of effort has gone into refining the geometric models of crop photosynthesis, coincident with greatly increased access to computers (e.g. de Wit, 1965; Duncan *et al.* 1967). The quantitative study of crop respiration has, on the other hand, been largely neglected. Crop geometry has been analysed almost entirely in relation to light penetration, with little concern for effects on carbon dioxide renewal, the microclimate of storage organs, or its relation to translocation patterns. Yield has been assumed to be limited primarily by photosynthesis and the supply of assimilates, but crop physiologists are now beginning to give more attention to other processes which may also limit yield capacity.

Translocation

Mason and Maskell (1928) wrote that their 'decision to investigate the method in which carbohydrates are distributed throughout the cotton plant was taken because of the conviction that such problems as the shedding of flower buds and young bolls, the condensation of sugar to form cellulose in the lint hair, and the variations in habit and development as a result of climatic and other causes could not be profitably attacked until a great deal more information was available relating to the machinery responsible for the transport of sugars and the factors determining the rate and direction of movement within the plant body'.

Experimental work in this area has, however, been overshadowed by that on crop photosynthesis, even in the period following the Second World War when increased access to radioisotopes transformed the ease and scope of translocation experiments. Much descriptive work has been done on the patterns of translocation, but how these are controlled and whether they limit crop yield, is not clear.

We have now reached the point where these questions must be answered because of the requirements for dynamic modelling of crop growth. The modelling of crop photosynthesis is essentially a geometric problem with a static crop canopy. But in order to simulate crop growth and yield development, not only must respiration be better understood, so that net photosynthesis can be derived, but the principles governing the partitioning of the assimilate and its storage or investment in new growth must also be known.

The techniques of dynamic modelling developed by Forrester were first applied by Brouwer and de Wit (1968) in their Elementary Crop Growth Simulator (ELCROS) for maize. They have also been applied to sugar beet, cotton and other crops, as may be seen from some of the chapters that follow.

Another necessary input for the dynamic modelling of crop growth is information on the potential growth rates of plants and of their organs as influenced by such factors as temperature and light, as well as by nutrient levels. It is here that the data accumulated on the growth of crop plants in controlled environments, particularly in phytotrons since the Earhart Laboratory was opened in 1949, will prove invaluable.

Storage

It was pointed out earlier that a predominant feature of the evolution of many crop plants has been the increased storage capacity of the harvested organs. Even so, this storage capacity may continue to limit yields, and the analysis of its components should have been a major theme in crop physiology. In fact it has been relatively neglected, although Engledow and Wadham of Cambridge emphasized this approach in their 1923 paper. It did not flourish, partly because growth analysis seemed a more promising approach at the time and partly because, under the suboptimal agronomic regimes used, negative correlations between the individual yield components were often evident, implying that it was the supply of assimilate rather than the storage capacity that was limiting.

With the increasing realization of the extent of feed-back relations between the rates of photosynthesis, transport and storage, and with the demands of dynamic modelling for more insight into the mechanisms that control the rate and duration of storage, there is likely to be more emphasis on this facet of crop physiology in future. Vernalization and photoperiodism were recognized as processes of the most fundamental importance in the control of reproduction in plants in the same decade which saw the establishment of crop physiology. In the intervening period their mechanisms have been explored and most emphasis has been given to their adaptive role in control of the timing of the reproductive cycle. However, they also affect, to a considerable extent, the components of the storage systems of crops, and provide a powerful tool for varying and perhaps increasing these.

Ideal plant type

The challenge to crop physiology in the future is that it should be able to specify to plant breeders an ever more comprehensive and complex set of characteristics which lead to continued increases in crop productivity, in yield reliability, and in quality of composition. In the past, as Balls suggested was likely, there has been much retrospective explanation of what plant breeders had already achieved, but not merely that, as consideration of the concept of ideal plant type reveals.

Engledow and Wadham, in 1923, made explicit what many plant breeders had probably always had in mind, that 'the procedure should be to find out the plant characters which control yield per acre and by a synthetic system of hybridizations to accumulate into one plant-form the optimum combination of yield-controlling factors'. In the intervening years the list of desirable characters has grown, with the emphasis shifting from readily observable, morphological characters to physiological ones. Rice breeding in Japan offers the outstanding example of this approach.

One of the first tasks set for the National Agricultural Experiment Station in Japan was to find ways to prevent lodging in rice grown under high fertilizer regimes, and the consensus arrived at by plant breeders was to develop varieties with short stems combined with high tillering and high panicle number (Baba, 1954). The advantage of a wide range of adaptability was also emphasized. Monsi and Saeki's analysis of light relations in plant communities then led to emphasis on the advantages of small, erect, thick leaves and of a compact tillering habit (Tsunoda, 1959; Baba, 1961; Murata, 1961). This concept of an ideal plant type for rice was very much in the Japanese traditions of close observation, miniaturization and idealization of their plants. It was subsequently transplanted to the tropics and applied with considerable success by plant breeders of the International Rice Research Institute to the development of rice varieties adapted to tropical lowland culture under high fertilizer applications (Jennings, 1964).

In both the Mexican wheat and IRRI rice programmes the aim was to develop varieties with broad adaptation, to be productive in a wide range of environments. This led to increased pressures to specify a generalized optimum plant type, but it also led to emphasis on other physiological characteristics, namely insensitivity to day-length for control of flowering and a shortening of the life cycle. These characteristics are not universally desirable, however. Wheat at high latitudes requires some sensitivity to day-length to delay ear development until the risk of frosts has passed; similarly, floating rice must delay its flowering until the monsoon floods recede, or the grain cannot be harvested.

Not so many years ago there was perhaps too much emphasis on close local adaptation in plant breeding: in Taiwan alone there were more than 2000 local varieties of rice. Now the pendulum may have swung too far the other way in the emphasis on transworld adaptation and optimum plant type. Our concept of the latter depends on agronomic practice. Social pressures to reduce the use of nitrogen fertilizers or herbicides and insecticides could lead to shifts in optimum plant type, just as further physiological insight could do. Plant breeders are more aware than physiologists of shifts in agronomic practice, and have provided much of the initiative for change, but physiological insight into the basis of yield has also contributed. Emphasis in the past has been particularly on growth habit and crop geometry insofar as they affect crop photosynthesis, and other factors influencing photosynthetic rate are currently receiving much attention. Those influencing translocation patterns, storage mechanisms and capacity, and protein and lipid synthesis should receive more attention in the future. They should provide ample opportunities for crop physiologists to point the way to further advances in crop productivity.

REFERENCES

Anderson, E. (1954). *Plants, Man and Life*. Andrew Melrose, London. 208 pp.

Autret, M. (1970). World protein supplies and needs. In *Proteins as Human Food*, ed. R. A. Lawrie. Butterworths, London, pp. 3–19.

Baba, I. (1954). Breeding of a rice variety suitable for heavy manuring. *Jap. J. Breeding* **4**, 167–184.

Baba, I. (1961). Mechanism of response to heavy manuring in rice varieties. *Intl. Rice Commission Newsletter* **10**(4), 9–16.

Balls, W. L. (1917). Analyses of agricultural yield. III. The influence of natural environmental factors upon the yield of Egyptian cotton. *Phil. Trans. Roy. Soc. Ser. B*, **208**, 157–223.

Balls, W. L. and Holton, F. S. (1915*a*). Analyses of agricultural yield. I. The spacing experiment with Egyptian cotton, 1912. *Phil. Trans. Roy. Soc. Ser. B*, **206**, 103–180.

Balls, W. L. and Holton, F. S. (1915*b*). Analyses of agricultural yield. II. The sowing date experiment with Egyptian cotton, 1913. *Phil. Trans. Roy. Soc. Ser. B*, **206**, 403–480.

Bardach, J. E. (1968). Aquaculture. *Science* **161**, 1098–1106.

Blackman, V. H. (1919). The compound interest law and plant growth. *Ann. Bot.* **33**, 353–360.

Boysen-Jensen, P. (1932). *Die Stoffproduktion der Pflanzen*. Fisher, Jena, 108 pp.

Briggs, G. E., Kidd, F. and West, C. (1920). A quantitative analysis of plant growth. *Ann. appl. Biol.* **7**, 103–123, 202–223.

Brouwer, R. and de Wit, C. T. (1968). A simulation model of plant growth with special attention to root growth and its consequences. In *Root Growth*, ed. W. J. Whittington. Butterworths, London, pp. 224–244.

Bunting, A. H. (1971). Productivity and profit, or is your vegetative phase really necessary? *Ann. appl. Biol.* **67**, 265–272.

Calder, N. (1967). *The Environment Game*. Secker and Warburg, London. 240 pp.

Clark, C. (1967). *Population Growth and Land Use*. Macmillan, London, 406 pp.

Concon, J. M. (1966). The proteins of opaque-2 maize. In *Proc. High Lysine Corn Conference*, eds E. T. Mertz and O. E. Nelson, Corn Refiner's Assoc. pp. 67–73.

Darlington, C. D. (1969). The silent millennia in the origin of agriculture. In *The Domestication and Exploitation of Plants and Animals*, eds P. J. Ucko and G. W. Dimbleby. Duckworth, London, pp. 67–72.

Darwin, C. (1868). *The variation of plants and animals under domestication.* John Murray, London, 2 vols.

Davidson, J. L. and Philip, J. R. (1958). Light and pasture growth. In *Proc. UNESCO Sympos. Climatology and Microclimatology, Canberra 1956. Arid Zone Research* **11**, 181–187.

Dore, J .(1959). Response of rice to small differences in length of day. *Nature, Lond.* **183**, 413.

Duncan, W. G., Loomis, R. S., Williams, W. A. and Hanau, R. (1967). A model for simulating photosynthesis in plant communities. *Hilgardia* **38**, 181–205.

Enebo, L. (1970). Single-cell protein: in *Evaluation of novel protein products*, eds A. E. Bender, R. Kihlberg, B. Lofquist and L. Munck. Pergamon, Oxford, pp. 93–103.

Engledow, F. L. and Wadham, S. M. (1923). Investigations on yield in the cereals. Part 1. *J. agric. Sci.* **13**, 390–439.

Evans, L. T. (1973). The effect of light on plant growth, development and yield. In *Plant Response to Climatic Factors.* UNESCO, Paris, pp. 21–35.

Evans, L. T. and Dunstone, R. L. (1970). Some physiological aspects of evolution in wheat. *Aust. J. biol. Sci.* **23**, 725–741.

Farmer, B. H. (1969). Available food supplies. In *Population and Food Supply*, ed. J. Hutchinson. Cambridge University Press, pp. 75–95.

Fisher, R. A. (1920). Some remarks on the methods formulated in a recent article on 'The quantitative analysis of plant growth'. *Ann. appl. Biol.* **7**, 367–372.

Gavin, W. (1951). The way to higher crop yields. *J. Min. Agric.* **58**, 105–111.

Gregory, F. G. (1917). *Third Ann. Rep., Exptl. and Research Station, Cheshunt*, pp. 19–28.

Harlan, J. R. (1967). A wild wheat harvest in Turkey. *Archaeol.* **20**, 197–201.

Harlan, J. R. (1971). Agricultural origins: centers and non-centers. *Science* **174**, 468–474.

Harlan, J. R. and de Wet, J. M. J. (1965). Some thoughts about weeds. *Econ. Bot.* **19**, 16–24.

Harland, S. C. (1961). William Lawrence Balls, 1882–1960. *Biograph. Mem. Fellows Roy. Soc.* **7**, 1–16.

Hatch, M. D. and Slack, C. R. (1970). Photosynthetic CO_2-fixation pathways. *Ann. Rev. Pl. Physiol.* **21**, 141–162.

Hawkes, J. G. (1969). The ecological background of plant domestication. In *The Domestication and Exploitation of Plants and Animals*, eds P. J. Ucko and G. W. Dimbleby. Duckworth, London, pp. 17–29.

Heath, O. V. S. and Gregory, F. G. (1938). The constancy of the mean net assimilation rate and its ecological importance. *Ann. Bot. N.S.* **2**, 811–818.

Helbaek, H. (1950). *Tollund mandens sidste maaltid.* Aarbøger for Nordisk Oldkyndighed og Historie.

Helbaek, H. (1966). Commentary on the phylogenesis of *Triticum* and *Hordeum*. *Econ Bot.* **20**, 350–360.

Hesketh, J. D. (1963). Limitations to photosynthesis responsible for differences among species. *Crop Sci.* **3**, 493–496.

Hesketh, J. D. and Moss, D. N. (1963). Variation in the response of photosynthesis to light. *Crop Sci.* **3**, 107–110.

Jennings, P. R. (1964). Plant type as a rice breeding objective. *Crop Sci.* **4**, 13–15.

Jimenez, J. R. (1966). Protein fractionation studies of high lysine corn. In *Proc. High Lysine Corn Conf.*, eds E. T. Mertz and O. E. Nelson. Corn Refiner's Assoc., pp. 74–79.

Johnson, M. (1967). Growth of microbial cells on hydrocarbons. *Science* **155**, 1515–1519.

Kaplan, L. (1965). Archaeology and domestication in American *Phaseolus* (beans). *Econ. Bot.* **19**, 356–368.

Kellogg, C. E. and Orvedal, A. C. (1968). World potentials for arable soils. *War on Hunger* **2**(9), 14–17.

Kortschak, H. P., Hartt, C. E. and Burr, G. O. (1965). Carbon dioxide fixation in sugar cane leaves. *Plant Physiol.* **40**, 209–213.

Ludwig, L. J., Saeki, T. and Evans, L. T. (1965). Photosynthesis in artificial communities of cotton plants in relation to leaf area. *Aust. J. biol. Sci.* **18**, 1103–1118.

Mangeldsdorf, P. C. (1966). Genetic potentials for increasing yields of food crops and animals. *Proc. Natl. Acad. Sci.* **56**, 370–375.

Mason, T. S. and Maskell, E. J. (1928). Studies on the transport of carbohydrates in the cotton plant. I. A study in diurnal variation in the carbohydrates of leaf, bark and wood, and of the effects of ringing. *Ann. Bot.* **42**, 189–253.

Matsuo, T. (1959). *Rice culture in Japan.* Min. Agr. For., Japan, 128 pp.

McPherson, A. T. (1965). Synthetic food for tomorrow's billions. *Bull. Atom. Sci.* **21**, 6–11.

Mertz, E. T., Bates, L. S. and Nelson, O. E. (1964). Mutant gene that changes protein composition and increases lysine content of maize endosperm. *Science* **145**, 279–280.

Monsi, M. and Saeki, T. (1953). Uber den Lichtfaktor in den Pflanzengesellschaften und seine Bedeutung für die Stoffproduktion. *Jap. J. Bot.* **14**, 22–52.

Murata, Y. (1961). Studies on the photosynthesis of rice plants and its culture significance. *Bull. Natl. Inst. Agric. Sci. Japan Ser. D* **9**, 1–169.

Njoku, E. (1959). Response of rice to small differences in length of day. *Nature, Lond.* **183**, 1598.

Pickersgill, B. (1969). The domestication of chili peppers. In *The Domestication and Exploitation of Plants and Animals*, eds P. J. Ucko and G. W. Dimbleby. Duckworth, London, pp. 443–450.

Pirie, N. W. (1970). Complementary ways of meeting the world's protein needs. In *Proteins as Human Food*, ed. R. A. Lawrie. Butterworths, London, pp. 46–61.

President's Science Advisory Committee (1967). *The World Food Problem.* The White House, 3 vols.

Prokes, B. and Zahradnik, J. (1968). *Ann. Rep. Lab. Algol. Trebon*, Czechoslovakia, pp. 139–142.

Pyke, M. (1970). *Synthetic Food.* John Murray, London, 145 pp.

Reed, H. S. (1942). A short history of the plant sciences. *Chron. Bot.*, Waltham, Mass., 320 pp.

Revelle, R. (1966). Population and food supplies: the edge of the knife. *Proc. Natl. Acad. Sci.* **56**, 328–351.

Ryther, J. H. (1969). Photosynthesis and fish production in the sea. *Science* **166**, 72–76.

Saeki, T. (1960). Interrelationships between leaf amount, leaf distribution and total photosynthesis in a plant community. *Bot. Mag., Tokyo* **73**, 55–63.

Schwanitz, F. (1966). *The origin of cultivated plants.* Harvard University Press, Cambridge, 175 pp.

Scrimshaw, N. S. (1966). Applications of nutritional and food science to meeting world food needs. Prospects of the World Food Supply. *Proc. Natl. Acad. Sci.* **56**, 352–359.

Smartt, J. (1969). Evolution of American *Phaseolus* beans under domestication. In *The Domestication and Exploitation of Plants and Animals*, eds P. J. Ucko and G. W. Dimbleby. Duckworth, London, pp. 451–462.

Solheim, W. G. (1972). An earlier agricultural revolution. *Sci. Amer.* **226**(4), 34–40.

Tamiya, H. (1957). Mass culture of algae. *Ann. Rev. Plant Physiol.* **8**, 309–334.

Tarchevskii, I. A. and Karpilov, Y. S. (1963). On the nature of products of short term photosynthesis. *Soviet Plant Physiol.* **10**, 183–184.

Thomas, M. D. (1965). Photosynthesis (Carbon Assimilation): Environmental and Metabolic Relationships. In *Plant Physiology: A Treatise*. ed. F. C. Steward. Academic Press N.Y. vol. IVa, pp. 9–202.

Tsunoda, S. (1959). A developmental analysis of yielding ability in varieties of field crops. II. The assimilation system of plants as affected by the form, direction and arrangement of single leaves. *Jap. J. Breeding* **9**, 237–244.

Vavilov, N. I. (1951). The origin, variation, immunity and breeding of cultivated plants. Transl. K. S. Chester. *Chron. Bot.* **13**, 1–366.

Watson, D. J. (1947). Comparative physiological studies on the growth of field crops. I. Variation in net assimilation rate and leaf area between species and varieties, and within and between years. *Ann. Bot. N.S.* **11**, 41–76.

Wit, C. T. de (1965). Photosynthesis in leaf canopies. *Agr. Res. Rept.* **663**. Centre Agr. Publ. Document., Wageningen.

Zohary, D. (1969). The progenitors of wheat and barley in relation to domestication and agricultural dispersal in the Old World. In *The Domestication and Exploitation of Plants and Animals*, eds P. J. Ucko and G. W. Dimbleby. Duckworth, London, pp. 47–65.

2

Maize

W. G. DUNCAN

Maize (*Zea mays* L.) is the only species usually included in the genus *Zea*, of the tribe Maydeae (or Tripsaceae) of the family Gramineae. There are only eight genera in the Maydeae, five of which are oriental and three American. The other two American genera, *Euchlaena* and *Tripsacum*, are much more like maize than are the oriental genera, and some authorities favor redesignation of *Euchlaena* as *Zea*. Relationships among the genera are discussed in considerable detail by Weatherwax and Randolph (1955). While crosses between maize and *Tripsacum* are possible, and have been produced under special conditions, there is little reason to think such crossing has influenced the evolution of maize. Maize crosses freely with teosinte (*Euchlaena mexicana*), however, and the hybrids are fertile. As a result most, if not all, races of maize can be shown to have teosinte as a part of their genetic background. This introgression by teosinte and the ability to recover it by inbreeding was responsible for earlier hypotheses that maize originated as a hybrid of teosinte and one or more other grasses.

The origin of maize poses a mystery because no wild ancestor from which it could have originated has ever been found. The mystery is heightened because no known variety of maize is capable of existing for more than two or three generations except under cultivation by man. Its kernels, attached to a cob and wrapped in husks, have no means for dispersal. Thus the ear that falls to the ground can only give rise to a clump of plants too crowded to produce viable kernels.

A part of the mystery of its origin was cleared in 1954 when Barghoorn *et al.* (1954) reported identification of maize pollen in a drill core taken 70 metres under Mexico City from strata established as being 80000 years old. This identification of the pollen has not gone unquestioned; but if accepted, it shows that the ancestor of maize was a wild maize and that it did not arise from hybridization of other species. This would also establish that it is American in origin and predated the coming of man to the continent. These conclusions have been supplemented by archaeological findings that have identified early maize remains (Mangelsdorf *et al.* 1964) and have established cultivated types dating back to 3000 to 5000 B.C. The origin of maize is discussed by Mangelsdorf (1965), Weatherwax (1954), and Galinat (1971). Finan (1950) has reviewed the mentions of maize in the early herbals.

Maize ranks after wheat and rice as the third most important crop in the world (cf. Table 1.1). It is used directly as food for human consumption and as a feed grain for animals. Its importance arises both because of its great productivity and because it can be grown over an extremely wide environmental range. It is grown without irrigation in regions with as little as 25 cm of rainfall and in areas with as much as 500 cm, and from sea level to altitudes of 4000 metres in the Andes.

Its usefulness is enhanced by its extreme diversity of form, quality and growth habit. In grain size alone there is more than a fifty-fold variation between the normal kernels of extreme varieties. The kernels themselves may range from hard and vitreous to soft and floury, and are of many colours. The height of mature plants may range from less than 60 cm with eight leaves on the main stalk to 7 metres with 48 leaves. The filled ears range in size from those of some dwarf popcorn varieties, little larger than a man's thumb, to the giant ears of the 'Jala' race that may be almost a metre in length. The time from planting to harvest may range from three months to more than a year in some high altitude locations.

The great diversity among varieties induced early botanists to attempt to divide the genus into numerous subspecies but these have not come into wide use. Under the leadership of the Rockefeller Foundation, the maize varieties of most of the important maize producing countries of South and Central America have been classified into groupings having common characteristics, called races, of which more than 250 have been described (e.g. Grant *et al.* 1963).

Presumably the different races of maize arose almost entirely as a result of selection by man over the past 10 000 years at most. A part of the process of improvement and diversification undoubtedly resulted from deliberate selection in the advanced pre-Columbian cultures of Central and South America, which were built largely around maize as a food source. The initial improvement from the wild ancestor, however, must have been achieved by primitive man, selecting by unknown criteria.

Whatever the wild form of maize may have been it probably originated and survived, like other cereals, in an area of alternating wet and dry seasons, as suggested by two of its characteristics. Maize does not tolerate shading so it seems unlikely that it evolved in forest shade; but neither is it adapted to dry conditions, so its period of active growth must have coincided with a time of ample rainfall. Such an alternation of wet and dry seasons is found in the part of Mexico generally assumed to be its most probable source of origin.

Most of the maize grown can be grouped for convenience into three classes, dent, flint, and soft or floury. Other lesser groups are waxy maize, sweet corn and popcorn. Dent is the most widely grown maize in the United States and northern Mexico (Berger, 1962). Anderson and Brown (1964) consider modern dent corn to be based on deliberate crosses between Northern Flint and Southern white dent varieties made mostly during the eighteenth and nineteenth centuries. Flint maize is predominant in the agriculture of Europe, Asia, Central and South America (Berger, 1962); soft or floury maize is grown to some extent in South America and South Africa; and sweet corn is grown primarily in the United States for human consumption. Sweet corn differs from dent maize by only one recessive gene (Neuffer *et al.* 1968) which interferes with the conversion of sucrose to starch in the kernels. Popcorn is almost entirely restricted to the American continent. It is not a type of maize but describes any kind with kernels that will 'pop' when heated. It is used entirely for human consumption. Waxy maize is a variety grown in East Asia and as an industrial raw material in the United States. It is characterized by a much higher proportion of amylopectin in the endosperm which gives useful properties to starch made from it (Neuffer *et al.* 1968).

The germinating seed

According to Sass (1951) most of the maize varieties he has examined have five leaves in embryonic form in the seed. This then is the minimum number of leaves on the main stalk of the developed plant. This count does not include the modified first leaf or scutellum which acts as an organ for absorbing growth materials from the endosperm. The attachment of the scutellum to the endosperm is such that succeeding true leaves are oriented at right angles to the longest lateral axis of the seed.

The germination of maize seed is similar to that of many grasses except for scale differences resulting from the relatively large endosperm and embryo. There is no inhibition of germination so that under wet conditions kernels may germinate immediately after maturity even while still attached to the plant. Kernels developed under sterile conditions on agar by Poneleit and Hamilton (unpublished) germinated shortly after the seed dented. Several attempts have been made to plant maize after the onset of cold weather hoping for earlier germination in the spring but without known success, in spite of repeated observations that whole ears buried at the same time will germinate almost completely when the soil warms in the spring.

The large size of endosperm and embryo permits emergence from considerable depths for some varieties. An extreme example is presented by a type grown since prehistoric times by the Hopi Indians in arid areas of the southwestern part of the United States which can emerge after being planted as deep as 45 cm (Collins, 1914). This is a highly specialized adaptation, however, and most dent varieties will not emerge if planted deeper than 10 to 15 cm (Aldrich and Leng, 1966). The maximum depth from which a maize seedling can emerge is determined by the maximum enlargement potential of the mesocotyl in pushing the coleoptile to the surface of the soil. The first permanent roots emerge at the node joining mesocotyl and coleoptile. Thus the depth of the root crown is determined by the length of the coleoptile rather than by the depth of planting.

The root system

In the field the primary root and the seminal or seed roots are soon supplanted by others that form the permanent root system. This is not a fixed rule in all cases, however, as the initial roots can be made to remain functional for the life of the plant. They are actually capable of supplying all of the nutrition for the fully developed plant under special conditions (Ohlrogge, 1958). Internodes between the first nodes that develop above the mesocotyl elongate only slightly so the base of the plant comes to resemble an inverted cone as the diameter of each succeeding node increases. Each node gives rise to a ring of roots thus forming the underground root system. The diameter of each root is determined to a considerable extent by the circumference of the node from which it originates and since the roots have no cambial layer the permanent diameters of the roots are fixed by factors that affect early plant growth.

Aerial or 'brace' roots usually emerge from the lower above-ground nodes near the end of vegetative development. If they reach down to the surface of the soil they then assume the appearance and function of normal underground roots in subsequent growth. The number of aerial roots and the number of nodes from which they emerge is highly variable among varieties and is affected by rate of planting and nutrition within a variety. Most of the highly developed dent varieties have them on only the lower two or three nodes at usual planting rates. Some tropical races have aerial roots on nodes halfway or more to the top of the stalk. In all varieties they may form from almost any node if the stalk is placed in a horizontal position.

The shoot

The leaves of the maize plant are the first part to emerge from the soil after the tip of the coleoptile and remain the only above-ground plant part for a considerable time. The nodes from which the leaves arise develop in rapid succession above the mesocotyl but elongation of the internodal tissue is delayed, often until primordial development is complete.

The sequence of development of the shoot may be described most easily if one thinks of it as being formed of a repeating unit structure consisting of a leaf blade, a leaf sheath, a disk of insertion or node and an internode. This structural unit is repeated, with variations in relative dimensions of the component parts, to make up the entire vegetative shoot except the tassel and its stem. The number of such structural units that develop determines to a large extent the relative duration of vegetative development among varieties and their size determines plant height.

Each of the structual units develops in what may be described as a wave of growth and elongation that commences in the leaf blade and moves down so that sheath development follows that of the leaf blade and growth of the unit terminates with elongation of the internode. The absence of internode elongation in the first few structural units, taken with the delayed elongation of later developing internodes, results in the apical meristem of the shoot remaining at or below the surface of the soil until as many as ten leaves are visible.

The successive elongation of the lower internodes forms the stalk which rises through the tube formed by the leaf sheaths which developed earlier. The sheath serves an important function in preventing stalk breakage at the weak point of each internode created by its active intercalary meristem.

In most developed varieties primordial development is complete before there is any appreciable internode elongation. Thus all leaves, sheaths, nodes and internodes are present from the first completely developed unit to the most undeveloped one before stalk development starts. Subsequent growth results from the enlargement of structural units. As many as three internodes may be in active elongation at the same time although in different phases of their developmental cycle. The final episode of development for each structural unit is the formation of roots or root initials at the lowest part of the internode and the hardening of the meristematic tissues.

Tillers

All axillary buds are morphologically identical when initiated but some of the lowest ones may develop into tillers rather than ears. Maize cultivars vary widely in the number and nature of their tillers. Some form few if any tillers under any conditions, others form numerous tillers under almost all conditions. In some, tillers function as normal stalks almost indistinguishable from the primary shoot and may bear normal ears, whereas in others, tillers rarely bear a normal ear but often have 'tassel ears' that develop from functional complete flowers in the tassel. Within a cultivar the number of tillers that complete their development and approach the primary stalk in size is inversely related to the planting rate; the denser the stand, the fewer developed tillers.

Tillers are usually regarded as undesirable although there is little evidence to show that they reduce grain yield. Dungan (1931) and many others studied the effect of tillers on grain yield with generally inconclusive results. There is little current interest in the problem since present higher planting rates inhibit tiller development in any case. From a practical point of view, the grain produced in tassel ears is usually lost chiefly because it is not protected from birds and insects by husks. The tendency of a variety to tiller is probably related to the degree of introgression of the profusely tillering teosinte, according to Sehgal (1963).

There are races of maize, such as 'Tuxpeno' and 'Conico', that frequently have from one to eight large tillers, the number depending on spacing, fertility, moisture, and other environmental factors. These tillers bear ears almost indistinguishable from those of the main stalk in appearance, silking date, or maturity. Some effort has been given to developing high tillering strains of these races because of their flexibility in adapting to conditions of varying fertility over different planting densities. There is also some thought that plants with several tillers sharing a single root system might have advantages under drought conditions.

Stem height

The height of the final plant, the diameter of its stalk, and to some extent its yield potential, are strongly influenced by environmental conditions during stem elongation. Temperature and photoperiod may influence stalk height by affecting the number of internodes, as will be discussed. There are more direct effects, however, resulting from moisture stress, nutrition, temperature, and light quantity and quality.

It has often been observed in experiments involving different plant densities that maize plants are taller as mutual shading increases, although there is considerable varietal variation in this characteristic. Hozumi *et al.* (1955) noted that in closely spaced plants of different initial heights the shorter plants elongated more rapidly than the taller ones. They gave the phenomenon the name 'cooperative interaction' because smaller plants were able to catch up with taller ones by means of it and compete on more even terms. Although the shorter plants elongated more rapidly, their rate of dry weight gain was less, so their stalks were lighter and smaller in diameter. This characteristic of maize is at least partially responsible for the increase in susceptibility to lodging with increased planting rates.

Moisture stress also affects the length of internodes, probably by inhibiting the elongation of developing cells. Only the two or three internodes in the elongation phase during the period of moisture stress are affected. Temperature can also affect internode elongation although the mechanism involved is not clear. Thus the length of the internodes reflects the environment that existed when they were elongating and the final height of the plant is an integration of the factors that influenced the elongation of each of its internodes.

The relatively late elongation of internodes in relation to leaves and sheaths means that stalks are more affected than leaves by stresses or deficiencies that increase during vegetative plant development. One example of such a stress is the increase in mutual shading in a planting as leaf area increases with plant growth. Thus higher plant densities, which increase mutual shading, decrease the stalk:leaf ratio. This augments the effect of shading on stalk diameter mentioned earlier. Other environmental stresses that might increase with plant development are those due to loss of soil moisture, exhaustion of available nitrogen, or increasing seasonal temperatures. All may act to decrease the stalk:leaf ratio.

In addition to supporting leaves and grain the maize stalk also functions as a storage organ for soluble solids, consisting chiefly of sucrose, that may contribute to grain yield (Daynard *et al.* 1969; Hume and Campbell, 1972). The amount of soluble solids that can be stored in the stalk or other tissues is a function of the volume occupied by fluids. Clearly the cubic capacity of the stalk is affected by environmental stresses as mentioned earlier. Measurements of the specific gravity of maize stalks developed under either extreme drought stress or irrigation showed considerably higher values for the irrigated maize, indicating greater water content. If true for less extreme contrasts in environment, conditions during development may affect both cubic capacity and fluid volume.

The stalk of maize differs from the stems of many grasses in that it is filled with parenchymatous tissue, commonly called pith. The vascular strands of xylem and phloem elements surrounded by sclerenchyma sheaths are located in the pith with a tendency to concentrate near the periphery. The solid filling adds strength to the stalk by preventing its collapse under bending stresses. The pith of the lowest internodes is usually completely filled with fluids, thus providing maximum strength at the point of greatest strain. The vascular bundles with their strong sheaths probably increase the tensile strength of the stalk. Stalk rots, which may be a serious problem, cause lodging chiefly by destruction of pith, allowing stalk failure by collapse of the unsupported sides.

The leaf sheaths add appreciably to the strength of the upper part of the stalk after completion of vegetative development and aid in supporting the ears in an upright position. The leaf sheaths also function as storage sites for soluble solids produced in the leaves, as well as for nitrates and mineral nutrients, but there is little in the literature to indicate the quantities involved or their significance in terms of yield.

Leaves

Each maize leaf consists of a thin flat blade with a definite midrib and a thicker, more rigid, sheath with a smaller midrib. They meet at a definite collar at which the initial angle between leaf and stalk is established. Leaf inclination varies considerably among genotypes, from almost horizontal to almost vertical in one mutant with no collar at all. The angle between sheath and blade also varies with leaf position, lower leaves usually making a greater angle with the stalk.

The blade consists of an upper and lower epidermis between which is a mesophyll having a panicoid structure typical of grasses with the C_4-dicarboxylic acid pathway of photosynthesis. The upper epidermis consists mostly of cells elongated parallel with the veins. At intervals on the upper surface there are bands of bulliform cells which, by changes of turgor, cause leaves to roll and unroll. Some of the epidermal cells adjacent to the bulliform cells develop hairs and the upper surface of the leaves may range from glabrous to velvety pubescent; the lower epidermis is glabrous. Stomata are somewhat more numerous on the lower epidermis, but their density is quite variable. Kiesselbach (1949) reports approximately 7500 stomata per square centimetre on the upper and 9000 on the lower surface of leaves he examined.

The rolling of maize leaves under moisture stress is considered to reduce moisture loss by covering some stomata, but the presence of large numbers of stomata in the lower epidermis would reduce its effectiveness. Another effect of leaf rolling would be reduction of the exposed leaf area, and hence of the radiant energy intercepted.

Many of the lower leaves of maize plants are subject to early loss for various reasons. First leaves are torn off by roots that develop from the base of internodes directly above them. In dense plantings some lower leaves die, probably because they do not intercept enough radiant energy for maintenance. This effect is greatly increased by insufficient fertility, especially lack of nitrogen, and by drought. Leaf diseases and insects often cause progressive losses of leaf area after vegetative maturity. Lower leaves also senesce first as the plant matures.

REPRODUCTIVE DEVELOPMENT

The termination of vegetative development in maize is signalled when the apical meristem of the stem, which has a hemispherical form when leaves are being formed, begins to elongate and initiate the primordia of the staminate flower or tassel. Since no more leaf initials can be formed on the main stalk after the change in the apical meristem, the timing of the onset of reproductive development determines the number of leaves, nodes and internodes on the main stem and thus the relative time from emergence to vegetative maturity.

It is incorrect to say that reproductive development begins with the initiation of the tassel because the early initials of ears are visible as buds at the axils of the lower leaves before the tassel is differentiated. An axillary bud forms at each node of the stem up to the one which bears the uppermost ear. Bonnett (1940) believes that axillary buds which would otherwise form at the upper leaf axils are inhibited by the development of the tassel, but for whatever the reason the upper ear initial becomes the dominant one and soon becomes the largest. The next below, and possibly more, may develop depending on genotype and environment, but there is a clear hierarchy of dominance of each ear initial over the ones below.

The number of nodes that develop before the tassel is initiated is a function of the genotype, modified by temperature and day-length. Duncan and Hesketh (1968) showed that, in long days, decreasing temperature induced flowering at an earlier stage of development. Hesketh *et al.* (1969) showed that shorter days at the same temperature also decreased leaf number. Francis *et al.* (1970) found a similar effect of shortened nights on a wide variety of races of maize and in addition showed that there were differences among genotypes in their sensitivity to differences in the length of the dark period.

The development of the tassel proceeds as the internodes of the stalk elongate so it is almost fully developed when it emerges from the leaf whorl. From emergence of the tassel to its full development and shedding of pollen may take ten days or less during which time the supporting stalk elongates and vegetative plant growth is completed. The shedding of pollen by an individual tassel usually starts near the tip and proceeds both upward to the tip and down and out along the branches to their tips, although there may be considerable variation in this. The shedding process for an individual plant may last a week or slightly longer. Pollen grains are shaken from the extruded anthers by air movement. Thus they are usually retained until conditions are favourable for their dissemination. Kiesselbach (1949) estimated that an average-sized plant released 25 million pollen grains to fertilize approximately 1000 kernels per plant.

The axillary buds mentioned earlier develop into ear shoots with the greatest growth in the uppermost position. Several in lower positions may attain considerable size but those that do not reach the fertilization stage regress. The ear shoot starts development by enlargement of the prophyllum which eventually becomes a husk. This is accompanied by growth of the branch or shank which differs from ordinary stems in remaining slender with usually short internodes. This enables sheaths of the leaves that develop at each node to surround the ears as husks.

When all leaf or husk initials have formed the growing point of the ear shoot elongates to form the beginning of the ear. The development of the ear shoot then parallels the development of the main stalk, to the point that buds formed at the axils of husks may themselves develop into ear shoots that may silk but rarely actually bear grain.

Rapid development of the ear shoots appears to start at about the time the tassel emerges, and growth of the husks is well advanced at anthesis. The earliest development of the ear consists of the formation of a structure with rows of two-lobed protuberances from base to tip. Each of these lobes develops into a spikelet with two flowers, only one of which commonly persists. Since each spikelet produces a kernel, the kernels also occur in double rows and the ear always has an even number of rows of kernels. There are some cultivars in which both spikelet flowers form kernels, however, and rows of kernels are indistinguishable. Descriptions of the detailed morphological development of the flower parts are given by Bonnett (1953) and Kiesselbach (1949).

The pistil of the female flower, known as the silk, develops from the growing point of the flower. It elongates through the length of the husks propelled by the growth of an intercalary meristem located at its base. Each silk continues to grow until it is pollinated and fertilization takes place. If not pollinated, the silk may continue to elongate for ten days to two weeks and may extend 30 to 40 cm beyond the husks. When fertilization is accomplished, activity of the meristematic region quickly ceases, that short part of the silk shrivels up, and the balance slowly turns brown.

Development of the silk starts first at the basal spikelets but the silks from higher up the ear that start growth later may emerge first because of the shorter distance they have to traverse. Thus the first embryos fertilized are usually those near the middle of the ear. Silks from near the top of the ear may continue to emerge for some time after most of the embryos are fertilized, often long after no more pollen is available. Attempts to fertilize such late silks rarely cause formation of kernels.

Silks are receptive over ten days or two weeks or more and pollen may be released from each plant for up to a week so there is usually no difficulty in pollination, especially with open-pollinated varieties where there is considerable plant to plant variability in timing of both silking and pollen shed. Extreme heat or drought may damage a high percentage of tassels (Aldrich and Leng, 1966) or environmental stresses may delay silking until after all pollen is shed in some cases, particularly in single cross hybrids or inbreds with high uniformity.

Temperature and development

The rate of development of maize from planting to anthesis is a function of temperature rather than of photosynthesis. Thus plants from plots with widely different plant populations tassel at nearly the same time although the individual plants may differ greatly in weight due to differences in mutual shading. This is in agreement with observations by many workers but the idea was expressed clearly by Dobben (1962). Millerd and McWilliam (1968) and Brouwer et al. (1970) have shown that the temperature of the growing point of the stalk is controlling and thus that the rate of development from planting to anthesis depends almost entirely on the temperature experienced by the growing point over the whole period.

The growing point of maize remains below the surface of the soil for more than half of its period of vegetative growth, so development rate during that period is a function of soil temperature. After emergence, the growing point inside the leaf-sheath tube is at a temperature, influenced by both transpiration and radiation, that may differ by 5 °C or more from the temperature of the ambient air. Measurements have shown that growing point temperatures are usually lower than those of the ambient air during the day, but similar at night (Duncan, Davis and Chapman, 1973).

The independence of photosynthesis and development rate is a point of considerable importance in understanding the growth and yield of maize. Clearly, the cooler the average temperature the more days of sunlight a maize plant will experience between any two stages of vegetative development. As a result the maize plant will have more photosynthate for growth unless the reduced temperature causes a reduction in leaf photosynthetic rates large enough to offset the increase in the number of days of photosynthesis.

Both photosynthesis and development are very slow at 10 °C and both reach their maximum rates at 30–33 °C (Brouwer *et al.* 1970; Duncan and Hesketh, 1968). The important difference, however, is that photosynthesis is governed by leaf temperatures during daylight hours only, whereas development rate is a function of temperature over the whole day. Thus environments with lower night but similar day temperatures will have slower development rates, whereas higher night temperatures speed development, decreasing both the number of days of photosynthesis between developmental events and plant dry weight. In agreement with this, a recent experiment reported by Peters *et al.* (1971) showed large reductions in maize grain yield resulting from treatments in which plants in the field were kept warmer at night. Unfortunately they did not report plant weight but it is not unreasonable to assume that the reduction in grain yield was accompanied by a reduction in total plant weight.

To pursue this further, one would expect greatest maize growth in environments conducive to leaf temperatures of 30–33 °C during the day but with cool nights. Such conditions are characteristic of locations in regions that are arid, or at higher elevation. Conversely, warm humid environments at low elevations usually have less diurnal variation and might be expected to produce less total growth.

There is some confirmation for this conclusion in high maize grain yields reported from arid and high altitude locations in the USA and from the Cuzco Valley of Peru (Cuany *et al.* 1969; Duncan, Shaver and Williams, 1973; Grobman *et al.* 1962). Such locations are also characterized by unusually high radiation levels, which may be a strong contributing factor.

High altitude maize in tropical locations in South America develops very slowly due to low temperatures. In the Cuzco Valley of Peru, at 3000 metres elevation, maize may take 140 days from planting to anthesis (Grobman *et al.* 1962) and in Venezuela some maize races are reported to take as much as 13 months from planting to harvest (Grant *et al.* 1963). In contrast, they report that low altitude varieties take as few as four months from planting to harvest. Much of the maize in the USA is produced where the useful growing season is not more than 160 days.

In spite of temperatures often too low for efficient photosynthesis, maize at high altitudes in the tropics usually outyields that from low elevations for a single crop although it takes much longer from planting to harvest. There may be many factors involved, however, and the reasons for the superior yields of high altitude maize are not completely understood.

GRAIN YIELD

Two sequential steps are necessary for a maize plant to produce grain; a sink of pollinated kernels capable of further development must be created, and these must be supplied with photosynthate over the period of their subsequent development. Thus grain yield at harvest may be determined either by the kernel capacity established at pollination or by the quantity of photosynthate made available between pollination and maturity. Sink size may be thought of as the number of pollinated, developing kernels per plant multiplied by the potential kernel weight. The photosynthate available for filling the kernels is the integration of the rate of crop photosynthesis over the filling period, minus respiration losses, but plus any labile carbohydrate reserves accumulated earlier in the plant structure and available during the grain filling period.

Kernel numbers

In maize plants the number of kernels per ear and the number of ears that can develop is established at or shortly after pollination and no more can develop later. The kernels of most varieties are limited as to their maximum weight so the number of kernels available may limit yield regardless of how favourable growing conditions may be subsequent to pollination. Thus it is vital to any understanding of maize grain yield to know more about the physiological processes that determine ear and kernel numbers. Unfortunately there is little or no published information known to the author, so we are faced with a subject too important to neglect but about which little experimental information is available.

The question of how many potential kernels can develop on an ear, and how this number might be affected by environmental influences, has not been answered satisfactorily. Bonnett (1948) writes that the ear primordium is formed by an indeterminate meristem, but cites no evidence to support his statement. If the ear is formed by an indeterminate meristem, one would expect that the number of potential kernels per ear within a variety would vary with differences in the environment, but there is no evidence to show that it does. Siemer (1964) found differences among varieties but no differences associated with planting rates or years. We have made numerous counts of potential kernels within varieties but have found no meaningful differences associated with ear position on plants or with treatments.

More attention has been paid to the number of rows of grain per ear, usually called row number, which is variable within and among varieties. Anderson and Brown (1948) have related row number within a variety to certain characteristics of their tassels. We have observed an inverse relationship between row number and potential kernels per row such that the kernels per ear remains nearly the same. It is as if the ear were a helical structure of given length such that the greater the diameter of the helix the less its height.

The helical or spiral structure of the ear has been pointed out by Reeves (1950). Reeves states that the cob structural unit consisting of two spikelets each with one kernel is hexagonal, and each unit is fitted half its diameter above the units on either side. Thus a spiral can be traced around the cob that rises half as many unit widths as there are double rows of kernels.

The axial bud that forms at each of the lower nodes is capable of developing into either a tiller or an ear shoot, but ordinarily only those on nodes below ground develop into tillers. Even though no axial buds form above the node on which the dominant ear eventually appears, this still leaves half or more of the nodes with an axial bud that may develop into a normal ear under some conditions.

As internode elongation commences, the lowest axial buds are larger than those above but the uppermost ones develop more rapidly so that by the time the tassel emerges the upper ear shoots are much larger whereas the lowest ear primordia may have changed very little since stalk elongation commenced. The general result is as if the development of each ear primordium contributes to the inhibition of all ear primordia below it on the stalk so that at anthesis the topmost ear is largest and those below are progressively smaller. The size of the ears at anthesis and the number that undergo appreciable enlargement is strongly influenced by the light and soil environment.

The number of ears at silking is related to the maximum number of kernels per ear of the genotype. In general, a cultivar with fewer kernels per ear will have more ears, other conditions being the same. The earliest maize specimens found indicate that each plant had a large number of small ears, but selection through the ages, until very recently, has been toward plants with fewer and larger ears. In the following discussion, total grain potential per plant will be used with the understanding that this may be a single ear or divided among two or more depending on the ear size potential of the particular variety. Varieties with a strong tendency to have more than one ear at low populations are termed 'prolific', but it is not a precise designation.

Grain yield potential at anthesis is a function of the prior growth of the plant. Thus plants grown under less competition have higher potential yields than those from dense plantings.

The potential yield at silking as defined here may be reduced immediately after pollination because the emergence of a silk while pollen is available does not ensure that a fertilized embryo will develop into a growing kernel. For example, there may be a hormonal mechanism preventing kernel development even when all of the elements needed are present and viable. It has been observed that even though fertilization is accomplished the fertilized embryos may never start development. As Daynard and Duncan (1969) have shown, fertilized kernels develop a black layer at the base of the nucellus whether the embryo actually develops or not. Thus the presence of a black layer at the base of an undeveloped kernel at maturity shows that it was fertilized. In all ears with undeveloped kernels at the tip that I have examined, pollinated but undeveloped kernels were present. Thus part of the drop in yield potential after pollination results from failure of development of these fertilized embryos.

If only one ear is present, the undeveloped embryos will be at the tip with rare exceptions. If more than one ear is present, the upper ear is usually well filled with most kernels developed. In an experiment with plants with two ears Bauman (1960) showed that removal of leaves, with consequent decrease in photosynthesis per plant, caused the second ear to have fewer developed kernels at harvest. The more leaves he removed the fewer developed kernels there were on the second ear but the upper ear maintained approximately the same number of kernels. It thus appears that the yield potential of a plant is satisfied, in order, first by kernels from the base to the tip in the dominant ear, and then by kernels at the base of the second ear, and so on until the yield potential is satisfied.

The fertilized but undeveloped kernels pose several other questions. What prevented the embryo from developing as in the other fertilized kernels? When was development arrested? For how long after pollination do these embryos retain their potential for further development? There are no experimental data to answer any of these questions but there are observations of later developing kernels on the upper ends of ears. This suggests that such kernels might lie dormant for a period of time and then start development under the influence of more favourable environmental conditions.

On many ears there are also very small kernels just above the normal kernels, that have obviously undergone some development and then been arrested. This suggests that there is some mechanism for inhibiting kernels even after development has started. This might occur if conditions deteriorated shortly after grain filling commenced. There are also cases where kernels on the upper half or more of the ear are shrivelled but this is usually explained by severe late drought or leaf destruction or nutrient deficiency, especially of potassium. There appears to be a definite priority in favour of the lowest kernels on the ear in almost every case where there are stresses of any kind.

In summary, the genetic yield potential of a maize plant is lost in a series of steps. Lower ears fail to develop, embryos on ears that do develop may not exsert silks in time for pollination, and pollinated kernels may not develop. Kernels that start growth may be aborted quite early or later if growth conditions deteriorate. The possibility that kernels might lie dormant for a period before commencing growth suggests that adjustment of yield potential may also be upward under some conditions.

Yield adjustment after pollination

Yield potential for a maize plant soon after pollination represents potential capacity that may or may not be translated into final grain yield. To do so requires a definite quantity of photosynthate. Production of a quantity of photosynthate requires a rate of photosynthesis operating over a period of time. The rate of photosynthesis of a maize canopy apparently tends to remain relatively constant over a considerable period of time in the absence of major foliage or climatic changes, as shown in work by Sayre (1948), Chandler (1960), and others.

The more uncertain element is the filling time or duration of active kernel development. Several studies have reported the number of days between silking and maturity, usually defined as the date the kernels attain their maximum dry weight. Shaw and Thom (1951) found that an interval of 51 days was required under Iowa conditions over a period of several years. In the same location Hallauer and Russell (1962) found a relatively constant 60-day interval for their material. Daynard *et al.* (1971), using black layer formation as an index of maturity, found differences of up to four days in effective filling period duration among three hybrids. In Kentucky, using the black layer index of maturity, I observed differences of as much as 15 days among 63 commercial hybrids in the time from silking to black-layer maturity. The consensus is that varieties differ in the period from silking to maturity but the lack of agreement among observers indicates a need for more study.

There is little published information on the difference in filling time for a single genotype among years or different planting dates in the same year. Peaslee *et al.* (1971) reported that there was no difference in filling time for the same variety planted at two different dates even though the average temperature during the filling period was greater for the second planting date. Similarly, Funnah (1971) found that for six varieties and three planting dates the filling period averaged six days longer for the latest planting, which experienced the warmest average temperature during the filling period.

These comparisons at single locations suggest that the length of the filling period is unaffected, or may even be lengthened, by higher temperatures but comparisons of data from different climatic regions tell a somewhat different story. Varieties judged to be of similar maturity based on degree days from planting to silking in Iowa and Florida had shorter filling periods in Florida (42 days) than in Iowa (62 days). In another comparison where the same variety was grown at latitudes 38 and 44 degrees north (Iowa and Kentucky) the filling periods were 1200 and 1400 degree days and 67 and 58 calendar days, respectively. Clearly this important varietal characteristic needs much more attention than it has received to date.

One possible determinant of the length of the filling period could be the relationship between photosynthetic rate and sink capacity. If kernels mature when they reach some final weight, the time required might vary with sink capacity relative to plant photosynthesis. Daynard (unpublished) investigated this possibility by using various means to reduce sink capacity without otherwise affecting the plants. He bagged ears so that only a few kernels were pollinated, he removed parts of young ears in various patterns, and he removed the upper ear of plants with two ears soon after pollination. He then observed the time of black-layer maturity for the treated plants as compared with similar untreated plants for the same plot, and found little or no difference in filling time, concluding that it was not closely related to the sink capacity.

The use of filling time, defined as the time elapsed between the appearance of the first silks and grain maturity, has the disadvantage that it includes two periods of slow development and uncertain duration. The first is the period after silking lasting until kernels are established and reach their full growth rate. This is followed by a period of growth that is almost linear under favourable conditions, and then by a slowing down period to final maturity, as shown in Figure 2.1. The duration of the period from silking to maturity is easily established by observation but most of the grain increase takes place during the period of linear growth. Thus in attempts to improve grain yield the interest is in increasing the duration of the linear phase.

Hatfield and Ragland (1967) and Daynard *et al.* (1971) have measured an interval called the Effective Filling Period Duration, which is the time it would have taken for ear development had it proceeded at the linear rate of growth from pollination. This is computed by dividing the final yield per plant by the linear rate of grain growth obtained by successive weighings of kernels or ears. It is more difficult to determine but may have more significance in the determination of grain yield.

It is convenient and approximately correct to think of maize grain yield as a product of the rate of photosynthesis during the grain filling period multiplied by the duration of the grain filling period, plus the change in labile reserves, with grain sink capacity as an upper limit. This is not quite correct because it evades the problem of the early and late periods of slow development. A question of major interest in any particular case is, which of these three parameters, rate, time, or kernel capacity, limited the final grain yield? Unfortunately there have been few attempts to answer it.

The photosynthate used for kernel growth may come from both current photosynthesis and from labile carbohydrates accumulated earlier in stalk, leaf sheaths, and possibly roots. During vegetative growth there is normally no accumulation of labile carbohydrates in the stalk, probably because of

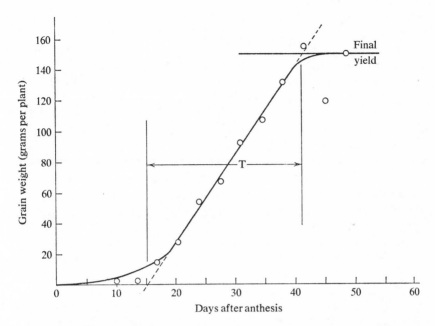

Figure 2.1. Grain growth in maize. T is the effective filling period. (Unpublished data of P. R. Goldsworthy, CIMMYT.)

the high rate of utilization of photosynthates by the growing plant. Refractometer readings of stalk sap by the author have rarely exceeded 5.0 % during active vegetative growth. Measurements by Campbell and Hume (1970) indicate that such refractometer readings, usually referred to as percent Brix, are a measure of various soluble solids, with sucrose from 31 to 39 % of the total. As vegetative growth ceases, however, there is a period before kernels start their rapid growth when little growth activity is apparent in the plant. There is no corresponding reduction in photosynthesis, however, as indicated by the uninterrupted linear dry weight increase of the plants, so it seems reasonable to assume that the carbohydrates resulting from photosynthesis are accumulating in storage tissues in the plant. This assumption is supported by work of Campbell and Hume (1970), whose data show that soluble solids in dried corn stalks can weigh more than the insoluble matter. Hume and Campbell (1972) have shown that soluble solids lost from the stalk can equal 20 % of the weight of grain ultimately produced, under their climatic conditions.

When I removed leaves from maize plants just before the dent stage and left the ears attached, stalk Brix readings of the lower internodes decreased from 5.0 to 1.0 % within nine days. This indicated that labile material from the stalk was being utilized by the still developing ears, since similarly treated plants with ears removed showed no measurable decrease. This was also shown in an experiment by Duncan *et al.* (1965) in which ears continued to increase in weight after all leaves were removed and the entire plant was wrapped in foil. Daynard *et al.* (1969) also reported large increases in grain weight after leaves were killed by early frost, presumably due to transfer of labile carbohydrates from stalks and sheaths.

Clearly, transfer of carbohydrates can take place from stalks and leaf sheaths to ears under extreme conditions. Hume and Campbell (1972) have shown transfer from stalks to be a normal occurrence under their conditions. It seems a reasonable assumption that sheaths also function as reservoirs of soluble materials that are augmented when photosynthesis exceeds utilization, and depleted when the reverse conditions prevail.

If this is indeed the case, the level of soluble materials accumulated in the stalk sap might furnish useful information about the yield limiting factor in a particular crop. If sink size is limiting, soluble materials should accumulate during the period of active ear filling as photosynthesis exceeds utilization. Conversely, where photosynthesis is limiting, solubles should decline as utilization exceeds supply.

Figure 2.2 illustrates the relative dry weights of the plant parts as well as their developmental timing. The increase in weight of the stalk after anthesis and the subsequent decrease with the approach of maturity show the storage and utilization of soluble solids.

Within a field there is usually considerable variation among plants in Brix readings of stalk sap. Among plants selected at random from an experiment in Kentucky, stalk sap Brix readings from lower internodes ran from 3.0 to 11.0 %, probably indicating that some plants had inadequate sink capacity, others inadequate photosynthesis. Thus stalk Brix readings could furnish diagnostic indications as well as point to reasons for yield limitations of varieties and locations.

Any dissolved carbohydrates remaining in maize stalks at grain maturity represent energy fixed by photosynthesis but not converted into grain, and hence potential grain yield not realized. This simple picture is complicated to a degree by the fact that resistance to stalk rots, serious diseases of maize, is positively correlated with sugar content of stalk and roots (Mortimore and Ward, 1964). Thus selection of varieties for resistance to stalk rots operates against complete utilization of photosynthate for grain. Incomplete utilization of photosynthate for grain would also occur in varieties whose grain matures before leaf senescence.

The maize canopy

Plant canopies intercept light with varying degrees of efficiency associated chiefly with their leaf area index (LAI), with the aspect of their leaves in

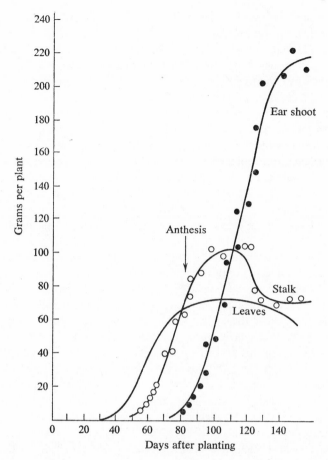

Figure 2.2. Dry weight change in irrigated maize plants in Ohio. (Unpublished data of R. B. Curry.)

relation to the sun and, to a lesser extent, with the spatial arrangement of their leaves, properties collectively referred to as canopy architecture. Other factors influencing interception are the reflectivity of leaf surfaces and soil which together with the canopy architecture determine the albedo of the crop surface. The efficiency of interception of incident light, combined with the efficiency of the photochemical reactions of the leaves, determines the efficiency of the canopy in utilizing radiant energy per unit of land area. Several characteristics of maize contribute to highly efficient foliage canopies.

The leaf area produced on the main stem of maize plants does not decrease in inverse proportion to an increase in plant density. Therefore, when there is little tendency to form tillers, as is the case with most improved varieties, the LAI of a maize canopy can be controlled within wide limits by the density of planting. In this, maize differs from many other crop plants. Loomis *et al.* (1968) obtained mature maize canopies from LAI 3.5 to 8.5 with planting densities from 17500 to 125000 plants per hectare. Donald (1968) cites maize as close to ideal as a plant for building efficient canopies.

Maize leaves have the highly efficient C_4 pathway of photosynthesis and utilize intercepted radiant energy with high efficiency even under intense light. The photosynthetic characteristics of maize leaves under field conditions have been described by Hesketh and Musgrave (1962).

Maize plants have other characteristics that generate an efficient plant canopy. Leaves are well separated on the stalk, ensuring ventilation within the canopy and a minimum of close overlapping. Leaves are attached to the stalk at angles that give generally favourable light exposure and there is enough genetic variability in this to allow improvement by breeding techniques. It is also possible to orient leaves directionally by appropriate placing of the seed.

There are two characteristics of maize plants, however, that reduce the potential efficiency of maize canopies to a degree. The most serious is the growth habit of many leaves attached to a single stalk. Maximum efficiency requires many vertical leaves (Duncan, 1971) but this is not possible in maize because clumping against the stalk would cause excessive mutual shading. There is some work with maize mutants having near vertical leaves, called liguleless, but early reports do not appear promising probably because of excessive mutual shading and the absence of more horizontal lower leaves. Another characteristic impairing efficiency of maize canopies is the presence of tassels at the top of the stalks. After anthesis these die and shade the plants below. This shading effect was estimated by Duncan *et al.* (1967), who calculated maximum shading of 19 % at a density of 100000 plants per hectare. Hunter *et al.* (1969) measured the effect of tassel shading on grain yield and found significant reductions.

Control of the maize canopy

The establishment of an efficient maize canopy requires attention to both planting rate and the proper distribution of plants over the surface. These are both affected by the genotype involved. Genotypes with less leaf area per plant require more plants per hectare; shorter plants require narrower rows than taller varieties for efficient light interception. The efficiency of a canopy design is affected by the intensity of solar radiation, latitude, and date of planting. Because of the complexity of the problem, as well as its importance in terms of yield, many experiments involving rates and patterns of planting have been reported. An intriguing aspect of the problem is that grain yield rises with planting rate to some maximum value and then declines. The rate that produces a maximum yield varies with genotype, environment, fertility,

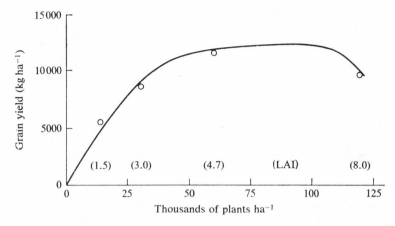

Figure 2.3. Relation between plant density, LAI, and grain yield for an irrigated crop of hybrid maize.

and planting pattern. Yoshida (1972) and Duncan (1973) think that the decline in yield after a maximum is reached is chiefly due to barren plants. Increase in grain yield with increase in planting rate normally ceases, however, before there are significant numbers of barren stalks. This yield plateau presumably occurs when light interception by the canopy is essentially complete so that little further increase in photosynthesis per unit area is possible.

Figure 2.3 illustrates the relation between plant density, LAI, and grain yield of irrigated maize under favourable cultural conditions. Under less favourable conditions of light, moisture, or fertility the yield decline associated with the onset of barrenness would occur at lower LAI values. If the yield decrease started at LAI values at or below those required for nearly complete light interception, the yield-plant density curve would have a sharp maximum rather than a plateau as shown.

The mathematical relationship between plant density and grain yield has been given in slightly different equations by Fery and Janick (1971) and by Duncan (1958). Both give functions that express the density-yield relationship in linear form so that grain yields for any density within the limits given can be estimated if yields for two densities are known.

PHYSIOLOGICAL SYNTHESIS

The work reviewed here has been essentially analytic in character. The maize plant has been reduced to parts or processes simple enough to study experimentally or to describe. Until recently there was little more that could be done. We have been like a watchmaker's apprentice, examining and admiring the parts of a fine watch but with no understanding of how to put them together to create a timepiece.

Within the last decade, however, high speed digital computers have become available and several crop physiologists are attempting to apply the power of such computers to the problems of physiological synthesis. It would leave this chapter on maize incomplete if I did not give brief mention to work now in progress with such synthetic models of maize.

Models for studying maize fall into two categories; those intended to show the operation of subsystems, and models to simulate the growth and development of whole maize plants through time. Duncan *et al.* (1967) developed a subsystem model to compute the effects of leaf area and angle in simulations of photosynthesis in a plant canopy. With it one can describe the positions and characteristics of leaves in space, and their radiation environment, and from this information compute an estimate of the rate of photosynthesis. By changing different elements one at a time one can ascertain the relative importance of each. De Wit (1965) has published a similar model, and Stewart and Lemon (1969) have constructed a more comprehensive model, SPAM, by adding other elements of the canopy environment such as windspeed, carbon dioxide, thermal radiation, and transpiration. DeMichele and Sharpe (1973) have developed an elegant mathematical model of the operation of stomata that contributes to our understanding of this maize subsystem.

De Wit and Brouwer (1968) have published a model of maize growth through the vegetative phase, called ELCROS. This is the first attempt to model the growth of a maize crop through time, and it is to be hoped they will continue their work to develop a model capable of simulating maize development from emergence to maturity.

The author has developed a simulation model of maize growth and development, called SIMAIZ, that others have used for various purposes. It is considerably less elaborate than some other models under development but it does combine much of what is known about the responses of maize to environmental influences in simulating development, growth and grain yield. The basic philosophy underlying any attempt to simulate plant growth is that plants respond to environmental factors according to rules simple enough to be stated in mathematical language. These rules change as the plants grow but in ways that can also be expressed mathematically.

Plant growth results from the influence of discontinuous and variable daily inputs into a system whose response is continually changing. The end result is extremely difficult to describe satisfactorily by any single equation. It is relatively easy, however, to estimate the effect of a day's inputs if a suitable description of the plant at the beginning of the day is available. Computer simulation is nothing more than making a series of such daily estimates and making suitable modifications in the plant being simulated. The simulation by computer is no better than could be obtained by using the same information and methods to calculate day-by-day growth by hand except that the results are obtained without computational errors and in seconds.

A useful simulation requires not only that the plant growth be described with reasonable accuracy but also that the results be arrived at by steps or processes that parallel those known or inferred to occur in the plant. In many cases the plant processes are not known in a quantitative sense, if at all. To develop a model, quantitative input–output relationships must be used for each process to be simulated regardless of how little we may know about the details or principles involved. For example, there is little information about the plant mechanism that initiates the change from vegetative to reproductive development in maize but the number of degree-days from planting to anthesis for the variety to be simulated may well be documented for its area of adaptation. If simulation is to be useful, it is essential that the date of anthesis be predicted with reasonable accuracy. In a model this can be done by computing degree-day accumulation without any understanding of the underlying process. If we knew the temperature–day-length relationship for floral induction it might be possible to improve the prediction for a wider range of latitude but we can build a useful model without doing so.

Unfortunately there are processes and controls in the growth process that little or nothing is known about. In attempting to build a model one discovers steps or controls that must be present but which have not been commented on at all in the physiological literature. For example, maize plants can survive periods of shading with no apparent sign of injury, indicating the presence of available reserves of some sort. These reserves must have accumulated in many cases in competition with other growth needs of the plant. By what process does the plant assign photosynthate to reserves when other demands exist and with what sort of priority?

Obviously it is possible to produce almost any desired result by manipulating several input–output relationships or controls. The usefulness of a simulator, however, depends on how accurately the assumed relationships imitate the real but unknown ones. The problem then becomes one of discovering the best approximation of unknown real functions and of proving that they are satisfactory.

If experimental data are available for a variety over a wide range of environmental conditions, one test for validity is given by the comparisons between simulated and experimental observations over the whole range of input data. Adequate experimental data are rare, however, so other means must be found in many cases although even limited amounts of good data can be extremely useful. Another approach is to vary input parameters such as rate or date of planting or time of irrigation to see if the simulator predicts the type of response expected. Experiments reported in the literature can be simulated, where sufficient information is available, to see if the results predicted agree with those actually found.

The real potential for accuracy in a simulator is that it can continue to evolve. Every comparison with experimental data presents a test of accuracy and an opportunity to improve the model. Every experimental discovery of quantitative physiological relationships within the plant system permits the improvement of accuracy.

In SIMAIZ, photosynthesis is considered as a process that produces a quantity of net photosynthate each day as determined by leaf area, temperature, and radiation. This photosynthate is divided among plant parts depending on their state of development, based on experimental observations found in the literature. The computer not only does the computations but keeps a record of the daily growth of all plant parts through time. The output includes a print-out of these weights for each day so that the simulated plant can be compared in detail with any type of observations made on the experimental plants grown under the conditions being simulated.

The initial information supplied includes the known or inferred characteristics of the particular variety being simulated, the moisture characteristics of the soil to rooting depth, and management details such as planting rate and date and information about irrigation policy if water is to be applied. Environmental information includes the daily weather parameters for the growing season comprised of radiation, maximum and minimum temperatures, rainfall, and pan evaporation. As the computation proceeds, moisture in the soil is modified by the computed transpiration losses and by rainfall received. Moisture tension in the soil is estimated and its effect on rate of photosynthesis and growth assessed. Each day's growth is added, and the plant is modified accordingly.

When time and temperature summations indicate that the variety should terminate vegetative development, growth of ears is speeded up and kernels are initiated. Their number depends on the growth of the plant to vegetative maturity and on conditions at silking. Since the actual rules regulating kernel establishment are entirely unknown, different hypotheses have been used and those giving the closest consistent approximation to experimental values have been retained. It is only when yield is limited by kernel numbers that this estimate affects yield.

After kernels have been initiated their subsequent growth, as in real plants, results from the transfer of photosynthate from leaves and from labile reserves. When the amount of photosynthate produced exceeds the capacity of the kernels to utilize it, the excess is stored in the stalk to be recovered and translocated to kernels when and if their potential utilization rate exceeds current photosynthesis. Maturity is assumed to terminate kernel growth at a time set by a temperature accumulation computed on a different basis from that used to estimate vegetative development.

Simulation models are built to serve different purposes and will differ greatly in detail and complexity, but they must meet certain general criteria.

They must truly mimic the physiological processes that are known to occur or that may be inferred in the growth and development of the plant simulated. Similarly control mechanisms that modify the physiological processes must be simulated in a reasonable manner. Feed-back mechanisms must be linked to plant processes, for example. This is necessary to meet another criterion, that the simulator must be able not only to predict end results but also to describe all intermediate steps and products in experimentally verifiable ways.

For example, one can predict maize yields by correlation methods if sufficient past yield and weather history is available, but this gives little information about why the yields varied. A simulation model should predict grain yields, given the same weather information, but in addition it should describe the state of plant at any date of the growing season. More importantly, one should be able to estimate the results of any changes in the variables involved. One could predict the consequences of earlier or later planting, or of irrigation at any time, or of the use of a variety with different characteristics. An important use of almost any simulator is to answer the question, 'what would result if . . .?'

Good crop simulators will greatly enlarge our understanding of the mechanisms controlling the development of maize crops. Experiments that take years in the field can be done in minutes in a computer. One can assess the values of ideotypes such as Donald (1968) has proposed by 'planting' them in the computer years before they could be produced by plant breeders. If the necessary weather records are available for use one can 'grow' any described variety of maize at any location in the world. With a simulator and historical weather records one can learn what would have resulted over past years from the use of new practices or varieties, thus accumulating valuable 'experience' without loss of time.

One of the important and often overlooked features of simulators is that they can evolve toward any desired degree of perfection. The limitations are imposed by our understanding of how plants function and by our ability to describe it to the computers, not by the capacity of the machines. Thus simulators can continue to improve as long as the needs justify.

REFERENCES

Aldrich, S. R. and Leng, E. R. (1966). Modern Corn Production. *The Farm Quarterly*, Cincinnati, Ohio, U.S.A.

Anderson, E. and Brown, W. L. (1948). A morphological analysis of row number in maize. *Missouri Bot. Gard. Ann.* **35**, 323–336.

Anderson, E. and Brown, W. L. (1964). Origin of corn belt maize and its genetic significance. In *Heterosis*, ed. J. W. Gowan. Hafner, New York.

Barghoorn, E. S., Wolfe, M. K. and Clisby, K. H. (1954). Fossil maize from the Valley of Mexico. *Botanical Museum Leaflets, Harvard University* **16**, 229.

Bauman, L. P. (1960). Relative yields of first (apical) and second ears of semi-prolific southern corn hybrids. *Agron. J.* **52**, 220–222.

Berger, Joseph (1962). *Maize production and the manuring of maize.* Centre d'etude de L'azote. Geneva, Switzerland.

Bonnett, O. T. (1940). Development of inflorescences of sweet corn. *J. Agr. Res.* **60**, 25–37.

Bonnett, O. T. (1948). Ear and tassel development in maize. *Mo. Bot. Gard. Ann.* **35**, 269–287.

Bonnett, O. T. (1953). Developmental morphology of the floral shoots of maize. *Bull. Univ. Ill. Agr. Expt. Sta.* **568**.

Brouwer, R., Kleinendorst, A. and Locher, J. Th. (1970). Growth response of maize plants to temperature. In *Plant response to climatic factors.* UNESCO, Paris, pp. 169–174.

Campbell, D. K. and Hume, D. J. (1970). Evolution of a rapid technique for measuring soluble solids in corn stalks. *Crop Sci.* **10**, 625–626.

Chandler, W. V. (1960). Nutrient uptake by corn in North Carolina. *Tech. Bull. of North Carolina Agr. Expt. Sta.* **143**.

Collins, G. N. (1914). A drought resisting adaptation in seedlings of Hopi maize. *J. Agric. Res.* **1**, 293–302.

Cuany, R. L., Swink, J. F. and Shafer, S. L. (1969). *Corn performance tests. Colorado State Univ. Expt. Sta. General Series* **904**.

Daynard, T. B. and Duncan, W. G. (1969). The black layer and grain maturity in corn. *Crop Sci.* **9**, 473–476.

Daynard, T. B., Tanner, J. W. and Duncan, W. G. (1971). Duration of the grain filling period and its relation to grain yield in corn (*Zea mays* L.). *Crop Sci.* **11**, 45–48.

Daynard, T. B., Tanner, J. W. and Hume, D. J. (1969). Contribution of stalk soluble carbohydrates to grain yield in corn (*Zea mays* L.). *Crop Sci.* **9**, 831–834.

DeMichele, D. W. and Sharpe, P. J. H. (1973). An analysis of the mechanisms of guard cell motion. *J. Theoret. Biol.* **41**, 77–96.

Dobben, W. H. van (1962). Influence of temperature and light conditions on dry-matter distribution, development rate, and yield in arable crops. *Neth. J. Agr. Sci.* **10** (5), 377–389.

Donald, C. M. (1968). The breeding of crop ideotypes. *Euphytica* **17**, 193–211.

Duncan, W. G. (1958). The relationship between plant population and yield. *Agron. J.* **50**, 82–84.

Duncan, W. G. (1971). Leaf angles, leaf area, and canopy photosynthesis. *Crop Sci.* **11**, 482–485.

Duncan, W. G. (1973). Plant spacing, density, orientation, and light relationships as related to different corn genotypes. *Proc. 27th Ann. Corn and Sorghum Res. Conf. ASTA*, 159–167.

Duncan, W. G., Davis, D. R. and Chapman, W. A. (1973). Developmental temperatures in corn. *Florida Soil and Crop Sci. Soc.* **32**, 59–62.

Duncan, W. G., Hatfield, A. L. and Ragland, J. L. (1965). The growth and yield of corn. II. Daily growth of corn kernels. *Agron. J.* **57**, 221–223.

Duncan, W. G. and Hesketh, J. D. (1968). Net photosynthetic rates, relative leaf growth rates and leaf numbers of 22 races of maize grown at eight temperatures. *Crop Sci.* **8**, 670–674.

Duncan, W. G., Loomis, R. S., Williams, W. A. and Hanau, R. (1967). A model for simulating photosynthesis in plant communities. *Hilgardia* **38**, 181–205.

Duncan, W. G., Shaver, D. N. and Williams, W. A. (1973). Insolation and temperature effects on maize growth and yields. *Crop Sci.* **13**, 187–190.

Duncan, W. G., Williams, W. A. and Loomis, R. S. (1967). Tassels and the productivity of maize. *Crop Sci.* **7**, 37–39.

Dungan, G. H. (1931). An indication that corn tillers may nourish the main stalks under some conditions. *Agron. J.* **23**, 662–669.

Fery, F. L. and Janick, J. (1971). Response of corn (*Zea mays* L.) to population pressure. *Crop Sci.* **11**, 220–224.

Finan, J. J. (1950). *Maize in the Great Herbals.* Chronica Botanica Co. Waltham, Mass. USA.

Francis, C. A., Sarria, V. D., Harpstead, D. D. and Cassalott, D. C. (1970). Identification of photoperiod insensitive strains of maize (*Zea mays* L.). *Crop Sci.* **10**, 465–468.

Funnah, S. M. (1971). Association of grain yield in corn (*Zea mays* L.) with duration of actual grain filling period and degree day accumulation. Thesis, University of Florida.

Galinat, W. C. (1971). The origin of maize. *Ann. Rev. Genet.* **5**, 447–478.

Grant, U. J., Hatheway, W. H., Timothy, D. H. Cassalott, D. C. and Roberts, L. M. (1963). *Races of Maize in Venezuela. Publ. Nat. Res. Council, Nat. Acad. Sci., Washington*, No. 1136.

Grobman, A. W., Salhuano, W. and Sevilla, R. in collaboration with Mangelsdorf, P. C. (1962). *Races of Maize in Peru. Publ. Nat. Res. Council, Nat. Acad. Sci., Washington*, No. 915.

Hallauer, A. R. and Russell, W. A. (1962). Estimates of maturity and its inheritance in maize. *Crop Sci.* **2**, 289–294.

Hatfield, A. L. and Ragland, J. L. (1967). New concepts in corn growth. *Plant Food Review* **12**(1), 2.

Hesketh, J. D., Chase, S. S. and Nanda, D. K. (1969). Environmental and genetic modification of leaf numbers in maize, sorghum, and Hungarian millet. *Crop Sci.* **9**, 460–463.

Hesketh, J. D. and Musgrave, R. B. (1962). Photosynthesis under field conditions. IV. Light studies with individual corn leaves. *Crop Sci.* **2**, 311–315.

Hozumi, K., Koyami, H. and Kira, T. (1955). Interspecific competition among higher plants. IV. A preliminary account of the interaction between adjacent individuals. *J. Inst. Polytechnics, Osaka City University Ser. D* **6**, 121–130.

Hume, D. J. and Campbell, D. K. (1972). Accumulation and translocation of soluble solids in corn stalks. *Can. J. Plant Sci.* **52**, 363–368.

Hunter, R. B., Daynard, T. B., Tanner, J. W., Curtis, J. D. and Kannenberg, L. W. (1969). The effect of tassel removal on grain yield of corn (*Zea mays* L.). *Crop Sci.* **9**, 405–406.

Kiesselbach, T. A. (1949). The Structure and Reproduction of Corn. *Res. Bull. Univ. Nebr. Agr. Expt. Sta.* **161**.

Loomis, R. S., Williams, W. A., Duncan, W. G., Dovrat, A. and Nunez, A. F. (1968). Quantitative description of foliage display and light absorption in field communities of corn plants. *Crop Sci.* **8**, 352–356.

Mangelsdorf, P. C. (1965). The evolution of maize. In *Essays on Crop Plant Evolution*, ed. J. Hutchinson. Cambridge Univ. Press, Cambridge, pp. 23–49.

Mangelsdorf, P. C., MacNeish, R. S. and Galinat, W. C. (1964). Domestication of corn. *Science* **143**, 538–545.

Millerd, A. and McWilliam, J. R. (1968). Studies on a maize mutant sensitive to low temperature. I. Influence of light on the production of chloroplast pigments. *Plant Physiol.* **43**, 1967–1972.

Mortimore, C. G. and Ward, G. M. (1964). Root and stalk rot of corn in southwestern Ontario. III. Sugar levels as a measure of plant vigor and resistance. *Can. J. Plant Sci.* **44**, 451–457.

Neuffer, M. G., Loring Jones and Zuber, M. S. (1968). *The Mutants of Maize.* Crop Sci. Soc. of America, Madison, Wis. USA.

Ohlrogge, A. J. (1958). How roots tap a fertilizer band. *Plant Food Review* **4**, (2 & 3) 4–6.

Peaslee, D. E., Ragland, J. L. and Duncan, W. G. (1971). Grain filling period of corn as influenced by phosphorus, potassium, and the time of planting. *Agron. J.* **63**, 561–563.

Peters, D. B., Pendleton, J. W., Hageman, R. H. and Brown, C. M. (1971). Effect of night air temperature on grain yield of corn, wheat, and soybeans. *Agr. J.* **63**, 809.

Reeves, R. G. (1950). Morphology of the ear and tassel of maize. *Amer. J. Bot.* **37**, 697–704.

Sass, J. E. (1951). Comparative leaf number in the embryos of some types of maize. *Iowa State College J. Sci.* **25**, 509–512.

Sayre, J. D. (1948). Mineral accumulation in corn. *Plant Physiol.* **23**, 267–281.

Sehgal, S. M. (1963). *Effects of teosinte and tripsacum introgression in maize.* The Bussey Institution of Harvard University.

Shaw, R. H. and Thom, H. C. S. (1951). On the phenology of field corn. *Agron. J.* **43**, 541–546.

Siemer, E. G. (1964). Major developmental events in maize – their timing, correlation, and mature plant expression. Ph.D. Thesis, Univ. Ill.

Stewart, D. W. and Lemon, E. R. (1969). *The energy budget at the earth's surface: a simulation of net photosynthesis of field corn.* Tech. Rept. ECOM 2-68, I-6, US Army Electronics Command, Arizona.

Weatherwax, P. (1954). *Indian Corn in Old America.* Macmillan Co., N.Y.

Weatherwax, P. and Randolph, L. F. (1955). History and origin of corn. In *Corn and Corn Improvement*, ed. G. F. Sprague. Academic Press, N.Y.

Wit, C. T. de (1965). Photosynthesis of Leaf Canopies. *Versl. Landbouwk. Onderz. Nederl.* **663**.

Wit, C. T. de and Brouwer, R. (1968). Über ein dynamisches Modell des vegetativen Wachstums von Pflanzenbeständen. *Zeitschr. Angew. Bot.* **42**, 1–12.

Yoshida, S. (1972). Physiological aspects of crop yield. *Ann. Rev. Plant Physiol.* **23**, 437–464.

3

Sugar cane

T. A. BULL AND K. T. GLASZIOU

The original sugar cane (*Saccharum officinarum* L. 2*n* = 80) probably evolved in New Guinea from strains of the wild species *S. spontaneum* L. (Brandes, 1958). *S. spontaneum* (2*n* = 40 to 128; Panje and Babu, 1960) is indigenous to India. Penetration into South East Asia, Melanesia, the Middle East and parts of Africa has been accompanied by polyploidization and hybridization. Some *S. spontaneum* strains became isolated in New Guinea where altered selection pressures and introgression from related genera including *Miscanthus* are thought to have given rise to a new species, *S. robustum* 'Brandes et Jesuit ex Grassl.', and eventually to *S. officinarum*. Selection has involved a definite pressure for increased sugar storage which may have been associated with tiller survival (Bull and Glasziou, 1963) or due largely to selection as chewing canes by village communities (Brandes, 1958).

Trading and local wars caused the highly prized clones of *S. officinarum* to become dispersed throughout Polynesia and South East Asia. Satellite centres of diversity developed along the paths of migration, and natural hybridization with *S. spontaneum* in Northern India resulted in a further species, *S. sinense* 'Roxb.', which was widely grown until recently in both Northern India and China (Parthasarathy, 1947).

Up to the end of the nineteenth century, only a few clones of *S. officinarum* had been used to establish the major portion of the world sugar cane industry. Genetic variability was increased to some extent by importing new varieties and breeding within *S. officinarum*. As the industry expanded most regions were increasingly troubled with diseases and pests. Early this century the industry was boosted by the discovery of natural resistant hybrids between *S. spontaneum* and *S. officinarum* in Java. Further controlled hybridization, initially in Java and then mainly in India, led to varieties which quickly replaced *S. officinarum*, and are still important in the complex ancestry of modern sugar cane cultivars.

VEGETATIVE GROWTH

Propagation

The crop is produced from stalk cuttings called setts. Each node has an axillary bud, and a band of root primordia, and is capable of giving rise to a

[51]

new plant. The germinating bud is initially dependent on the sett and sett roots for nutrients and water, but develops its own root system after about three weeks in favourable conditions. The shoot roots arise from underground nodes, and the axillary buds at these nodes give rise to tillers. As many as 144 stalks have been recorded in a stool arising from a one bud sett (Shamel, 1924). A detailed review of work on the development of the cane plant from germinating setts is given by van Dillewijn (1952).

Root growth and nutrient uptake

Sett roots supply the germinating bud with water until shoot roots are formed. It is doubtful whether sett roots contribute significant amounts of mineral nutrients. In water culture, radioactive inorganic phosphate in the bathing solution did not move readily from old roots to new shoots (Ann. Rep. H.S.P.A., 1961).

Experiments on root growth in long plastic pipes filled with either perlite–vermiculite or soil, and watered from the top, bottom or intermediate depths showed that root proliferation was most abundant wherever conditions of water availability and aeration were favourable. Sugar cane roots will penetrate downwards through soil of water potential less than -15 to -20 bars, if the main root mass is in a well-watered zone at the surface. Water may also be conducted to the leaves from a well-watered root mass by relatively few main roots traversing a 2–3 metre long column of dry soil of water potential less than -20 bars (Bull, Farquhar, Waldron, unpublished results).

In controlled experiments, root growth in soils was affected by temperature, the volume of soil available and the geometry of the available soil volume. For the same soil volume, root dry matter production was greater in shallow containers. During the first six months of growth over a temperature range of 18 to 30 °C, cultivar 'Pindar' showed little difference in total dry matter production for the whole plant, but the shoot:root ratio increased about nine-fold as the temperature increased because of reduced root growth at the high temperature. Restriction of water supply decreased the shoot:root ratio (Bull, unpublished data) as did increasing the photoperiod (Whiteman *et al.* 1963). However, it seems likely that considerable varietal differences in responses of root growth to environmental factors would be generated by vastly different selection pressures prevailing at the numerous cane breeding and selection stations. For example, root growth in Hawaiian varieties was adversely affected by cooling the roots (Mongelard and Mimura, 1972).

Tillering, canopy development and crop growth

Growth analysis of sugar cane crops in South Africa (Gosnell, 1968), Guyana (McLean *et al.* 1968), Hawaii (Borden, 1946) and Australia (Bull, unpublished) have revealed similar growth patterns, but with some differences

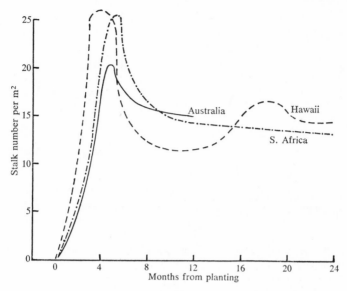

Figure 3.1. Changes in stalk numbers during the growth of sugar cane in Hawaii (Nickell, 1967), South Africa (Gosnell, 1968), and Australia (Bull, unpublished).

arising from the use of one-year to two-year cropping cycles. During the closing-in stages of crop growth, there is an overproduction of stalks (Figure 3.1). Peak numbers are attained about three to five months from planting but up to 50 % of these die and a stable stalk population is achieved before nine months. Further stalk production occurs at 15 to 17 months in Hawaii because variety selection has favoured continued crop growth in the second year. Although this pattern of growth appears inefficient, it has probably been indirectly selected because it provides rapid early growth which quickly achieves a full canopy, competes with weeds, and provides an insurance against adverse conditions.

Leaf area index is not much affected by the rather drastic loss of stalks (Figure 3.2). In general, maximum leaf area index is achieved about six months from planting and then slowly declines, but this may be affected by both variety and conditions of growth.

Phytotron studies show that leaf area expansion rates in sugar cane are more closely related to air temperature than to solar radiation during the first few months of growth (Figure 3.3a). Subsequently leaf area expansion is more responsive to solar radiation but expansion at low temperatures (16 °C) is too slow to be greatly affected. The prolonged juvenile growth phase which occurs at low temperatures results in higher relative leaf area expansion rates during periods of high radiation receipt (Figure 3.3b). However, lower solar radiation inputs in autumn greatly reduce the effect of

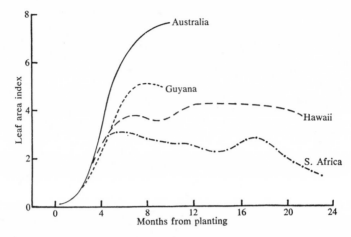

Figure 3.2. Changes in leaf area index during the growth of sugar cane in Hawaii (Borden, 1946), Guyana (McLean *et al.* 1968), South Africa (Gosnell, 1968) and Australia (Bull, unpublished).

temperature on relative expansion rates. Field measurements of relative expansion rates failed to detect the marked response to daily fluctuations in maximum temperature reported for temperate crops (Bull, 1968).

Total dry matter produced by sugar cane may average 40 g m^{-2} day^{-1} and exceed 150 t ha^{-1} year^{-1} under exceptional circumstances. These values were calculated from recorded fresh weight yields of about 250 t cane stalks ha^{-1} year^{-1} (Ann. Rep. H.S.P.A., 1967; Ham, 1970) by assuming that the stalks account for about 50 % of the total production (van Dillewijn, 1952). In general the ability of sugar cane to achieve high yields can be traced to its extended growing season. Comparison of average sugar cane crops with a temperate zone crop, kale, shows that until the kale reached maturity its yield was not greatly different from that of sugar cane (Figure 3.4). This is reflected in the crop growth rates which reach a similar maximum in both kale and cane, but do not decline as rapidly in cane (Figure 3.5).

Light penetration

Most of the light in a closed-in canopy is intercepted by the top six fully expanded leaves in cultivar 'Pindar', which is a fairly typical variety (Waldron *et al.* 1967). The older leaves receiving low light intensity tend to senesce. The number of green leaves on a stalk varies between about 6 and 12, with fewer leaves being maintained during dry or cold conditions. The top three fully opened leaves of nearly all cultivars tend to be erect, but droop more or less at the tips. The older leaves may be erect or droopy (planophile) depending on variety and environmental conditions. Under controlled environment,

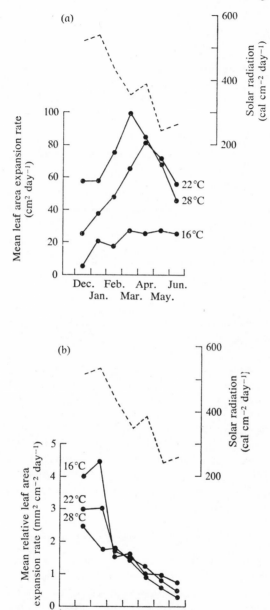

Figure 3.3 The effect of temperature on (*a*) mean leaf area expansion rates and, (*b*) mean relative leaf area expansion rates in sugar cane in Australia (Bull, unpublished).

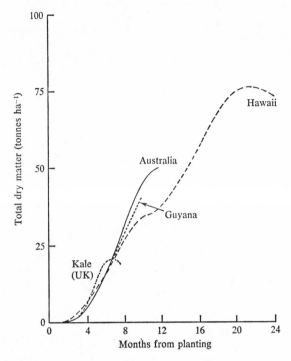

Figure 3.4. Total dry matter production per unit area in sugar cane in Hawaii (Borden, 1946), Guyana (McLean *et al.* 1968), and Australia (Bull, unpublished) compared with kale in U.K. (Goodman, 1968).

on high water and nutrient regimes and simulating normal field population density of stalks, no significant differences were observed in total dry matter production between erect and droopy leaf commercial varieties (Bull, unpublished). However, an experiment in which growth was compared at several planting densities indicated that erect leaved varieties gave significantly increased yields at high stalk densities.

Transpiration and water use

Controlled experiments in Hawaii and Queensland have shown that the linear relationship between total dry matter production and water consumption (7 to 9 g l⁻¹) is not greatly affected by temperature or the stage of growth of the plant (Mongelard and Mimura, 1971; Bull, unpublished). Maximum water use efficiency for sugar cane in Hawaii occurs with an effective pan ratio of 0.8 (Chang *et al.* 1963).

Water use by irrigated cane is reported to be from 200 to 240 cm year⁻¹ in Hawaii (Chang, 1961) and 200 cm year⁻¹ in Queensland (Ham, 1970). The ratio of evapotranspiration to class A pan evaporation reached a

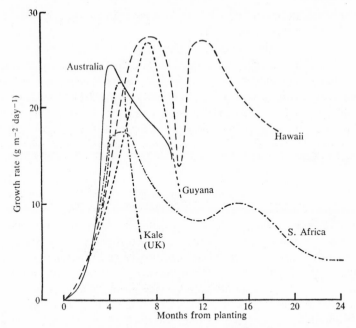

Figure 3.5. Growth rates in sugar cane and kale as a function of time. See Figure 3.4 for sources.

maximum value of 1.2 at both locations. This high ratio is associated with the increased roughness of a tall cane crop (Chang, 1961).

A large proportion of the cane crop is not irrigated and is usually subject to alternating wet and dry seasons under tropical and sub-tropical conditions. Provided prolonged periods of water stress do not occur during the wet season the potential yield is roughly 2.4 t cane stalks ha^{-1} per 2.5 cm of evapotranspiration. It is thus possible to use evaporation and water budget figures to estimate potential yield and so guide research into factors limiting yields.

Control of the rate of photosynthesis and transpiration under mild water stress situations resides mainly in the boundary layer and stomatal resistance (Bull, 1969). However, drought resistance is a much more complex phenomenon and many mechanisms exist whereby cultivars may avoid or cope with the consequences of drought periods. Commonly, during the initial stages of a water stress period, the bulliform cells of the upper leaves lose turgor and the lamina curls inward to reduce greatly the exposed leaf area. The older leaves are more resistant to curling and the effective light intercepting surface is thus lower in the canopy where turbulence is reduced. Consequently canopy water use is also reduced whilst an effective but lower level of photosynthesis is maintained. If stress continues, water loss is further reduced by leaf senescence although in many varieties the transpiration rate per unit of green leaf area remains unaltered.

Some varieties maintain tight control over stomatal resistance during drought to ensure virtually no water loss and the retention of a green leaf canopy (Naidu and Bhagyalakshmi, 1967). In other varieties the leaf canopy rapidly collapses and becomes senescent early in the drought, but is capable of rapid regeneration when the drought breaks. If the droughts are severe during the growing season the rapidity of resumption of crop growth becomes of major importance, and is not necessarily associated with retention of a green leaf canopy.

Respiration, photosynthesis and photorespiration

Dark respiration of young sugar cane plants about two metres tall accounts for about 20 % of daily photosynthesis (Waldron *et al.* 1967). Stalk respiration increases in the absence of carbon dioxide by a factor of about three compared with normal atmospheric carbon dioxide levels (Bull, unpublished). Data from steady state gas flow studies are given by Gayler and Glasziou (1972), and range up to about 20 μg CO_2 g^{-1} h^{-1} for stalks of intact plants, which amounts to about 15 % of the total dark respiration for a large stalk of about 2 kg. The respiration rates for stalks had no apparent relationship to their hexose content.

Dark respiration is highly temperature dependent (Waldron, 1966). It is also affected by the preceding photoperiod, being increased following days of intensive photosynthetic activity (Bull, unpublished).

The products of short term photosynthesis using $^{14}CO_2$ were first studied at the Hawaiian Sugar Planters' Experiment Station in the late 1940s. Malate was found to be the principal labelled compound, and not 3-phosphoglyceric acid as reported at that time by Calvin and co-workers for algae and some other plants. In pulse-chase experiments, label from malate moved into 3-phosphoglyceric acid, the time course being that expected for precursor-product relations. Early results are briefly reviewed by Burr *et al.* (1957). It took some time for this work to gain general acceptance. Recently the sugar cane pathway (C_4 or dicarboxylic acid pathway) of photosynthesis has been found to be widespread in the plant kingdom. Besides specific biochemical characteristics (Kortschak *et al.* 1965; Hatch and Slack, 1966) certain anatomical and physiological characteristics appeared to be uniquely associated with the pathway, at least while extreme types were under study. This work is reviewed by Hatch and Slack (1971).

Sugar cane is one of the extreme types in that it can have extraordinarily high rates of photosynthesis, no stimulation of photosynthesis in an oxygen-free atmosphere, virtually zero carbon dioxide compensation point at high light intensity, and the ability to utilize light intensity up to full sunlight with only slight fall off in the curve relating photosynthetic rate to light intensity. Anatomically, sugar cane and other extreme dicarboxylic acid pathway plants have their chloroplasts located in a layer of mesophyll and an adjacent layer of bundle sheath cells which surrounds the vascular bundle. The mesophyll chloroplasts have well developed grana, and do not normally accumulate starch. The chloroplasts of the bundle sheath cells have either no or very much reduced grana, and accumulate starch soon after sunrise. These features are characteristic, and are different from Calvin cycle (C_3) plants, but there are some plants which are intermediate at least in some respects. Even in extreme types conditions can occur in which the uniqueness of any distinguishing character can be lost. For example, in cultured parenchyma tissue from sugar cane, the C_3 pathway was dominant (Kortschak and Nickell, 1970).

There are at least two views on the roles of the mesophyll and bundle sheath cells in the biochemistry of carbon fixation of C_4 plants. One view proposes a co-operative role between the two cell types in the synthesis of sugars (Hatch, 1971; Edwards and Black, 1971), while the second view assigns this role to the mesophyll, and considers that the bundle sheath chloroplasts act mainly as amyloplasts (Baldry *et al.* 1971; Laetsch and Kortschak, 1971; Coombs and Baldry, 1972). Because of the high concentration of microbodies (peroxisomes) and mitochondria in bundle sheath cells, it is also suggested that these are possible sites of photo-oxidation (Laetsch and Kortschak, 1971).

Canopy respiration appears to be inhibited in the light. Kong, quoted by Burr (1970), pre-labelled cane leaves with $^{14}CO_2$, swept the system free of unused $^{14}CO_2$, then pumped the air passing over the plant through tubes of sodium hydroxide. The label leaking from sugar cane leaves in the light either at 300 ppm or zero carbon dioxide was about 5 % of the dark respiration measured in the same way. For geranium, a C_3 species, more labelled carbon dioxide was released in light than in darkness. For sugar cane leaves at 40 °C, Burr (1970) found that dark respiration was 25 % of the maximum photosynthetic rate, and equal to the photosynthetic rate if the carbon dioxide level was reduced to 100 ppm, whereas the carbon dioxide compensation point at the light level being used was about 4 ppm. Enhancement of photosynthesis by reducing the oxygen concentration in the atmosphere is normal for C_3 but not C_4 plants. However, Bull (1969) has observed slightly enhanced carbon dioxide fixation in the light in the absence of oxygen in leaves of mature cane.

Ribulose diphosphate carboxylase appears to be a major regulator of photorespiration in the soybean, carbon dioxide fixation by the purified enzyme being competitively inhibited by oxygen (Ogren and Bowes, 1971). In the presence of oxygen, it is suggested the reaction products may be 3-phosphoglycerate and phosphoglycolate, the latter being a presumed intermediate in photorespiration. This viewpoint had been deduced many years earlier from quite different evidence and is discussed by Burr (1970), who also suggests that there may be several forms of ribulose diphosphate carboxylase. If so, product control of forms of the enzyme could explain observations of photorespiration in mature sugar cane (Bull, 1969; Irvine, 1970).

The adaptive function of the specialized C_4 pathway appears to be the ability to retain carbon dioxide entering the plant in daylight under virtually any conditions (Laetsch and Kortschak, 1971), an adaptation which could have advantages under many stress situations. For sugar cane, the ability to carry out very high rates of photosynthesis at high light intensities may be of secondary importance (Bull, 1971). Hawaiian workers (Tanimoto, personal communication) found that the first fully opened leaf of mature stalks had rates of photosynthesis well below the same leaf type from young plants. This observation was investigated in detail as being a possible feedback effect on photosynthesis as a result of sink limitations (Glasziou and Bull, 1971). No conclusive evidence was obtained to support the existence of direct end-product repression of photosynthesis, and in all cases investigated any reduction in photosynthesis was accompanied by increased stomatal resistance.

Bull (1971) showed that high photosynthetic rates were characteristic of young sugar cane plants grown under summer light conditions. Leaves of mature plants and leaves of young winter-grown plants gave maximum rates of photosynthesis similar to those of C_3 plants. The low rates were partially attributable to increased stomatal resistance in leaves that developed under low light regimes. Growth analysis showed that maximum rates of dry matter production occurred at a time of the year when maximum rates of photosynthesis per unit leaf area were similar to those of C_3 plants. The high crop growth rates attributed to sugar cane are at least partially due to an extended growing season, rather than to an inherently superior photosynthetic pathway. This conclusion would not necessarily apply to other C_4 plants, nor under all growth conditions.

Photosynthetic rates per unit leaf area showed more than a two-fold range between different sugar cane varieties when estimated from the fixation of $^{14}CO_2$ under field conditions (Irvine, 1967). However, when a wide range of phenotypes from *Saccharum* was grown under controlled environment conditions and steady state gas flow methods used to measure carbon dioxide exchange, only small differences in maximum photosynthetic rates per unit leaf area were observed (Bull, 1971). Total dry matter production in these phenotypes varied over a two-fold range after a five-month growth period and was correlated with the rate of leaf area production. Thus factors other than maximum photosynthetic capacity figure in determining yield potential.

REPRODUCTIVE DEVELOPMENT

Except for *S. edule*, which has an abortive inflorescence, all sugar cane species flower and produce fertile seed, as do most F_1 and more complex interspecies hybrids. Work on inflorescence structure is reviewed by van Dillewijn (1952), and the ontogeny of the inflorescence has been described by Moore (1971).

Inflorescence development is of little economic importance in sugar cane crops, and selection programs are biased against heavily flowering varieties. In two-year crops in Hawaii, chemical control of flowering is sometimes practised to reduce potential losses incurred due to the subsequent mortality of flowered stalks. On the other hand sugar cane breeders are vitally concerned in floral induction procedures so that they can synchronize flowering in various crosses.

Most *Saccharum* varieties will not flower on day-lengths longer than about 13 hours or shorter than about 12 hours, nor if given light in the middle of the dark period. There are exceptions, some of the *S. spontaneum* varieties behaving as long-day plants. Generally, a day-length of 12.5 hours and night temperatures between 20 and 25 °C will induce floral initiation if enough inductive cycles are given. To obtain floral initation on the minimum number of inductive cycles, perhaps about 10, very much more precise control is required. Extensive factorial experiments combining day-length and day and night temperature treatments are necessary to determine the optimum condition but this actual optimum varies considerably between varieties (Daniels *et al.* 1965). The expense of providing such an exact specification is scarcely warranted for most practical purposes.

Many varieties are particularly sensitive to water stress during the inductive period. Overhead irrigation is the method of choice to overcome this problem, provided that water-logging of the root system is avoided (Glasziou, unpublished data).

Under field conditions, the number of inductive cycles received by the plant decreases at higher latitudes because of the more rapid seasonal change in day-length, and flowering becomes more sensitive to environmental stresses such as water deficiency and cold nights. Spring initiation does not usually occur, which may be attributed to a number of variables including age of plant, lower night temperatures, harsher environments preceding the period of inductive day-lengths, and the possibility that, at least for some varieties, increasing day-length is less favourable than the diminishing day-lengths occurring during the autumn. Autumn flowering may be partially due to a promoting effect of shortening days on floral development and emergence rather than on induction and initiation. However, under controlled environment, free flowering varieties will initiate and develop a normal inflorescence regardless of whether increasing, decreasing or constant photoperiods are used (Daniels *et al.* 1965).

Provided sufficient inductive cycles are given, some interruptions by interposing non-inductive treatments can be tolerated. Apparently a quantitative amount of a final stimulus must be produced and accumulated before differentiation of a floral primodium will occur (Coleman, 1968).

Much of our present knowledge of environmental effects on flowering in cane was passed around by oral communication over a period of many years. Burr, Coleman, Davis and workers from the Sugarcane Breeding Institute, Coimbatore, India, contributed substantially to this fund of knowledge. Methods of co-ordinating flowering for breeding purposes by regulating post-inductive photoperiods have been published by Moore and Heinz (1971), and effects of photoperiod, light intensity and temperature have been described by James and Smith (1969), Paliatseas (1972) and others.

YIELD DETERMINING PROCESSES

Translocation

Using $^{14}CO_2$ as tracer, translocation of assimilates has been studied extensively at the Hawaiian Sugar Planters' Experiment Station, the most recent publication being by Hartt (1972). Similar work has been carried out in the authors' laboratory. Most of the results can be explained in terms of a pressure flow hypothesis, movement in the phloem being in the direction of the turgor pressure gradient within the conducting elements, and movement of water being in the direction of the water potential gradient. Major unknown factors which limit interpretation of data on conduction and distribution of assimilates are the mechanisms of loading and unloading of phloem. The sieve tubes and companion cells of major vascular bundles in the lamina and all bundles in the mid-rib are not in direct contact with the nearest chloroplast-containing cells in the bundle sheath, so that assimilate must traverse at least one layer of more or less lignified cells in order to move from sites of photosynthesis to conducting cells.

It seems likely that there would need to be a metabolically-mediated energy-utilizing mechanism to concentrate sugar into phloem in the leaves. The rate of loading is probably enhanced in the light. Unloading of phloem could be by diffusion in the direction of a concentration gradient, but it is more likely to be regulated, reversible loading and unloading in the non-photosynthetic tissues.

Hatch and Glasziou (1964) used [^{14}C]sucrose, labelled only in the fructose moiety, together with mixtures of labelled and unlabelled sugars to demonstrate that sucrose moved into and along the phloem without breakdown. This work poses a problem of how the fructose-labelled sucrose actually entered the phloem without randomization of the label as happens when it is transported against an apparent gradient across storage parenchyma cell membranes. The anatomy of vascular tissue in sugar cane is such that phloem loading mechanisms are intractable to a concise experimental approach.

A quantitative treatment of the partitioning of assimilates and mobilization of reserves is also intractable to an experimental approach because flow rates in phloem, xylem, storage and other tissues, and turnover rates of metabolic pools cannot be measured precisely, or in some cases even approximately.

Storage processes

Sucrose levels reach values of about 55 % of the total stalk dry matter in the hybrid variety 'Pindar' (Waldron, 1966), and can exceed 70 % in *S. officinarum* (Bull and Glasziou, 1963). The sugar in the stalk is capable of being mobilized and transported elsewhere in the plant to support growth requirements in excess of that provided by current photosynthesis. Under fairly extreme conditions, the sugar content in basal internodes fell from about 16 % fresh weight to 6.5 % fresh weight in 35 days when the plant was forced into high growth rates by a sharp temperature increase (Glasziou *et al.* 1965). The mechanism by which sugar is mobilized and transported from the base to the top of the plant is unknown. Hawker (1965) found sucrose in xylem and suggested xylem flow may be one pathway. From measurements of transpiration rates and the xylem sugar concentration, it can be calculated that most of the stalk sugar would move to the leaves in a single day. Sampling of individual xylem elements showed wide variation in sugar concentration and it was concluded that those xylem elements conducting water to the transpiring leaves were not likely to contain much sugar (Bull *et al.* 1972). This conclusion is supported by measurements of water and sugar flux across xylem using isotopic tracers.

Conditions which favour sugar storage in stalks include maturity of the plant and a check to internode elongation rates in the stalk by cold, nutritional, or water stress, provided rates of photosynthesis are maintained at a sufficient rate. However, actual sugar storage has to be differentiated from dehydration effects under water stress conditions. Spurious results may be obtained because the loss of water causes a shift in the basis of reference. The problem may be overcome by expressing results per unit length of cane for non-growing portions of the stalk, but not by volume measurements because in some varieties the stalk diameter changes under water stress, and the intercellular air volume can also change. The maturation and growth check effects which give apparent ripening are partially due to a change in the ratio of immature to mature portions of the stalk, and do not necessarily mean the rate of sugar deposition per unit area of crop has increased. In fact, it may decrease during a period when sugar as percent weight of cane has increased sharply. The concept of ripening in the cane crop is largely derived from mill juice results, and can be quite misleading in terms of rate of sucrose storage in the crop or amount of sugar per unit area of crop.

Work on sugar movement and storage in cane has been reviewed recently (Glasziou and Gayler, 1972). The picture that has emerged is one of a dynamic flux of sugar to or from storage in the stalks as well as relatively slow turnover of sucrose but rapid turnover of reducing sugar pools within mature storage parenchyma cells. Conclusive evidence has accumulated showing that sucrose is hydrolysed either by extracellular or intercellular invertase prior to accumulation in the vacuole, and that hexose transport across the outer cell membrane is energy-coupled and does not appear to involve phosphorylation of the hexoses. Sucrose phosphate, sucrose phosphate synthetase and a specific sucrose phosphatase are present in the cells. Sucrose synthetase has not been detected in the storage parenchyma cells. Tracer studies indicate that sucrose phosphate is released into storage as sucrose without breaking of the glucose–fructose link, whereas sucrose is broken down to the hexoses and re-synthesized during the storage process. We consider the evidence indicates that movement of sugar from the cytoplasm to the vacuole is an energy-coupled system capable of transport against a concentration gradient. Sucrose phosphate is the probable substrate.

Isolated plant vacuoles act as osmometers. Some workers in the field of ion and other solute fluxes across the vacuolar membrane favour the view that the membrane is differentially permeable to water and solutes, net transport being in the direction of activity or electrochemical potential gradients, and energy-coupled transport occurring only at the outer cell membrane. However, an explanation is needed as to how the vacuole of a highly vacuolated cell can be formed and maintained if this concept is true. The problem does not arise if there are energy-coupled carrier transport systems moving solutes into the vacuole.

Hormonal effects and source–sink relations

Hormonal effects on partitioning of photosynthate *in vivo* are poorly understood. Certainly applied gibberellic acid can stimulate internode elongation and, with a careful choice of conditions, stalk weight and sugar production can be increased (Robinson, Farquhar, Arvier, unpublished data; Ann. Rep. H.S.P.A., 1968, 1970). Bull (1964) obtained spectacular increases in dry matter production in young plants. These results appear to mean that in some conditions, rates of photosynthesis are limited by sink availability. However, no changes in photosynthesis of single leaves occurred when other leaves on the same stalk were shaded or removed (Bull, unpublished) so that any source–sink control of photosynthesis appears to have a slow response time.

Responses to auxin application may be obtained in the germinating sett (reviewed by van Dillewijn, 1952), but are difficult to obtain on intact or decapitated plants (Bull, unpublished). Kinetin also has interesting effects on sett germination. It will induce rapid germination, and changes in pigmentation and hair groups (Bull, 1969). Combinations of gibberellic acid and kinetin give rapid germination and early growth, but tend to give very thin stalks until the effects are lost. The treatments do not appear to have agronomic value.

Nitrogen metabolism

Low nitrogen supply during the maturation phase promotes cane ripening, but the mechanism is not adequately understood. No nitrate was found in the stalk tissue of Hawaiian varieties when ^{15}N was supplied to roots (Ann. Rep. H.S.P.A., 1963), and most of the label was in the amide fraction (Takahashi and Tanimoto, personal communication). Maretzki and De la Cruz (1967) found nitrate reductase in both roots and leaves of sugar cane. By sampling of xylem contents, Waldron (unpublished) found that most of the nitrogen assimilated by the roots is reduced to amide and amine forms prior to transport to the upper parts of the plant in mature cane under field conditions when fertilizer was applied at planting time. For potted plants supplied with high nutrient levels, nitrate nitrogen in xylem exceeded amino nitrogen by a factor of 10, but on low nitrogen, amide and amino nitrogen became the dominant forms. An interesting observation is that whereas all xylem vessels in the stalk contain organic nitrogen compounds, they are virtually undetectable in xylem of transpiring leaves, but are found in xylem of the developing leaves enclosed by the spindle. Specialized transfer cells are found in xylem and phloem parenchyma in many plants, including the complex nodes of monocotyledons (see review, Pate and Gunning, 1972). These specialized cells were not detected in the nodes or intercalary meristems of sugar cane stalks (Pate, personal communication). In sugar cane stalks the structure of the vascular bundles must change as they pass through the intercalary meristem region, since they tend to fall apart in that region in standard retting procedures used to study vascular anatomy. Also the intercalary meristem region, even of basal nodes, can expand from about 2 mm to more than 1 cm on one side of the stalk as a geotropic response. This observation appears to require that the vascular bundles in the intercalary meristem should be capable of undergoing extension growth and it seems to be the most likely zone in which solutes could be removed from the xylem.

Glutamine and asparagine are both found in the xylem, glutamine being the dominant amino-nitrogen compound in stalk xylem, but asparagine being dominant in root xylem (Waldron, unpublished).

Hardiness, adaptability and heterosis

Given optimal conditions the original cultivars of *S. officinarum* may yield as well as modern commercial hybrids. However, current sugar cane growing areas and agronomic methods are far from favourable for *S. officinarum*. Many areas are subject to pests, diseases, floods, droughts, frosts and other environmental stresses which severely restrict production in *S. officinarum* varieties. Modern hybrids, derived predominantly from *S. officinarum* × *S. spontaneum* parentage, are well adapted to these conditions and maintain economic production.

Other than introducing genetic variation for disease resistance, it is by no means certain what specific characters from the wild species contribute to the greater yield potential of hybrid commercial cultivars. Genetic analysis of hybrid populations is exceedingly difficult because the results of observations on many small plots of varieties having a wide range of growth habits has then to be extrapolated to how the varieties would perform in large fields with no intervarietal competition. The problem of energy budgets in small plots is discussed in the following section.

Roach (1968) has analysed a large F_1 population from *S. officinarum* × *S. spontaneum* crosses. Heterosis was observed in most crosses for early growth, stalk length, yield of cane per hectare, percent flowering, pollen production and percent reducing sugars. Yield of sucrose per hectare exceeded mid-parent value, but was below the *S. officinarum* parent. Stalk thickness, flowering time, percent total sugars and percent sucrose approximated mid-parent values. Fibre content and erectness of hybrids were below mid-parent values. The *S. officinarum* parents averaged about 14 t sucrose ha^{-1}, the *S. spontaenum* parents 1 t, and the F_1 population 9 t. Sucrose content in stalks of *S. officinarum* parents averaged 16.6.% of fresh weight, of *S. spontaneum* parents 4.1 %, and of F_1 hybrids 9.8 %.

The main problem facing the breeder is to try to maintain the advantage in the F_1 population in respect to cane yield whilst getting the sucrose content up to acceptable levels. Longer stalks and greater stooling ability relative to the *S. officinarum* parents are the main factors affecting yield of cane per hectare. In a further genetic analysis on a range of hybrid material, Brown *et al.* (1969) found negative correlations between sucrose percent dry weight and yield per plot, stalks per plot, and stalk length, illustrating the difficulty facing the plant breeder.

MODELLING AND GROWTH ANALYSIS

Limits and limitations to yield

Photosynthetic efficiency can be measured as grams carbohydrate (CH_2O) fixed per calorie of incident light in the photosynthetic region under conditions in which the only limiting factor is light. The highest photosynthetic efficiency recorded by Bull (1969) for single leaves was 2.4×10^{-5} g CH_2O per cal light in the photosynthetic waveband. Allowing an average photosynthetic radiation of 200 cal cm^{-2} day^{-1} in a wet tropical region, maximum dry matter production would be about 180 t ha^{-1} year^{-1}, or about 600–700 t wet weight.

Respiration losses during the night are about 15% of net photosynthesis in four-month-old cane (Waldron, 1966). Using an estimated respiration loss of 20% of gross photosynthesis for a one-year crop cycle, and half the photosynthate being utilized for leaf and root production, the expected stalk yield would be 240–280 t wet weight ha^{-1}.

Recorded yields of irrigated 12-month-old cane of 250 t cane stalks ha^{-1} (Ann. Rep. H.S.P.A., 1967; Ham, 1970) are quite remarkable. They are not likely to be exceeded since they represent a photosynthetic quantum efficiency of about 40% (allowing 8 quanta per molecule of CO_2 reduced) over the whole crop cycle. Few areas attain these high productivity levels, and in those areas where good agricultural practices are used and an adequate variety selection program is maintained, the most frequent limiting factor is too much or too little water at various times of the year.

Approaches to yield improvement

Any approach to yield improvement must have the aim of optimizing economic yield, which is unlikely to coincide with maximum yield. The optimal variety for the cane grower will seldom be the optimal variety for milling and refining of the product. Simulation modelling and systems analysis using computer techniques are capable of assisting to define goals, but only rarely will the ideal solution be immediately practical.

The time taken for an original seedling to get through a selection program, and become a significant component of the commercial crop is seldom less than 12 to 15 years. Few varieties released for commercial cropping last more than 10 years. Because of the rapidity of changes in the technology of cultivation, harvesting, transporting and milling of cane, attempts must be made to ensure that criteria imposed in variety selection programs will have validity under the probable economic and technical environment 10 to 20 years hence.

The ancestors of cultivars now being used under mechanized agriculture were originally cultivated with implements drawn by horse or buffalo. It is scarcely surprising that row spacing experiments show that optimum yields are obtained with row width corresponding approximately to the width of these animals, or that our farm machines tend to be analogues of the same animals.

There has been no recurrent breeding and selection program for different planting densities nor has the machinery been evolved to cope with some of the drastic possible changes required for cultivation and harvest. Adequate testing of alternative ideotypes will require much time and effort before pronouncements can be made about the optimum ideotypes for particular environments.

Data from Chang (1961) indicate that two-year Hawaiian varieties achieve full ground cover at about five months of age. Row spacing of Hawaiian crops is about 1.5 m. Data from the Annual Report of the South African Sugar Association Experiment Station (1964) show that full ground cover was attained in 14 weeks on 0.45 m row spacing, but had not been attained at 24 weeks with the more usual 1.4 m row spacing. Bull (unpublished data) obtained full ground cover in 12 weeks or less with 0.45 m row spacing of three varieties, but full cover was delayed by 4 weeks or more with 1.4 m row spacing. None of the varieties used was specifically selected for close row spacing, having been previously grown and selected on wide row spacing as were their parents. If water supply is not limiting during the early weeks of growth, there appears to be a gain to be made by attempting to get a full leaf canopy as early as possible in the crop cycle, particularly for short term crops. The advantage would be lost if quick close-in of the canopy was due to excessive tillering, combined with subsequent reduction of tiller number through competition.

Modelling

An outline of various approaches to improving agricultural productivity, including the role of simulation modelling, is given by Loomis *et al.* (1971). It is exceedingly difficult to lay down general guidelines for this type of research on the sugar cane crop, since there are so many potential limiting factors the importance of which varies enormously between areas. Insects and other pests and diseases, too much or too little water, frequent cyclones, labour availability and costs, organizational difficulties associated with co-ordinating cane supply and mill economics, marketing problems, and fluctu-ating prices must all be taken into account if we are to be realistic.

Much of the knowledge to commence a reasonable approach to modelling the sugar cane crop is already available, but some will need refining. As a first approach we have modelled the radiation budget over the year for experimental plots of varying size, and as a function of the height of the canopy surrounding the plots plus whether rows are oriented east–west or north–south (Tovey *et al.* 1973). The results are somewhat disturbing in relation to the plot sizes in current use for field experimentation. The effect of increments or decrements in total radiation received by a plot are to some extent reduced, and less commonly amplified, by the effects of advective energy. Despite good agreement between replicated small plots of a single variety, it can be forecast that the trial results will not necessarily extrapolate accurately to large plots. The geometry of absorption of advective energy within plots is now being studied, and should help in designing better procedures for variety selection work and for agronomic and genetic experiments.

The concept of randomized plot design introduced by statisticians is wrong unless each variety or treatment plot is sufficiently large to nullify energy budget differences between the experimental plot and the real crop. Since the expense of using sufficiently large plots to obtain valid results for crop plants as large as sugar cane is exorbitantly high, different trial designs are needed using a pre-sorting of treatments or varieties so that each plot is surrounded by a sufficient area (fetch) of plots having similar performance in respect to height, leaf canopy structure, and stalk density. Ideally the cane height, roughness parameter, and zero plane displacement of adjacent plots should be similar throughout the whole trial period. Attention to this problem will mean a big improvement in the information yield from genetic studies as well as from the more usual field experiments, and variety selection programs. The potential value of crop modelling is not confined to defining the ideal ideotype for a particular environment, and may indeed have as much or perhaps more value in helping to define how to get there.

REFERENCES

Ann. Rep. Hawaiian Sug. Planters' Assoc. Exp. Sta. (1961). Old roots lose activity after ratooning, p. 6.

Ann. Rep. Hawaiian Sug. Planters' Assoc. (1963). How fast does the cane plant take up nitrogen? A comparison of nitrate and ammonium, pp. 20–21.

Ann. Rep. Hawaiian Sug. Planters' Assoc. (1967). Transplanting-spacing trials, p. 14.

Ann. Rep. Hawaiian Sug. Planters' Assoc. (1968). Field trials with gibberellic acid, p. 39.

Ann. Rep. Hawaiian Sug. Planters' Assoc. (1970). Field trials with gibberellic acid, p. 42.

Baldry, C. W., Bucke, C. and Coombs, J. (1971). Progressive release of carboxylating enzymes during mechanical grinding of sugar cane leaves. *Planta* **97**, 310–319.

Borden, R. J. (1946). A search for guidance in the nitrogen fertilization of the sugarcane crop. *Hawaiian Planters' Record* **50**, 3, 4.

Brandes, E. W. (1958). In *Sugarcane*. USDA Handbook No. 122, Washington, pp. 307.

Brown, A. H. D., Daniels, J. and Latter, B. D. H. (1969). Quantitative genetics of sugarcane. II. Correlation analysis of continuous characters in relation to hybrid sugarcane breeding. *Theoret. and Appl. Genetics* **39**, 1–10.

Bull, T. A. (1964). The effects of temperature, variety and age on the response of *Saccharum* spp. to applied gibberellic acid. *Aust. J. Agric. Res.* **15**, 77–84.

Bull, T. A. (1968). Expansion of leaf area per plant in field bean (*Vicia faba* L.) as related to daily maximum temperature. *J. Appl. Ecol.* **5**, 61–68.

Bull, T. A. (1969). Photosynthetic efficiencies and photorespiration in Calvin cycle and C_4-dicarboxylic acid plants. *Crop. Sci.* **9**, 726–729.

Bull, T. A. (1969). Temperature effects on the development of hair groups and stalk coloration in *Saccharum* L. *Crop Sci.* **9**, 390–392.

Bull, T. A. (1971). The C_4 pathway related to growth rates in sugarcane. In *Photosynthesis and Photorespiration*, eds M. D. Hatch, C. B. Osmond and R. O. Slatyer. John Wiley, Inc., pp. 68–75.

Bull, T. A. and Glasziou, K. T. (1963). The evolutionary significance of sugar accumulation in *Saccharum*. *Aust. J. Biol. Sci.* **16**, 737–742.

Bull, T. A., Gayler, K. R. and Glasziou, K. T. (1972). Lateral movement of water and sugar across xylem in sugar cane stalks. *Plant Physiol.* **49**, 1007–1011.

Burr, G. O. (1970). *Photosynthesis via the PGA and malic acid pathways. Report of the Government Sugar Experimental Station, Taiwan* No. 6.

Burr, G. O., Hartt, C. E., Brodie, H. W., Tanimoto, T., Kortschak, H. P., Takahashi, D., Ashton, F. M. and Coleman, R. E. (1957). The sugarcane plant. *Ann. Rev. Plant Physiol.* **8**, 275–308.

Chang, Jen-Hui (1961). Microclimate of sugarcane. *Hawaiian Planters' Record* **56**, 195–223.

Chang, Jen-Hui, Cambell, R. B. and Robinson, F. E. (1963). On the relationship between water and sugarcane yield in Hawaii. *Agron J.* **55**, 450–453.

Coleman, R. E. (1968). Physiology of flowering in sugarcane. *Proc. Intern. Soc. Sugar Cane Technol.* **13**, 795–812.

Coombs, J. and Baldry, B. W. (1972). C-4 Pathway in *Pennisetum purpureum*. *Nat. New Biol.* **238**, 268–270.

Daniels, J., Glasziou, K. T. and Bull, T. A. (1965). Flowering in *Saccharum spontaneum*. *Proc. Intern. Soc. Sugar Cane Technol.* **12**, 1027–1032.

Dillewijn, C. van (1952). *Botany of Sugar Cane*. Chronica Botanica Co., Waltham, Mass.

Edwards, G. E. and Black, C. C. (1971). Photosynthesis in mesophyll cells and bundle sheath cells isolated from *Digitaria sanguinalis* L. leaves. In *Photosynthesis and Photorespiration*, eds M. D. Hatch, C. B. Osmond and R. O. Slatyer. John Wiley, Inc., pp. 153–168.

Gayler, K. R. and Glasziou, K. T. (1972). Physiological functions of acid and neutral invertase in growth and sugar storage in sugar cane. *Physiol. Plant.* **27**, 25–31.

Glasziou, K. T., Bull, T. A., Hatch, M. D. and Whiteman, P. C. (1965). Physiology of sugarcane. VII. Effects of temperature, photoperiod duration, and diurnal and seasonal temperature changes on growth and ripening. *Aust. J. Biol. Sci.* **18**, 53–66.

Glasziou, K. T. and Bull, T. A. (1971). Feedback control of photosynthesis in sugarcane. In *Photosynthesis and Photorespiration*, eds M. D. Hatch, C. B. Osmond and R. O. Slatyer. John Wiley, Inc., pp. 82–88.

Glasziou, K. T. and Gayler, K. R. (1972). Storage of sugars in stalks of sugarcane. *Botan. Rev.* **38**, 471–490.

Goodman, P. J. (1968). Physiological analysis of the effects of different soils on sugar beet crops in different years. *J. Appl. Ecol.* **5**, 339–357.

Gosnell, J. M. (1968). Some effects of increasing age on sugarcane growth. *Proc. Intern. Soc. Sugar Cane Technol.* **13**, 499–513.

Ham, G. J. (1970). Water requirements of sugar cane. *Report of Water Research Foundation of Australia* No. 32.

Hartt, C. E. (1972). Translocation of carbon 14 in sugarcane plants supplied with or deprived of phosphorus. *Plant Physiol.* **49**, 569–571.

Hatch, M. D. (1971). Mechanism and function of the C_4 pathway of photosynthesis. In *Photosynthesis and Photorespiration*, eds M. D. Hatch, C. B. Osmond and R. O. Slatyer. John Wiley, Inc., pp. 139–152.

Hatch, M. D. and Glasziou, K. T. (1964). Direct evidence for translocation of sucrose in sugarcane leaves and stems. *Plant Physiol.* **39**, 180–184.

Hatch, M. D. and Slack, C. R. (1966). Photosynthesis by sugarcane leaves. A new carboxylation reaction and the pathway of sugar formation. *Biochem. J.* **101**, 103–111.

Hatch, M. D. and Slack, C. R. (1971). The C_4 dicarboxylic acid pathway of photosynthesis. In *Progress in Phytochemistry*, **2**, 35–106, eds L. Reinhold and Y. Liwschitz. Interscience, London.

Hawker, J. S. (1965). The sugar content of cell walls and intercellular spaces in sugarcane stems and its relation to sugar transport. *Aust. J. Biol. Sci.* **18**, 959–969.

Irvine, J. E. (1967). Photosynthesis in sugarcane varieties under field conditions. *Crop. Sci.* **7**, 297–304.

Irvine, J. E. (1970). Evidence for photorespiration in tropical grasses. *Physiol. Plant.* **23**, 607–612.

James, N. L. and Smith, G. A. (1969). Effect of photoperiod and light intensity on flowering in sugarcane. *Crop Sci.* **9**, 794–796.

Kortschak, H. P., Hartt, C. E. and Burr, G. O. (1965). Carbon dioxide fixation in sugarcane leaves. *Plant Physiol.* **40**, 209–213.

Kortschak, H. P. and Nickell, L. G. (1970). Calvin-type carbon dioxide fixation in sugarcane stalk parenchyma tissue. *Plant Physiol.* **45**, 515–516.

Laetsch, W. M. and Kortschak, H. P. (1971). Chloroplast structure and function in tissue cultures of C_4 plants. *Plant Physiol.* **49**, 1021–1023.

Loomis, R. S., Williams, W. A. and Hall, A. E. (1971). Agricultural Productivity. *Ann. Rev. Plant Physiol.* **22**, 431–468.

McLean, F. C., McDavid, C. R. and Singh, Y. (1968). Preliminary results of net assimilation rate studies in sugarcane. *Proc. Intern. Soc. Sugar Cane Technol.* **13**, 849–858.

Maretzki, A. and De la Cruz, A. (1967). Nitrate reductase in sugarcane tissues. *Plant & Cell Physiol.* **8**, 605–611.

Mongelard, J. C. and Mimura, L. (1971). Growth studies on the sugarcane plant. I. Effects of temperature. *Crop Sci.* **11**, 795–800.

Mongelard, J. C. and Mimura, L. (1972). Growth studies on the sugarcane plant. II. Some effects of root temperature and gibberellic acid and their interactions on growth. *Crop Sci.* **12**, 52–58.

Moore, P. H. (1971). Investigations on the flowering of *Saccharum*. I. Ontogeny of the inflorescence. *Can. J. Bot.* **49**, 677–682.

Moore, P. H. and Heinz, D. J. (1971). Increased post-inductive photoperiods for delayed flowering in *Saccharum* spp. hybrids. *Crop Sci.* **11**, 118–121.

Naidu, K. M. and Bhagyalakshmi, K. V. (1967). Stomatal movement in relation to drought resistance in sugarcane. *Current Sci.* **36**, 555–556.

Nickell, L. G. (1967). Agricultural aspects of transplanting and spacing. *1967 Reports, Hawaiian Sugar Technologists*, pp. 147–155.

Ogren, W. L. and Bowes, G. (1971). Ribulose diphosphate carboxylase regulates soybean photorespiration. *Nature New Biol.* **230**, 169–170.

Paliatseas, E. D. (1972). Flowering of sugarcane with reference to induction and inhibition. *Proc. Intern. Soc. Sugar Cane Technol.* **14**, 354–364.

Panje, R. R. and Babu, C. N. (1960). Studies in *S. spontaneum*. Distribution and geographical association of chromosome numbers. *Cytologia* **25**, 152–172.

Parthasarathy, N. (1947). The probable origin of North Indian sugarcanes. *Indian Bot. Soc. J., Silver Jubilee Session*, pp. 133–150.

Pate, J. S. and Gunning, B. E. S. (1972). Transfer cells. *Ann. Rev. Plant Physiol.* **23**, 173–196.

Roach, B. T. (1968). Quantitative effects of hybridization in *Saccharum officinarum* × *Saccharum spontaneum* crosses. *Proc. Intern. Soc. Sugar Cane Technol.* **13**, 939–954.

Shamel, A. D. (1924). *Hawaiian Planters' Record* **28**, 400–428.

Tovey, D. A., Glasziou, K. T., Farquhar, R. H. and Bull, T. A. (1973). Variability in radiation received by small plots of sugarcane due to differences in canopy heights. *Crop Sci.* **13**, 240–242.

Waldron, J. C. (1966) Photosynthesis and sugar accumulation in sugar cane. M.Sc. Thesis, University of Queensland.

Waldron, J. C., Glasziou, K. T. and Bull, T. A. (1967). The physiology of sugar cane. IX. Factors affecting photosynthesis and sugar storage. *Aust. J. Biol. Sci.* **20**, 1043–1052.

Whiteman, P. C., Bull, T. A. and Glasziou, K. T. (1963). The physiology of sugar cane. VI. Effects of temperature, light and water on set germination and early growth of *Saccharum* spp. *Aust. J. Biol. Sci.* **16**, 416–428.

4

Rice

Y. MURATA AND S. MATSUSHIMA

Grain production in rice can be expressed as the product of total dry weight at harvest and the 'harvest index' (Donald, 1962), which shows considerable variation among rice varieties. This expression indicates the final result in the simplest form. However, actual grain yield is determined by a complex chain of developmental processes, which take place in a definite sequence, and a change in any one of these may influence all the following processes. Physiological and ecological research in the last two or three decades has elucidated this sequence of processes in rice to a considerable degree.

GROWTH OF VEGETATIVE ORGANS

Root growth

Except for the seminal root which emerges at germination, all roots are initiated at the stem nodes, keeping a definite relationship with the emergence and development of the leaves; when leaf n is developing, roots emerge simultaneously at node $n - 3$ of the same stem (Fujii, 1961).

In this way, the number and weight of roots increase, as shown in Figure 4.1, with increase in tiller number, attaining their maximum values at the time of heading and anthesis. Thus, root growth is more accentuated in early stages as compared with shoot growth. The absorption of nutrients such as nitrogen, potassium, sulphur and phosphorus is usually most rapid between tillering and early panicle formation (Takahashi and Murayama, 1953; Inada, 1967).

According to Okajima (1960), the emergence of a root is closely correlated with the nitrogen content of the stem base, active emergence of roots taking place only when it is above 1 %.

Tillering

Tillering begins at the four- to five-leaf stage. Emergence of tillers is closely linked to that of leaves, according to Katayama (1951), the primary tiller emerging from the axil of leaf $n - 3$ when leaf n of the main stem elongates. Secondary and tertiary tillers emerge in the same way. Thus, all tillerings are synchronized with the development of leaves on the main stem.

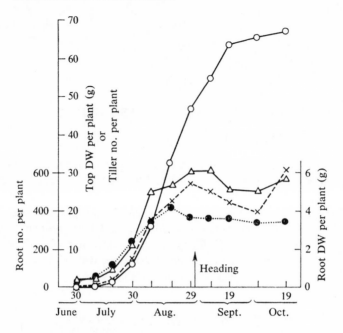

Figure 4.1. Increase in root number and dry weight, tiller number and shoot weight in a Japanese rice crop (Inada, 1967): (○) top dry weight; (△) root number; (×) root dry weight; (●) tiller number.

Tillers initially depend on the nutrient supply from the mother stem, but become autotrophic when they have three leaves and four to five roots (Ishizuka and Tanaka, 1963). The number of tillers attains its highest value about one month after transplanting, decreasing thereafter because of death of some of the last tillers to emerge as a result of their failure in competition for light and nutrients (Matsushima, 1957; Ishizuka and Tanaka, 1963). Most of the nutrients in the tillers which fail are translocated to other tillers (Ishizuka and Tanaka, 1963).

Tillering capacity is one of the most important characteristics of a variety. The initiation of tiller primordia is free from the influence of environment (Yamazaki, 1960), but their emergence and development are greatly influenced by such factors as nitrogen supply, solar radiation, and temperature. The most important of these is nitrogen content of the plant (Kumura, 1956); a concentration above 3.5 % is necessary for active tillering, at 2.5 % tillering stops, and below 1.5 % death of tillers takes place (Ishizuka and Tanaka, 1963).

Phosphorus level is also closely correlated with tillering, no tillering taking place when the phosphorus content of the mother stem is below 0.25 % (Honya, 1961). Climatic factors are also important. The number of emerging

Table 4.1. *Correlation between the number of tillers of rice plants grown at various localities in Japan and the average daily solar radiation during the six weeks after transplanting* (*Murata and Togari*, 1972)

Year	Correlation coefficient with solar radiation	
	Maximum tiller number	Number of bearing tillers
1967	0.436	0.690*
1968	0.093	0.714**
1969	0.270	0.677*

Significant at 5 % level (*) and 1 % level (**).

tillers is highest when water temperature is 15–16 °C at night and 31 °C during the day, but the optimum temperature for tiller development is 31 °C both day and night (Tsunoda, 1964; Matsushima *et al.* 1964*b*). With rice plants cultured under favourable conditions in the field at various localities in Japan, in the 'Maximal growth rate experiment' of IBP, the average daily solar radiation during the six weeks after transplanting (when tiller number reaches its maximum) did not show any significant correlation with the maximum number of tillers per square metre, but correlated highly with the number of ear-bearing tillers (Table 4.1).

Leaf growth

The total leaf area per unit ground area, i.e. the leaf area index (LAI), increases according to the compound interest law, reaches its highest value a little before heading, and decreases thereafter due to withering of lower leaves (Table 4.2).

The increase in LAI is caused by two factors, increase in tiller number and in the size of successive leaves. It is the former in a heavy-tillering variety, and the latter in a light-tillering variety, that mainly contributes to their LAI (Tanaka, A. *et al.* 1964).

Of various environmental factors, nitrogen fertilizer has the most marked effect on LAI, by increasing both the components. This effect is greatest for nitrogen applications just before panicle initiation (cf. Tables 4.2, 4.3, and 4.4).

As to climatic factors, it has been shown by the 'Seedling experiment' of IBP in Japan that increasing air temperature has a major positive influence on leaf area development, while increasing solar radiation has a slight, negative influence. Also it has been shown by the previously mentioned 'Maximal growth rate experiment' of IBP that the LAI of rice plants at

Table 4.2. *Changes in leaf area index according to growth stage* (*Murata*, 1961)

Nitrogen fertilizer applied kg ha $^{-1}$	Tillering stage	Max. tiller number stage	Young-ear-formation stage	Heading stage	Milk-ripe stage
0	0.27	0.51	2.06	2.58	1.87
75	0.30	0.77	3.23	3.38	1.87
112	0.38	1.15	4.03	4.67	3.19
150	0.31	1.17	4.05	4.94	4.42

Variety, 'Norin 29'.

heading in various localities is closely correlated with the heat unit accumulation from transplanting to heading. The same tendency has been observed in a monthly sowing experiment carried out near Bangkok (Osada *et al.* 1972).

The influence of solar radiation on LAI may be explained as an adaptation of the plants to develop thin, large leaves under weak light. However, increase in LAI is also limited by too low a level of solar radiation (Murata, 1961).

Dry matter production in vegetative phase

A close, positive correlation between the photosynthetic capacity of leaves and the relative growth rate of very young rice plants has been observed (Murata, 1964b). However, in rice crops with small LAI, the most important factor for dry matter production is generally LAI, the effect of photosynthetic capacity being scarcely recognizable (Murata, 1961). As LAI increases, its effect decreases.

According to a recent review by Yoshida (1972), many workers have reported values of 4 to 7 as the optimum LAI of rice stands, assuming that gross photosynthesis increases asymptotically with the increase in LAI, while respiration increases more or less linearly. In contrast to this, Yoshida *et al.* (1973) found that in 'IR-8', the high-yielding variety bred at the International Rice Research Institute, the amount of dry matter produced approaches an asymptotic plateau, because respiration increases asymptotically with the increase of LAI.

With increasing LAI, the extinction coefficient of the stand and the photosynthetic activity of the leaves play an increasingly important role in the photosynthesis of the plant community, as indicated by Monsi and Saeki (1953). Their equation showed that the distribution of light within a plant community is determined by its LAI and extinction coefficient, which in turn is mainly determined by leaf angle. The smaller the extinction coefficient, the more even is the light distribution within a community, and the less there is light saturation of the photosynthesis of individual leaves. As photosynthesis of rice leaves reaches light saturation at around 50 klx, about half full sunlight (Yamada *et al.* 1954), the effect of extinction coefficient is so great that it has a major influence on varietal differences in dry matter production, in spite of marked differences in other characteristics (Hayashi and Ito, 1962; Hayashi, 1969, 1972).

According to T. Tanaka *et al.* (1969), a rice stand whose leaves were made droopy by attaching weights to their tips was reduced in photosynthetic rate and dry matter production.

The extinction coefficient which has been recognized in this way as one of the most important morphological characteristics related to light utilization of a variety, shows some variation according to cultural and other conditions. It changes with planting density (Hayashi, 1966), depth of ploughing, growth stage (Murata *et al.* 1966), and time of nitrogen topdressing (Matsushima and Tanaka, 1963). However, much of this variation may be due to adaptations to changing LAI.

Tsunoda (1959 *a* and *b*, 1962, 1964) has pointed out that thickness and spatial distribution of leaves are also important, postulating a plant type which has small, thick, erect leaves arranged in a 'gathered form' as an ideal type for rice. At equal LAI, stands with many, small leaves are superior to those with few, large leaves (Matsushima *et al.* 1964 *a*). As to culm length, longer ones are advantageous in terms of light utilization (Murata, 1961; Hayashi, 1972), but disadvantageous in terms of lodging and in having greater respiration not directly related to grain yield (Tanaka, A. *et al.* 1966). The optimum culm length is a compromise between these three factors (Yoshida, 1972).

At high LAI, differences in photosynthetic rate might be reflected in the dry matter production of a crop stand. T. Tanaka (1972) demonstrated an increase in photosynthesis under strong light of rice stands having an LAI of 4.8, by nitrogen topdressing at the booting stage. In general such data are difficult to obtain because, between varieties, the variation of plant type is more conspicuous than that in photosynthetic rate, while within varieties the variation in photosynthetic rate is comparatively small during the vegetative growth phase.

GENERATIVE GROWTH

Differentiation and development of a young panicle

The generative growth of rice plants starts at the differentiation of young panicles. Matsushima (1957, 1966 *a*) divided panicle development into 21 stages.

Panicle initiation starts by the initiation of the first bract which later develops into the neck-node of a panicle.

About three or four days later, the primordium of a rachis-branch appears near the top of the developing panicle, followed by the differentiation of other primary rachis-branches in rapid, upward succession.

As the primary rachis-branches begin to grow, the secondary rachis-branch primordia appear on their dorsal side alternately in two rows, the primary rachis-branch located nearest to the tip of the developing panicle being higher than the tip of the growing point of the panicle. This means that the order of growth of primary rachis-branches is the reverse, at this stage, of the order of differentiation, i.e. the later a primary rachis-branch differentiates, the earlier it begins to grow.

About 10–12 days after panicle (neck-node) initiation, the differentiation of spikelets begins on the secondary branch primordia. During this stage the order of differentiation of secondary branch primordia and the order of their growth are reversed, as occurs earlier with the primary branches.

Structure of the spikelet

The spikelet is borne on a pedicel, of which the apex below the empty glumes (sterile lemmas) is expanded into a lobed facet. Thus, the enlarged, cup-like apex is homologous with a pair of true glumes, and is referred to as the rudimentary glumes. A spikelet contains a minute axis (rachilla) on which a single floret is borne in the axils of 2-ranked bracts which are termed empty glumes. The upper bracts or flowering glumes, the lemma and palea, together with the enclosed flower, form the floret. The flower proper consists of six stamens and one pistil which contains an ovule. The short style bears the bifurcate, plumose stigma. At anthesis the lodicules at the base of the flower become turgid and thrust the lemma and palea apart.

Pollination and fertilization

Pollination takes place almost simultaneously with the opening of flowers under normal conditions. When pollen falls on the stigma, the pollen tube emerges from the germ pore in one to three minutes, and grows very rapidly under normal conditions. Many of the pollen tubes which have entered the tissue of the stigma elongate downwards, pass through the space between the pericarp and integument, and finally enter the embryosac through the micropyle.

The optimum temperature for pollination is 31–32 °C, the minimum temperature 10–13 °C, and the maximum about 60 °C (Sasaki, 1919; Goto, 1931). Drought, as well as low temperatures, may have a detrimental effect on pollination.

As soon as the pollen tube has passed through the micropyle and entered the embryosac, within 30 minutes after flowering, the tip of the pollen tube ruptures, and its contents which include two male gametes cover the top of the egg cell. The fusion of one male gamete with the two polar nuclei to form the tripoid primary endosperm nucleus occurs 2.5–3 h after flowering. This nucleus then continues dividing to form an endosperm.

The diploid fertilized egg cell, however, only begins division to form the embryo after a resting stage of 6–8 h. Thus, the double fertilization of both the egg and polar nuclei is completed 2.5–3 h after flowering (Cho, 1956). The minimum temperature for fertilization is said to be 15 °C.

Growth of the caryopsis

Twenty-four hours after flowering, about 70 endosperm nuclei can be counted and the egg cell has divided into two to four cells.

The endosperm nuclei continue dividing and as soon as the embryosac is lined with one layer of these nuclei, their cell walls are formed. These cells then divide and multiply towards the interior of the embryosac, gradually forming cell layers. Four days after flowering the embryosac is filled with endosperm cell tissue and the formation of starch grains begins. These accumulate steadily and in 10 days the grain becomes so compact that it is difficult to section.

The outermost layer of the endosperm tissue gradually begins to change its form, to become an aleurone layer about seven to nine days after flowering. The caryopsis completes its growth in length by the sixth day, in width by 10–12 days, and in thickness by about 15 days, after flowering.

The dry weight of the caryopsis increases rapidly up to 26–30 days after flowering under temperate conditions. However, in tropical areas caryopsis weight seems to reach its highest value considerably earlier, judging from the number of days required for ripening there (Tanaka, A. and Vergara, 1967).

YIELD-DETERMINING PROCESSES

Determination of yield capacity

As there is only one floret in each spikelet in rice,

yield capacity (Murata, 1969)

$$= \begin{pmatrix} \text{number of} \\ \text{ears per m}^2 \end{pmatrix} \times \begin{pmatrix} \text{number of spike-} \\ \text{lets per ear} \end{pmatrix} \times \begin{pmatrix} \text{potential size} \\ \text{of a grain} \end{pmatrix}.$$

Whereas in wheat and barley there is considerable scope for variation in grain size (Thorne, 1965; Evans, 1973), this is not so in rice, as shown by Matsushima (1957). Because the grain is so rigidly enclosed by the outer and inner glumes, whose size is determined as early as five days before anthesis, it cannot grow to a size greater than that permitted by the hull. This was shown by placing a small stone or piece of plastic inside the hull at anthesis, and finding that at maturity the grain was indented with the shape of the enclosed object.

According to Matsushima (1957), the number of ears per square metre is determined during the period up to about 10 days after maximum tiller number is reached, and is greatly influenced by both nitrogen supply and the level of solar radiation at tillering. The number of spikelets per ear, on the other hand, is determined during the period from about 32 to 5 days before heading (in temperate rice) by the difference between the number of differentiated primordia and the number that degenerate. The former is strongly affected by nitrogen supply during panicle differentiation (*ca* 32–20 days before heading in temperate rice), and the latter by the level of solar radiation and other environmental factors around the reduction–division stage (*ca* 15–5 days before heading). The size of hulls is most profoundly influenced by the radiation level during the two weeks before anthesis.

Temporary storage of assimilates

After panicle initiation, the growth of vegetative organs, such as tillers, new leaves and roots, more or less slows down (cf. Figure 4.1), and as a result the accumulation of available carbohydrates – mostly starch and sugars – begins in the leaf sheath and culm base, sharply increases its amount during the two weeks before heading, and reaches its highest value at anthesis (Figure 4.2).

After anthesis, the amount rapidly decreases, reaching its lowest value about three weeks later (cf. Yoshida, 1972). Changes in the dry weights of leaves and stems run roughly parallel to those of the stored carbohydrates and, moreover, the increase in ear dry weight is about equal to the sum of the increase in total dry weight and the decrease in stored carbohydrates after heading. These facts suggest that the carbohydrates stored before heading are efficiently translocated to the ear after anthesis. This has been proved by experiments using ^{14}C-labelled assimilates (Murayama *et al.* 1961; Oshima, 1966; Cock and Yoshida, 1972). However, not all the reserves are translocated: in the experiments of Cock and Yoshida, 68% was translocated to ears, 20% was respired away, and 12% remained in leaf sheaths and culms.

The contribution of this pre-heading storage to grain yield is variable, from 0 to 90% in extreme cases, but mostly between 20 and 40% (Matsushima and Wada, 1959; Yoshida, 1972); it is lower at higher levels of fertilizer application and higher with a longer-duration variety or culture (Yoshida and Ahn, 1968). It is also dependant on climatic conditions; Soga and Nozaki (1957) suggested a buffer action of the pre-heading storage on grain filling under unfavourable weather conditions. Further, it was shown by Kumura

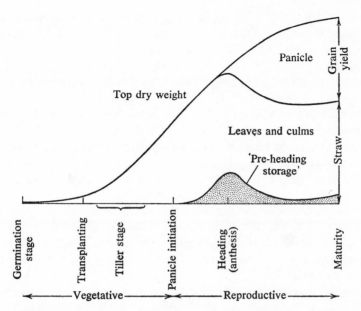

Figure 4.2. Changes in the amount of temporarily stored carbohydrates (pre-heading storage) and dry weight of various parts according to growth stages in rice (schematic illustration based on the data of Togari *et al.* 1954; Soga and Nozaki, 1957; Yamada *et al.* 1957; Murayama *et al.* 1961; Yoshida and Aha, 1968).

(1960), Kumura and Takeda (1962), and Tanaka, T. and Matsushima (1963) that an important role of pre-heading storage lay in preventing the occurrence of abortive grains around two to three weeks after flowering, when the growth of caryopsis was most active.

The rate of mobilization of the pre-heading reserves is greatest about one week after anthesis, reaching as high as twice the rate of total dry matter increase at that time (Monsi and Murata, 1970).

Partition, translocation, and storage of assimilates

The ultimate partitioning of dry matter between grain and vegetative parts is indicated by the grain:straw ratio or the 'harvest index', and one of the most fundamental factors affecting this is the ratio of duration of vegetative period to duration of reproductive period (Tsunoda, 1960).

The ratio of photosynthetic capacity of a rice crop to its growth capacity during panicle initiation and heading is important, because, when this ratio is high, pre-heading storage is abundant, and when low, reserves are small (Murata, 1969). The grain:straw ratio is influenced by the amount of assimilates translocated to the ear during grain filling which, in turn, depends on yield capacity, amount of photosynthesis, respiration rate, capacity of the conducting tissues, etc. It has been demonstrated by Murata (1969) that, if the capacity for translocation is not limiting, the grain:straw ratio is closely correlated with $P_0 . n_S / A_0$, where P_0 represents the average photosynthetic capacity of leaves during grain filling, n_S the number of spikelets per square metre, and $A_0 = $ LAI at heading. This is because P_0 is lower with a long-duration variety or culture than with a short-duration variety or culture, because A_0, which represents the capacity for vegetative growth and respiration, changes in the opposite direction to P_0 as duration of growth is extended, and because n_S can be taken as representing yield capacity so far as ordinary rice varieties are concerned.

If a large amount of starch and sugar remains in leaf sheaths and culms at harvest it is an indication that either translocation or storage was limiting (cf. Murata, 1969). For example, cool weather damage caused by temperatures below 17 °C at panicle initiation, reduction division, or flowering (Kondo, 1952; Satake *et al.* 1969; Hayase *et al.* 1969) or drought injury at the same stages (Wada *et al.* 1945), reduces yield capacity more than assimilate supply is reduced.

In *japonica* rice, the optimum average temperature for grain filling is said to be 20–22 °C (Matsushima, 1957; Aimi *et al.* 1959; Murata, 1964*a*; Munakata *et al.* 1967). Insufficient ripening at lower temperatures than the above is considered to be due to slowing down of translocation (Matsushima *et al.* 1957; Aimi *et al.* 1959), while on the other hand, the occurrence of incomplete grains at higher temperatures than the above is considered to be due to loss of assimilates caused by enhanced respiration (Yamamoto, 1954; Murata, 1964*a*), decrease of LAI (Murata, 1964*a*), ageing of grains (Matsushima *et al.* 1957; Aimi *et al.* 1959) or shortening of the grain-filling period (Tanaka, A. and Vergara, 1967).

As to translocation capacity, very little is known at present. Matsushima (1957) suggested that the diameter of the conducting tissue at the neck of an ear might have a correlation with the percentage of ripened grains. Nakayama (1969), on the other hand, has pointed out that the ageing of pedicels takes place before that of grains.

Dry matter production during grain filling, and the effect of photosynthetic capacity and leaf area upon it

All vegetative parts stop growing after flowering and, as a result, most of the net assimilates are translocated to the ear. Accordingly, the increase in total dry matter and carbohydrates during the grain-filling period usually shows a close correlation with grain yield (Matsushima and Wada, 1959; Murata and Togari, 1972).

The influence of LAI on the production of assimilates is marked in the later part of the grain-filling period when LAI is decreasing rapidly, but in

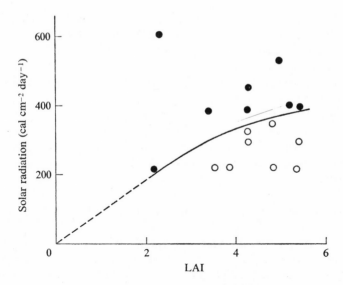

Figure 4.3. Effect of leaf area index (LAI) and solar radiation levels in 16 different experiments on the relation between photosynthetic activity of flag leaves and dry weight increase by the crop during grain filling: ($\bigcirc$) no significant correlation; ($\bullet$) significant correlations (Murata *et al.* 1966).

the earlier part, when it is high, LAI often shows a negative relation with grain filling. According to Murata and Osada (1958), the dry weight increase during grain filling in many varieties is positively correlated with the amount of photosynthesis when the level of solar radiation is high, but it is negatively correlated with LAI when radiation level is low. Furthermore, comprehensive examination of 16 field experiments differing in season and cultural conditions, made by Murata *et al.* (1966) to find out under what conditions the effect of photosynthetic rate on dry matter production would be revealed, has led to the conclusion that it can be recognized even in a stand whose LAI is beyond 3 when the radiation level is over 370 cals cm^{-2} per day, as is shown in Figure 4.3.

Moreover, the effect of photosynthetic rate is detectable even in grain yield among varieties differing in their responsiveness to nitrogen (Murata, 1964*b*). The effect of plant type is also recognizable (Hayashi and Ito, 1962; Hayashi, 1969; Tanaka, T. *et al.* 1969). However, in comparisons between varieties, plant type, photosynthetic rate, LAI and other characters may all vary, and at this stage the ear, which has little photosynthetic activity (Takeda and Maruta, 1956; Matsushima, 1957), obscures the effect of stand geometry. As a result, the effect of plant type is often not recognizable, while that of photosynthetic rate is variable, depending on conditions.

The nitrogen content of a leaf usually bears a close correlation with its photosynthetic rate, and nitrogen topdressing seldom fails therefore to bring about a promotion of the rate (cf. Murata, 1964b; Wada, 1969). The same correlation can also be observed among different varieties when they are at a comparatively young stage, but at later growth stages the rate falls due to ageing, especially in long-duration varieties, with the result that at anthesis and grain filling the photosynthetic rate is generally higher in short-duration varieties (Murata, 1957, 1961).

The photosynthetic rate during grain filling can be increased by the mid-summer drainage which is often carried out in poorly drained paddy fields to improve soil aeration (Koyama *et al.* 1962). In such fields, the soil becomes so severely reductive in summer, that toxic substances such as hydrogen sulphide and organic acids are produced, giving rise to root damage (Mitsui *et al.* 1951; Okajima and Takagi, 1959; Baba *et al.* 1957). By means of mid-summer drainage oxygen is supplied to the soil, leading to the recovery of root activity and of photosynthetic rate (Murata *et al.* 1965, 1966). The recovery of root activity may be the cause of the rise in photosynthetic rate. It has recently been concluded that in order to raise still further the grain yield of rice, it is necessary to keep the root system healthy by such methods as mid-summer drainage or intermittent irrigation (Agr. For. and Fish. Res. Council, 1971; Matsushima, 1973).

Nitrogen and assimilate supply

Nitrogen level influences grain yield through the determination of yield capacity during the vegetative phase and early reproductive phase, and through the production of assimilates by maintaining a high photosynthetic rate and LAI during the grain-filling period. Quantitatively, the nitrogen required for vegetative growth is far more than that required for reproductive development. A rice crop takes up more than 90% of the total nitrogen required for an average yield before the heading stage is reached (Ishizuka and Tanaka, 1953; Inada, 1967). However, for yields of more than seven tons per hectare of brown rice much more nitrogen is required for reproductive development of the crop, and 30–40% of the total nitrogen is absorbed during the grain-filling period (Shiroshita *et al.* 1962; Matsushima *et al.* 1966; Tsuno, 1968). Applications of nitrogen at appropriate times and amounts may therefore increase grain yield substantially. According to Matsushima and Manaka (1959), application of nitrogen at various growth stages gave the results shown in Table 4.3.

The number of spikelets per hill was largest when fertilizer was applied at panicle initiation (Treatments 3 and 7). The proportion of ripened grains, which usually shows an inverse correlation with the total spikelet number, was greatest for applications at heading (Treatments 6 and 8), perhaps due to promotion of photosynthetic activity. Weight per grain, on the other

Table 4.3. *Effect of nitrogen topdressing at various stages on grain yield and its components* (*Matsushima and Manaka*, 1959)

Time of N application	Number of spikelets per hill	Ripened grains (%)	Weight per grain (mg)	Grain yield per hill (g)
1 Transplanting	1290	75	26.8	37.5
2 Tillering stage	1350	72	26.4	37.0
3 Panicle-initiation stage	1490	66	26.1	37.7
4 Spikelet-differentiation stage	1420	69	26.8	40.0
5 Reduction-division stage	1360	71	27.2	39.7
6 Heading stage	1240	79	26.6	39.9
7 1/2 at Stage 3 + 1/2 at Stage 5	1530	67	26.7	39.4
8 1/2 at Stage 4 + 1/2 at Stage 5	1350	76	26.8	41.0

Amount of basal nitrogen, 38 kg ha^{-1} and that of topdressing nitrogen, 38 kg ha^{-1} or 60 kg ha^{-1}. The data are three-year averages.

hand, was greatest when nitrogen was given at the reduction-division stage, suggesting an increase in grain capacity. Thus, the highest grain yield was obtained in Treatment 8 where yield capacity and assimilate supply were considered to have been in good balance.

According to Hall *et al.* (1968) and Wells and Johnston (1970), maximum grain yields were obtained when nitrogen fertilizers were applied after the onset of internode elongation. In tropical areas, too, maximum grain yields were raised by applying three-quarters of the nitrogen fertilizer during land preparation and one quarter during panicle initiation (*IRRI*, 1970).

On the other hand, nitrogen applied in excess has an adverse effect on growth and especially on grain yield, the effect being greatest when it is applied at the end of the vegetative phase (Kumura, 1960; Matsushima and Manaka, 1961). Table 4.4 shows the response of rice plants to heavy applications, 151 kg nitrogen per hectare, at various growth stages.

The greatest yield reduction was observed when nitrogen was applied at panicle initiation (32–30 days before heading). In Treatment 3, the total spikelet number was high, the reserve of available carbohydrates was very low, as was resistance to lodging. On the other hand, LAI at heading was the largest and culm length at maturity the second highest.

Similar responses are observed when an excessive amount of nitrogen is applied at planting (Kumura and Takeda, 1962). This may be due to an excessive amount of nitrogen remaining in the soil at panicle initiation, causing excessive vegetative growth, LAI, and culm elongation, leading to a less favourable photosynthesis–respiration balance and greatly decreasing

Table 4.4. *Effect of a large dose of nitrogen fertilizer at various growth stages on grain yield, its components, and related parameters. (Matsushima and Manaka, 1961; Wada, 1969)*

Time of nitrogen application	Number of spikelets per hill	Ripened grains (%)	Grain yield per hill (g)	At heading stage			Culm length at maturity (cm)
				LAI	starch + sugars (%)	lodging resistance	
Transplanting	1720	63	26.4	5.4	9.0	4-5	94
Maximum tillering	1800	46	20.2	5.7	5.2	2-3	105
Panicle initiation	1780	30	13.5	7.4	2.5	1	96
Spikelet differentiation	1440	54	21.0	6.1	7.8	3	88
Reduction division	1300	74	25.2	4.5	10.8	4-5	86
Heading	1140	80	24.0	2.9	12.8	6	84

The amount of nitrogen applied, 151 kg/ha^{-1}. Lodging resistance is expressed in relative values, a larger figure indicating a larger resistance.

the carbohydrate reserves at anthesis (Kumura, 1960). Should low solar radiation and a high temperature be added to this situation, severe carbohydrate shortage would prevail.

On the other hand, because of the increase in spikelet number, serious competition for carbohydrates would take place between spikelets, and weak spikelets in the lower parts of panicles would fail to be fertilized, or would abort the grains (Togari and Kashiwakura, 1958; Kumura, 1960; Kumura and Takeda, 1962; Wada, 1969). Active elongation of culms under carbohydrate deficiency would limit the development of mechanical tissues, leading to lodging. Once lodging occurs, translocation and absorption of nutrients is hindered and the arrangement of leaves is disturbed, with the result that photosynthesis is decreased (Hidaka, 1968). Even if lodging did not occur, photosynthesis of lower leaves would be greatly reduced due to severe mutual shading, decreasing the physiological activity of the root and leading, in turn, to a decrease in photosynthetic activity of leaves (Murata, 1969). Assimilate supply to ears during grain filling will then be decreased, especially under low solar radiation, and grain filling is severely checked, resulting in low grain yield.

In this connection, a special effect of plant type on root activity should be noted. As rice plants are usually grown on flooded soil, their root system is under unfavourable conditions, and photosynthesis of lower leaves, which supply assimilates to the root system (Tanaka, A. 1958), plays an important role in maintaining root activity. Thus, plant types which allow the solar radiation to penetrate deep into the stand should have special merit in rice.

From the results of experiments mentioned above and others (Matsushima *et al.* 1965) in which it was found that restriction of nitrogen supply from the soil during the 20 days centred around panicle initiation greatly improved the percentage of ripened grains, Matsushima *et al.* (1969) have suggested a new rice-cultivation system which includes a temporary restriction of nitrogen supply, by mid-summer drainage and the use of nitrate-nitrogen which is leached from the soil more easily than ammonium-nitrogen.

Nitrogen-responsiveness and yielding ability of rice varieties

Heavy application of nitrogen fertilizers is a prerequisite for increase in the production of dry matter, and varietal characteristics must avoid excessive vegetative growth and increase assimilate supply to the ear. Because high yields are possible only under high levels of nitrogen supply, characteristics which confer high yielding ability are in many cases concerned with responsiveness to nitrogen. According to Yoshida (1972) the following characteristics are desirable.

Leaves should be small, thick, and erect, and stems should be of a 'gathered type' for favourable leaf photosynthetic capacity, stand geometry, and photosynthesis–respiration balance. Heavy tillering is desirable to compensate for missed hills and to give rapid leaf area expansion. Short, stiff culms are desirable to confer resistance to lodging. The grain:straw ratio should be high, and the proportion of ripened grains high under a high nitrogen level. High spikelet number per unit LAI may also be important to avoid excessive vegetative growth (Murata, 1969).

As to more physiological characteristics, the following may be mentioned: the promotion of photosynthetic activity by fertilizer nitrogen should be as large as possible, and that of respiration as small as possible (Osada and Murata, 1962, 1965; Osada, 1964, 1966); the plants should be resistant to root-rot (Baba, 1961) and they should have a relatively large amount of reserve carbohydrates in leaves and culms under high nitrogen levels (Takahashi *et al.* 1959). According to Nagato and Chaudhury (1969), 'Indica' varieties have a greater number of total spikelets in spite of their smaller carbohydrate reserves at heading as compared with 'Japonica' varieties, resulting in a greater proportion of non-fertilized, abortive, and partially filled grains.

It may be concluded that for a variety to be highly responsive to nitrogen fertilizer, its photosynthesis should increase with heavy applications, but little of the assimilate should be used in vegetative growth and respiration during the reproductive phase.

MAXIMUM LIMIT OF GRAIN YIELD AND LIMITING FACTORS

Estimation of maximum grain yield

The period of production of assimilates translocated to the ear extends from two weeks before heading to four weeks after it. In high yielding rice crops, however, the pre-heading storage is small but the grain filling period is extended (Murayama *et al.* 1955; Takahashi *et al.* 1955; Tsuno, 1968). Let us assume, therefore, that the period for the production of yield content is 40 days after heading, and calculate the maximum amount of net assimilates available. On this basis, the maximum grain yield of rice has been estimated with the following assumptions:

(i) The average daily solar radiation is assumed to be 400 cal cm^{-2}, of which 45 % is photosynthetically active (Loomis and Williams, 1963).

(ii) Of the photosynthetically active radiation, 5.5 % is lost by reflection at the canopy suface (Kishida, 1970), and another 10 % through absorption by inactive tissues (Loomis and Williams, 1963).

(iii) Eight photons are required to reduce one molecule of carbon dioxide, corresponding to an efficiency of 26 % in energy conversion (Evans, 1973).

(iv) The loss due to light saturation in upper leaves is 17 % at 400 cal cm^{-2} day^{-1} calculated from the data of Tsuno and Kitakado (1970).

(v) The conversion factor for dry matter is 3900 cal g^{-1} (Murata *et al.* 1968).

(vi) Respiration loss was calculated according to McCree (1970) as the sum of 1.5 % of dry weight per day (1.5 kg m^{-2} in this case) and 25 % of the gross photosynthesis.

Thus,

Maximum grain yield

$$= (400 \times 10^4 \times 0.45 \times 0.945 \times 0.90 \times 0.26 \times 0.83 \times 0.75$$
$$- 1500 \times 0.015 \times 3900) \times 40/3900$$
$$= 1640 \text{ g m}^{-2} \text{ (dry weight)}$$
$$= 19.1 \text{ t ha}^{-1} \text{ (at 14 \% moisture content).}$$

The efficiency for solar energy utilization in this case was 4.0 %.

In contrast to this estimate, the actual record yield of rice in the temperate zone is 10.052 t ha^{-1} of brown rice (14 % moisture) achieved by Y. Kudo, Japan in 1961, using 'Ootori'. In the tropical zone, 10.341 t ha^{-1} of rough rice was raised at *IRRI*, the Philippines, in 1967, and 10.67 t ha^{-1} of rough rice (8.52 t ha^{-1} of brown rice), at Battambang, Cambodia, by Hirano *et al.* (1968), both using 'IR-8'. The record yield is thus about one half of the estimated maximum yield, and we now examine what factors are responsible for the difference between them.

Limiting factors for grain yield

If no special problem is involved in the translocation processes, the actual yield of rice must be limited by either the yield capacity or the assimilate supply, and three cases are possible (Murata, 1969): (a) Yield capacity is limiting, the percentage of ripened grains is high, and assimilates in excess of the storage capacity remain in the straw. (b) Assimilate supply is limiting, and the proportion of ripened grains is low. (c) Assimilate supply and yield capacity are well-balanced, both the proportion of ripened grains and weight per grain are large, and little assimilate remains in the straw. Examples of these cases are shown in Figure 4.4.

(a) *Cases where yield capacity is limiting*

When the percentage of ripened grains is over 85 %, the crop falls into this category (Matsushima, 1966*b*), and close correlations are usually observed between grain yield and the number of spikelets per square metre as shown in Figure 4.4*a*. This is seen under such conditions as low fertilizer level, thin planting, etc., where vegetative growth is insufficient. According to Yoshida (1972) it is of universal occurrence in tropical rice. Crops with a close correlation between LAI at heading and grain yield are also considered to belong to this category, because of the correlation often found between

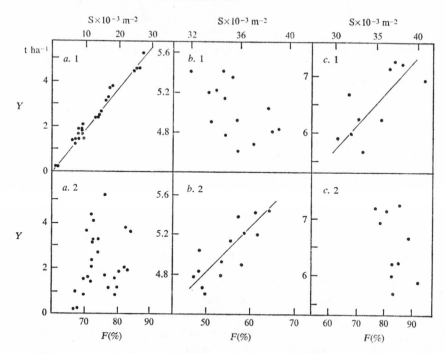

Figure 4.4. Relations of grain yield, *Y*, to spikelet number, *S*, and the proportion of ripened grains, *F*, in rice plants cultured under: (*a*), various planting densities combined with nitrogen topdressing (Wada and Matsushima, 1962); (*b*), various combinations of ploughing depth, planting density, and nitrogen level (Murata *et al.* 1966); and (*c*), different seasons with nearly optimum cultural conditions (Nakamura and Suzuki, 1969–1973).

LAI at heading and the number of spikelets per square metre (Murata, 1969), as in the IBP 'Maximal growth rate experiment' (Table 4.5). Here, LAI is also correlated with grain yield to a considerable degree, but the correlation is much reduced when the influence of spikelet number is excluded, indicating that the relationship between LAI and grain yield is not a direct one.

Even allowing for varietal differences, yield of rice in warmer areas is more likely to be limited by storage capacity than in cooler areas: in both areas there is usually close correlation between LAI at heading and the spikelet number, but the ratio of the latter to the former is smaller with rice plants grown in warmer areas (Murayama, 1971).

Thus, increase in yield capacity in order to increase yield in warmer-area rice crops is liable to lead to excessive vegetative growth and reduced assimilate supply. In cooler areas, on the other hand, tiller number can be increased without detriment to yield, partly due to low temperatures at tillering and partly to varietal characteristics; as a result yield capacity can be increased

Table 4.5. *Correlation coefficient of grain yield* (1) *with LAI at heading* (2) *and with the number of spikelets* (3) *in the IBP 'Maximal growth rate experiment' (Murata and Togari, 1972)*

Year	$r_{1,2}$	$r_{1,3}$	$r_{12,3}$
1967	0.600*	0.765**	−0.033
1968	0.787**	0.858**	0.579
1969	0.260	0.554	0.221
Pooled	0.492**	0.683**	0.203

(*) Significant at 5 % and (**) at 1 % level.

without excessive growth. This may be one of the reasons why heavy-tillering but non-elongating varieties are suitable in warmer areas.

One of the problems encountered when increase in yield capacity is sought, is competition for assimilates between vegetative and reproductive organs at the beginning of the reproductive phase. Because the time when elongation of culms begins coincides approximately with the stage of panicle differentiation and development (Arashi, 1960), the two organs are liable to compete under certain condition (Navasero and Tanaka, 1966). According to Wada (1969) about 60 % of the net assimilates produced at this stage are used for stem elongation, the rest being directed to the growth of leaves and young ears. As mentioned earlier (Table 4.4), a heavy topdressing of nitrogen at this stage greatly increases the number of spikelets per ear in spite of the reduced carbohydrate content, so there must be little competition for assimilates at the beginning of the reproductive phase.

The degeneration of spikelets and the size of hulls are particularly influenced by the level of solar radiation during the two weeks before heading (Matsushima, 1957), and this is the time when elongation of culms and growth of the flag leaf are active (Seko *et al.* 1956). Thus, in rice crops with excessive growth, especially under a spell of cloudy weather, competition for assimilates is quite likely to occur at the booting stage. A variety with small flag leaves may therefore be more suitable for heavy fertilization and dense planting conditions.

The following agronomic measures may increase yield capacity: to increase ear number – increased basal nitrogen application (Kumura, 1956; Yamada *et al.* 1957), dense planting (Yamada *et al.* 1961), exposure to low temperature and strong solar radiation at tillering (Matsushima *et al.* 1946b; Matsushima, 1957; Murata and Togari, 1972); to increase the number of spikelets per ear – nitrogen topdressing between spikelet differentiation and reduction–division stage (Matsushima, 1957; Matsushima and Manaka, 1959); and to enlarge hull size – exposure to strong solar radiation or nitrogen topdressing at the reduction–division stage (Matsushima, 1957; Wada, 1969).

Table 4.6. *Correlation coefficient of grain yield (1) with net assimilation rate during grain filling (4) and with average LAI during grain filling period (5) in rice plants grown at various localities in Japan* (*Murata and Togari*, 1972)

Year	$r_{1,4}$	$r_{14,5}$
1967	−0.198	0.419
1968	−0.118	0.723*
1969	−0.214	0.502
Pooled	−0.200	0.372*

(*). Significant at 5 % level

(b) *Cases where assimilate supply is limiting*
According to Matsushima (1966b) this is the case when the proportion of ripened grains is less than 80 % (cf. Figure 4.4b). The following measures are effective in increasing assimilate supply: promotion of photosynthetic activity of leaves and prevention of withering of lower leaves by nitrogen topdressing at booting or heading, or by deep layer placement of nitrogen fertilizer just before panicle initiation (Matsushima and Manaka, 1959, 1961; Murata, 1961; Matsuura et al. 1969; Heu and Ota, 1969; Lee and Ota, 1971); exposure to abundant solar radiation during grain filling (Matsushima, 1957; Murata, 1961, 1964a; Takeda, 1961; Munakata et al. 1967; Moomaw et al. 1967); and carbon dioxide fertilization (Matsushima et al. 1960; Yoshida, 1972; Yoshida et al. 1973).

(c) *Cases where yield capacity and assimilate supply are well-balanced*
In the IBP experiment cited above and partly shown in Figure 4.4c, it appeared at first that grain yield was limited by yield capacity. However, whereas net assimilation rate during grain filling shows no correlation with grain yield, as shown in Table 4.6, the correlation between them becomes significant at equal LAI levels.

Therefore, it is assumed that in this case not only yield capacity but also assimilate supply after heading was limiting to grain yield. Such cases have been also recognized in other experiments (Murata, 1969), and may be interpreted as supporting the existence of a 'sink–source' feedback.

Improvement of grain yield

Rice breeding during the past 50 years in Japan, and 10 years in the Philippines, has led to the development of erect-leaved, short-strawed, heavy-tillering varieties (Tanaka, A. et al. 1968; Ito and Hayashi, 1969; Chandler, 1969), of a plant type with good light utilization and little lodging. However, several aspects deserve further study. The contribution of ear photosynthesis to grain yield in rice is far less than in wheat and barley (Thorne, 1965), being less than 10 % at most (Takeda and Maruta, 1956; Tanaka, A. 1958). However, except for IR-8 and some others, most varieties have their ears at the top layer of the canopy, where they absorb much of the incident radiation, lowering the efficiency of light utilization (Murata *et al.* 1968). Thus, varieties which have their ears below the top of the canopy are more desirable (Yoshida, 1972). For this to be achieved, the topmost internode of the stem must be shortened, which has the further virtue of reducing assimilate competition.

By raising the carbon dioxide concentration in the air above 300 ppm, photosynthesis of rice is increased (Yamada *et al.* 1955; Murata, 1962), and also grain yield (Yoshida, 1972; Yoshida *et al.* 1973). But the real problem in field crops is how to deliver the additional carbon dioxide efficiently to them (Waggoner, 1969). Wind can increase photosynthesis of rice in the field (Yabuki and Kamotani, 1971), and rice crops may benefit from this more than upland crops because they suffer less from the enhanced transpiration.

In wheat the photosynthetic rate of the flag leaf has fallen whereas leaf area and grain size have increased in the course of evolution (Evans and Dunstone, 1970). In rice, however, the contribution of leaf photosynthesis to the grain yield is much greater than that of the ear and the importance of leaves may be greater than in wheat or barley. Varieties with thick leaves and a high nitrogen content have high photosynthetic activity, but less capacity to expand leaf area (Tsunoda, 1962). Photosynthetic capacity is therefore likely to be inversely correlated with that for leaf area expansion, as in wheat. In order to develop a variety which is highly responsive to nitrogen it is necessary to break this inverse correlation, but from the viewpoint of high yield responsiveness to nitrogen, a combination of high photosynthetic capacity and low growth capacity is necessary.

REFERENCES

Agriculture, Forestry & Fisheries Res. Council (1971). Studies on maximizing yield level of rice. *Min. Agr. and For., Res. Bull.* **49**, 1–252.

Aimi, R., Sawamura, H. and Konno, S. (1959). Physiological studies on the mechanism of ripening in crop plants. The effect of the temperature upon the behavior of carbohydrates and some related enzymes during the ripening of rice plant. *Proc. Crop Sci. Soc. Japan* **27**, 405–407.

Arashi, K. (1960). *Growth of rice plant and diagnosis of its 'autumn decline'.* Yokendo, Tokyo.

Baba, I. (1961). Mechanism of response to heavy manuring in rice varieties. *IRC Newsletter* **10**(4), 9–16.

Baba, I., Iwata, I. and Takahashi, Y. (1957). Studies on the nutrition of the rice plants with reference to *Helminthosporium* leaf spots. XII. *Proc. Crop Sci. Soc. Japan* **25**, 222–224.

Chandler, R. F. (1969). Plant morphology and stand geometry in relation to nitrogen. In *Physiological aspects of crop yield*, eds J. D. Eastin, F. A. Haskins, C. Y. Sullivan and C. H. M. van Bavel. ASA and CSSA, Madison, Wisconsin. pp. 265–285.

Cho, J. (1956). Double fertilization in *Oryza sativa* and development of endosperm with special reference to the aleurone layer. *Bull. Nat. Inst. Agr. Sci. Japan Ser. D* **6**, 61–101.

Cock, J. H. and Yoshida, S. (1972). Accumulation of ^{14}C-labelled carbohydrate before flowering and the subsequent redistribution and respiration in the rice plant. *Proc. Crop Sci. Soc. Japan* **41**, 226–234.

Donald, C. M. (1962). In search of yield. *J. Aust. Inst. Agr. Sci.* **28**, 171–178.

Evans, L. T. (1973). The effect of light on plant growth, development and yield. In *Plant Response to Climatic Factors*, pp. 21–35. UNESCO, Paris.

Evans, L. T. and Dunstone, R. L. (1970). Some physiological aspects of evolution in wheat. *Aust. J. Biol. Sci.* **23**, 725–741.

Fujii, Y. (1961). Studies on the regular growth of the roots in the rice plants and wheats. *Bull. Fac. Agr. Saga Univ* **12**, 1–117.

Goto, K. (1931). Physiological research on pollen with special reference to artificial germination of *Gramineae* pollen. *Memoir, Fac. Sci. Agr. Taihoku Imper. Univ.* **3**, 61–197.

Hall, V. L., Sims, J. L. and Johnston, T. H. (1968). Timing of nitrogen fertilization of rice. II. *Agron. J.* **60**, 450–453.

Hayase, H., Satake, T., Nishiyama, I. and Ito, N. (1969). Male sterility caused by cooling treatment at the meiotic stage in rice plants. II. *Proc. Crop Sci. Soc. Japan* **38**, 706–711.

Hayashi, K. (1966). Efficiencies of solar energy conversion in rice varieties as affected by planting density. *Proc. Crop Sci. Soc. Japan* **35**, 205–211.

Hayashi, K. (1969). Efficiencies of solar energy conversion and related characteristics in rice varieties. *Proc. Crop Sci. Soc. Japan* **38**, 495–500.

Hayashi, K. (1972). Efficiencies of solar energy conversion in rice varieties. *Bull. Nat. Inst. Agr. Sci. Japan Ser. D* **23**, 1–67.

Hayashi, K. and Ito, H. (1962). Studies on the form of plant in rice varieties with particular reference to the efficiency in utilizing sunlight. *Proc. Crop Sci. Soc. Japan* **30**, 329–334.

Heu, H. and Ota, Y. (1969). The effect of the deep placement of topdressing fertilizer and intermittent irrigation on physiological and ecological characters in rice plant. *Proc. Crop Sci. Soc. Japan* **38**, 501–506.

Hidaka, N. (1968). Experimental studies on the mechanism of lodging and of its effect on rice plants. *Bull. Nat. Inst. Agr. Sci. Japan Ser. A* **15**, 1–175.

Hirano, S., Shiraishi, K. and Tanabe, S. (1968). *Rice crop and water use in Southeast Asia. Symposium on World Rice* **2**, Agron. Soc. Japan, pp. 128–140.

Honya, K. (1961). Studies on the improvement of rice plant cultivation in volcanic ash paddy field in Tohoku district. *Tohoku Agr. Expt. Sta. Bull.* **21**, 1–143.

Inada, K. (1967). Physiological characteristics of rice roots, especially with the view-point of plant growth stage and root age. *Bull. Nat. Inst. Agr. Sci. Japan Ser. D* **16**, 19–156.

IRRI (1970). Soil fertility and fertiliser management. In *Annual report of Inter. Rice Res. Inst. for 1969*, The Philippines, pp. 118–119.

Ishizuka, Y. and Tanaka, A. (1953). Studies on the developmental processes in rice plants. III. *J. Sci. Soil and Manure, Japan* **23**, 159–165.

Ishizuka, Y. and Tanaka, A. (1963). *Studies on the nutrio-physiology of the rice plant.* Yokendo, Tokyo, 307 pp.

Ito, H. and Hayashi, K. (1969). The changes in paddy field rice varieties in Japan, *Proc. Symposium on Optimization of Fertilizer Effect in Rice Cultivation*, Agr. For. and Fisher. Res. Council, Min. Agr. and For., pp. 13–23.

Katayama, T. (1951). *Studies on the tillering of rice, wheat, and barley.* Yokendo, Tokyo, 117 pp.

Kishida, K. (1970). Balance sheet of photosynthetically active radiation in rice community. II. *Proc. General Meet. Soc. Agr. Meteor. Japan*, pp. 25–26.

Kondo, Y. (1952). Physiological studies on cool-weather resistance of rice varieties. *Bull. Nat. Inst. Agr. Sci. Japan Ser. D* 3, 113–228.

Koyama, T., Miyasaka, A. and Eguchi K. (1962). Studies on water management in the ill-drained paddy field. VII. *Proc. Crop Sci. Soc. Japan* 30, 143–145.

Kumura, A. (1956). Studies on the effect of internal nitrogen concentration of rice plants on the constitutional factor of yield. *Proc. Crop Sci. Soc. Japan* 24, 177–180.

Kumura, A. (1960). Yield analysis from the view-point of matter production. In *Experimental methods in crop science*, vol. II, eds Y. Togari, T. Matsuo, M. Hatamura, N. Yamada, T. Harada and N. Suzuki, pp. 195–274. Nogyo-gijutsu-kyokai, Tokyo.

Kumura, A. and Takeda, T. (1962). Analysis of grain production in rice plant. VII. *Proc. Crop Sci. Soc. Japan* 30, 261–265.

Lee, J. H. and Ota, Y. (1971). The role of root system of rice plant in relation to the physiological and morphological characteristics of aerial parts. V. *Proc. Crop Sci. Soc. Japan* 40, 217–222.

Loomis, R. S. and Williams, W. A. (1963). Maximum crop productivity, an estimate. *Crop Sci.* 3, 67–72.

McCree, K. J. (1970). An equation for the rate of respiration of white clover plants grown under controlled conditions. In *Prediction and measurement of photosynthetic productivity*, ed. I. Setlik, pp. 221–229. Centre for Agr. Publ. and Docum. Wageningen.

Matsushima, S. (1957). Analysis of development factors determining yield and yield prediction in lowland rice. *Bull. Nat. Inst. Agr. Sci. Japan Ser. A* 5, 1–271.

Matsushima, S. (1966a). *Crop science in rice.* Fuji Publ. Co., Tokyo, 365 pp.

Matsushima, S. (1966b). *Theory and practices in rice culture.* Yokendo, Tokyo, 302 pp.

Matsushima, S. (1973). *A method for maximizing rice yield through 'ideal plants'.* Yokendo, Tokyo, 393 pp.

Matsushima, S. and Manaka, T. (1959). Analysis of developmental factors determining yield and its application to yield prediction and culture improvement of lowland rice. LI. *Proc. Crop Sci. Soc. Japan* 27, 432–434.

Matsushima, S. and Manaka, T. (1961). *Ibid.* LVIII. *Proc. Crop Sci. Soc. Japan* 29, 202–206.

Matsushima, S., Manaka, T. and Tsunoda, K. (1957). *Ibid.* XXXIX and XL. *Proc. Crop Sci. Soc. Japan* 25, 203–206.

Matsushima, S. and Tanaka, T. (1963). *Ibid.* LXVI. *Proc. Crop Sci. Soc. Japan* 32, 44–47.

Matsushima, S., Tanaka, T. and Hoshino, T. (1964a). *Ibid.* LXIX. *Proc. Crop. Sci. Soc. Japan* 33, 44–48.

Matsushima, S., Tanaka, T. and Hoshino, T. (1964b). *Ibid.* LXX. *Proc. Crop Sci. Soc. Japan* 33, 53–58.

Matsushima, S., Tanaka, T. and Hoshino, T. (1965). *Ibid.* LXXIII *Proc. Crop Sci. Soc. Japan* 34, 25–29.

Matsushima, S., Tanaka, T. and Okabe, T. (1960). *Ibid.* LVI. *Proc. Crop Sci. Soc. Japan* **28**, 374–376.

Matsushima, S. and Wada, G. (1959). Relationships of grain yield and ripening of rice plants (2). *Nogyo-oyobi-engei* **34**, 303–306.

Matsushima, S., Wada, G. and Matsuzaki, A. (1966). Analysis of developmental factors determining yield and its application to yield prediction and culture improvement of lowland rice. LXXIV. *Proc. Crop Sci. Soc. Japan* **34**, 321–328.

Matsushima, S., Wada, G. and Matsuzaki, A. (1969). *Ibid.* LXXX. *Proc. Crop Sci. Soc. Japan* **38**, 11–17.

Matsushima, S., Wada, G., Tanaka, T. and Okabe, T. (1960). *Ibid.* LVII. *Proc. Crop Sci. Soc. Japan* **29**, 29–30.

Matsuura, K., Iwata, T. and Hasegawa, T. (1969). Studies on the effect of deep-layer application of fertilizers in rice plant. I. *Proc. Crop Sci. Soc. Japan* **38**, 215–221.

Mitsui, S., Aso, S. and Kumazawa, K. (1951). Dynamic studies on the nutrient uptake by crop plants. I. *J. Soil and Manure, Japan* **22**, 46–52.

Monsi, M. and Murata, Y. (1970). Development of photosynthetic system as influenced by distribution of matter. In *Prediction and measurement of photosynthetic productivity*, ed. I. Setlik, pp. 115–129. Centre for Agr. Publ. and Docum. Wageningen.

Monsi, M. and Saeki, T. (1953). Ueber den Lichtfaktor in den Pflanzen Gesellschaften und seine Bedeutung für die Stoffproduktion. *Jap. J. Bot.* **14**, 22–52.

Moomaw, J. C., Baldazo, P. and Lucas, L. (1967). Effects of ripening period environment on yields of tropical rice. *IRC Newsletter Special Issue*, pp. 18–25.

Munakata, K., Kawasaki, T. and Kariya, K. (1967). Quantitative studies on the effects of the climatic factors on the productivity of rice. *Bull. Chugoku Agr. Expt. Sta. A* **14**, 59–96.

Murata, Y. (1957). Photosynthetic characteristics of rice varieties. *Nogyo-gijutsu* **12**, 630–632.

Murata, Y. (1961). Studies on the photosynthesis of rice plants and its culture significance. *Bull. Nat. Inst. Agr. Sci. Japan Ser. D.* **9**, 1–169.

Murata, Y. (1962). The effect of carbon dioxide concentration in the air on the growth of crop plants. *Nogyo-oyobi-engei* **37**, 5–10.

Murata, Y. (1964a). On the influence of solar radiation and air temperature upon the local differences in the productivity of paddy rice in Japan. *Proc. Crop Sci. Soc. Japan* **33**, 59–63.

Murata, Y. (1964b). Photosynthesis, respiration, and nitrogen response. In *Mineral nutrition in rice plants, IRRI*, pp. 385–400. Johns Hopkins Press, Baltimore.

Murata, Y. (1969). Physiological responses to nitrogen in plants. In *Physiological aspects of crop yield*, eds J. D. Eastin, F. A. Haskins, C. Y. Sullivan and C. H. M. van Bavel, ASA and CSSA, Madison, Wisconsin, pp. 235–259.

Murata, Y., Iyama, J. and Honma, T. (1965). Studies on the photosynthesis of rice plants. XIII. *Proc. Crop Sci. Soc. Japan* **34**, 148–153.

Murata, Y., Iyama, J., Himeda, M., Izumi, S., Kawabe, A. and Kanzaki, Y. (1966). Studies on the deep-plowing, dense-planting cultivation of rice plants from the point of view of photosynthesis and production of dry matter. *Bull. Nat. Inst. Agr. Sci. Japan Ser. D* **15**, 1–53.

Murata, Y., Miyasaka, A., Munakata, K. and Akita, S. (1968). On the solar energy balance of rice population in relation to the growth stage. *Proc. Crop. Sci. Soc. Japan* **37**, 685–691.

Murata, Y. and Osada, A. (1958). Studies on the photosynthesis in rice plant. X. *Proc. Crop Sci. Soc. Japan* **27**, 12–14.

Murata, Y. and Togari, Y. (1972). Analysis of the effect of climatic factors upon the productivity of rice at different localities in Japan. *Proc. Crop Sci. Soc. Japan* **41**, 372–387.

Murayama, N. (1971). Nutritional characteristics of high-yielding rice plants. *Nogyo-oyobi-engei* **46**, 145–149.

Murayama, N., Oshima, M. and Tsukahara, S. (1961). Studies on the dynamic status of substances during ripening processes in rice plants. VI. *J. Sci. Soil and Manure, Japan* **32**, 261–265.

Murayama, N., Yoshino, M., Oshima, M., Tsukahara, S. and Kawarazaki, Y. (1955). Studies on the accumulation process of carbohydrates associated with growth of rice. *Bull. Nat. Inst. Agr. Sci. Japan Ser. B* **4**, 123–166.

Nagato, K. and Chaudhury, F. M. (1969). A comparative study of ripening process and kernel development in *japonica* and *indica* rice. *Proc. Crop Sci. Soc. Japan* **38**, 425–433.

Nakayama, H. (1969). Senescence in rice panicle. I. *Proc. Crop Sci. Soc. Japan* **38**, 338–341.

Nakamura, K. and Suzuki, M. (1969–1973). In *JIBP/PP-Photosynthesis Level I experiments. Report* II–VI. JIBP/PP-Photosynthesis, Local Productivity Group.

Navasero, S. A. and Tanaka, A. (1966). Low-light induced death of lower leaves of rice and its effect on grain yield. *Plant and Soil* **14**, 17–31.

Okajima, H. (1960). Studies on the physiological function of the root system in the rice plant, viewed from the nitrogen nutrition. *Tohoku Univ. Inst. Agr. Res. Bull.* **12**, 1–146.

Okajima, H. and Takagi, S. (1959). Harmful effect of hydrogen sulphide on the rice plant growth. *J. Sci. Soil and Manure, Japan* **29**, 577–582.

Osada, A. (1964). Studies on the photosynthesis of *indica* rice. *Proc. Crop Sci. Soc. Japan* **33**, 69–76.

Osada, A. (1966). Relationship between photosynthetic activity and dry matter production in rice varieties, especially as influenced by nitrogen supply. *Bull. Nat. Inst. Agr. Sci. Japan Ser. D* **14**, 117–188.

Osada, A. and Murata, Y. (1962). Studies on the relationship between photosynthesis and varietal adaptability for heavy manuring in rice plant. I. *Proc. Crop Sci. Soc. Japan* **30**, 220–223.

Osada, A. and Murata, Y. (1965). *Ibid.* III. *Proc. Crop Sci. Soc. Japan* **33**, 460–466.

Osada, A., Nara, M., Dhammanuvong, S., Chakurabanthu, H., Rahony, M. and Gesprosert, M. (1972). The effect of seasons on the growth of *indica* rice in tropical area. *Proc. Crop Sci. Soc. Japan* **41**, *Extra Issue* 2, 87–88.

Oshima, M. (1966). On the translocation of ^{14}C assimilated at various growth stages to grains in rice plants. *J. Sci. Soil and Manure, Japan* **37**, 589–593.

Sasaki, T. (1919). Effects of environmental factors on the germination of pollen. *Nogakukaiho* **207**, 921–944; **208**, 1033–1049.

Satake, T., Nishiyama, I., Ito, N. and Hayase, H. (1969). Male sterility caused by cooling treatment at the meiotic stage in rice plants. I. *Proc. Crop Sci. Soc. Japan* **38**, 603–609.

Seko, H., Samoto, K. and Suzuki, K. (1956). Studies on the development of various parts of paddy rice plant. I. *Proc. Crop Sci. Soc. Japan* **24**, 189–190.

Shiroshita, T., Ishii, K., Kaneko, J. and Kitajima, S. (1962). Studies on the high production of paddy rice by increasing the manurial effects. *J. Centr. Agr. Expt. Sta.* **1**, 47–108.

Soga, Y. and Nozaki, M. (1957). Studies on the relation between seasonal changes of carbohydrates accumulated and the ripening at the stage of generative growth in rice plant. *Proc. Crop Sci. Soc. Japan* **26**, 105–108.

Takahashi, J. and Murayama, N. (1953). Absorption of mineral nutrients by rice plants according to growth stage. *Nogyo-gijutsu* **8**, 27–30.

Takahashi, J., Murayama, N., Oshima, M., Yoshino, M. and Yanagisawa, M. (1955). Influence of the amount of application of nitrogenous fertilizer upon the composition of paddy rice plants. *Bull. Nat. Inst. Agr. Sci. Japan Ser. B* **4**, 85–122.

Takahashi, Y., Iwata, I. and Baba, I. (1959). Studies on the varietal adaptability for heavy manuring in rice. I. *Proc. Crop Sci. Soc. Japan* **28**, 22–24.

Takeda, T. (1961). Studies on the photosynthesis and production of dry matter in the community of rice plants. *Jap. J. Bot.* **17**, 403–437.

Takeda, T. and Maruta, H. (1956). Studies on CO_2 exchange in crop plants. IV. *Proc. Crop Sci. Soc. Japan* **24**, 181–184.

Tanaka, A. (1958). Studies on the physiological characteristics and significance of rice leaves in relation to their position on the stem. XI. *J. Sci. Soil and Manure, Japan* **29**, 327–333.

Tanaka, A., Kawano, K. and Yamaguchi, J. (1966). Photosynthesis, respiration, and plant type of the tropical rice plant. *IRRI Tech. Bull.* **7**, 1–46.

Tanaka, A., Navasero, S. A., Garcia, C. V., Parao, F. T. and Ramirez, E. (1964). Growth habit of the rice plant in the tropics and its effect on nitrogen response. *IRRI Tech. Bull.* **3**, 1–80.

Tanaka, A. and Vergara, B. S. (1967). Growth habit and ripening of rice plants in relation to the environmental conditions in the Far East. *IRC Newsletter Special Issue*, pp. 26–42.

Tanaka, A., Yamaguchi, J., Shimazaki, Y. and Shibata, K. (1968). Historical changes of rice varieties in Hokkaido viewed from the point of plant type. *J. Sci. Soil and Manure, Japan* **39**, 526–534.

Tanaka, T. (1972). Studies on the light-curves of carbon assimilation of rice plants. *Bull. Nat. Inst. Agr. Sci. Japan Ser. A* **19**, 1–100.

Tanaka, T. and Matsushima, S. (1963). Analysis of yield-determining processes and its application to yield prediction and culture improvement of lowland rice. LXIV. *Proc. Crop Sci. Soc. Japan* **32**, 35–38.

Tanaka, T., Matsushima, S., Kojo, S. and Nitta, H. (1969). *Ibid.* XC. *Proc. Crop Sci. Soc. Japan* **38**, 287–293.

Thorne, G. N. (1965). Physiological aspects of grain yield in cereals. In *The growth of cereals and grasses*, eds F. L. Milthorpe and J. D. Ivins, pp. 88–105. Butterworths, London.

Togari, Y. and Kashiwakura, Y. (1958). Studies on the sterility in rice plant induced by superabundant nitrogen supply and insufficient light intensity. *Proc. Crop Sci. Soc. Japan* **27**, 3–5.

Togari, Y., Okamoto, Y. and Kumura, A. (1954). Studies on the production and behavior of carbohydrates in rice plants. I. *Proc. Crop Sci. Soc. Japan* **22**, 95–97.

Tsuno, Y. (1968). Analysis of matter production in high-yielding rice. *Proc. Symposium on comparative studies on the primary productivity of various terrestrial ecosystems*, J.IBP. 1967, pp. 22–28.

Tsuno, Y. and Kitakado, K. (1970). Physiological and ecological studies on the high yield of rice. VII. *Proc. Crop Sci. Soc. Japan* **39**, *Extra Issue* 1, 11–12.

Tsunoda, K. (1964). Studies on the effects of water-temperature on the growth and yield in rice plants. *Bull. Nat. Inst. Agr. Sci. Japan Ser. A* **11**, 75–174.

Tsunoda, S. (1959a). A developmental analysis of yielding ability in varieties of field crops. I. *Jap. J. Breed.* **9**, 161–168.

Tsunoda, S. (1959b). *Ibid.* II. *Jap. J. Breed.* **9**, 237–244.

Tsunoda, S. (1960). High-yielding varieties viewed from their morphology and function. In *Morphology and function of the rice plant*, ed. T. Matsuo, Nogyo-gijutsu-kyokai, Tokyo, pp. 179–228.

Tsunoda, S. (1962). A developmental analysis of yielding ability in varieties of field crops. IV. *Jap. J. Breed.* **12**, 49–56.

Tsunoda, S. (1964). Leaf characters and nitrogen response. In *The mineral nutrition of rice plants, IRRI*, Johns Hopkins Press, Baltimore, Maryland, pp. 401–418.

Wada, E., Baba, I. and Furuya, T. (1945). Studies on the prevention of drought damage. I. *Nogyo-oyobi-engei* **20**, 131–132.

Wada, G. (1969). The effects of nitrogenous nutrition on the yield-determining process of rice plant. *Bull. Nat. Inst. Agr. Sci. Japan Ser. A* **16**, 27–167.

Wada, G. and Matsushima, S. (1962). Analysis of yield-determining processes and its application to yield prediction and culture improvement of lowland rice. LXI. *Proc. Crop Sci. Soc. Japan* **31**, 15–18.

Waggoner, P. E. (1969). Environmental manipulation for higher yields. In *Physiological aspects of crop yield*, eds J. D. Eastin, F. A. Haskins, C. Y. Sullivan and C. H. M. van Bavel, ASA and CSSA, Madison, Wisconsin, pp. 343–373.

Wells, B. R. and Johnston, T. H. (1970). Differential response of rice varieties to timing of mid-season nitrogen applications. *Agron. J.* **62**, 608–612.

Yabuki, K. and Kamotani, K. (1971). The effect of wind speed on the photosynthesis of a rice field, measuring carbon dioxide flux on the surface air layer over plant community with a newly devised instrument for the aerodynamical method. *JIBP/PP-P, Level III Experiments, Report 1970*, pp. 1–5.

Yamada, N., Murata, Y., Osada, A. and Iyama, J. (1954). Photosynthesis of rice plants. I. *Proc. Crop Sci. Soc. Japan* **23**, 214–222.

Yamada, N., Murata, Y., Osada, A. and Iyama, J. (1955). *Ibid.* II. *Proc. Crop Sci. Soc. Japan* **24**, 112–118.

Yamada, N., Ota, Y. and Kushibuchi, K. (1957). Studies on ripening of rice. I. *Proc. Crop Sci. Soc. Japan* **26**, 111–115.

Yamada, N., Ota, Y. and Nakamura, T. (1961). Ecological effects of planting density on growth of rice plant. *Proc. Crop Sci. Soc. Japan* **29**, 329–333.

Yamamoto, K. (1954). Studies on the maturity of rice plants. III. *Nogyo-oyobi-engei* **29**, 1161–1163, 1303–1304, 1425–1427.

Yamazaki, K. (1960). Studies on the morphogenesis of crop plants under different growing conditions. II. *Proc. Crop Sci. Soc. Japan* **28**, 262–265.

Yoshida, S. (1972). Physiological aspects of grain yield. *Ann. Rev. Plant Physiology* **23**, 437–464.

Yoshida, S. and Ahn, S. B. (1968). The accumulation process of carbohydrate in rice varieties in relation to their response to nitrogen in the tropics. *Soil Sci. Plant Nutrition* **14**, 153–162.

Yoshida, S., Cock, J. H. and Parao, F. T. (1973). Physiological aspects of high yields. In *Rice Breeding. IRRI*, Los Baños, pp. 455–468.

5

Wheat

L. T. EVANS, I. F. WARDLAW AND R. A. FISCHER

Wheat is among the oldest of man's crops, the most extensively grown, and the one produced in greatest amount (cf. Table 1.1). Several different species are still cultivated, but the modern bread wheat, the hexaploid *Triticum aestivum* L., is the most abundant by far, and the main subject of this chapter.

The first steps in the evolution of wheat as a crop plant occurred almost 10000 years ago in the Fertile Crescent of the Middle East (Harlan and Zohary, 1966). Archaeological investigations of village sites inhabited about 8000 B.C. have yielded plant remains almost exclusively of the wild diploid einkorn, *Triticum boeoticum*, along with wild barley, lentils and vetch, whereas remains at the village of Ali Kosh in southwest Iran, of 7500–6750 B.C. vintage, include a significant proportion of domesticated barley and wheat.

Even today there are extensive, though endangered, communities of wild einkorn, barley and oats, together with the wild diploid goatgrass, *Aegilops speltoides*, which hybridized with wild einkorn to give the wild tetraploid emmer wheat, *T. dicoccoides*. The domesticated forms of the diploid and tetraploid wheats differed from their wild progenitors mainly in having a tougher rachis, thereby retaining much more grain in the ear until threshing. Grain size was also much larger in the domesticated forms, and along with this effect of primitive selection there were larger leaves and more rapid seedling establishment. In some of the domesticated tetraploids, such as *T. dicoccum*, the grains were still invested by the glumes as in the wild progenitors, but in others, such as *T. durum*, the grains were free-threshing.

The next step in the evolution of wheat was the hybridization of a tetraploid wheat with another species of *Aegilops*, probably *A. squarrosa*, to yield the hexaploid wheat species, some with invested grains such as *T. spelta*, others with free-threshing grains such as *T. aestivum*. *A. squarrosa*, the donor of the D genome, is naturally distributed from the eastern edge of the Fertile Crescent to Kazakhstan and, as such, occupies a far wider range of environments than the other wheat progenitors. Zohary *et al.* (1969) argue, therefore, that the addition of the D genome conferred not only the protein characteristics so desirable for bread making, but also a greatly increased adaptive range, enabling wheat to become a crop of both sub-humid and semi-arid steppes. They suggest that this hybridization has been recurrent, and still continues. *T. aestivum* is known to have occurred in the Middle East at least as early as 5800 B.C. (Helbaek, 1966), and subsequently spread from there around the Mediterranean, and through Central Europe to higher latitudes and more humid environments (Waterbolk, 1968).

Cytogenetic aspects of wheat evolution have been discussed by Riley (1965), and physiological aspects by Evans and Dunstone (1970). With increase in ploidy, and with the shift from wild to cultivated forms, there has been a parallel increase in grain and leaf size, over a 20-fold range, along with a fall in the light-saturated rate of photosynthesis. Grain filling has become of longer duration, associated with progressively more delayed senescence of the upper leaves. The protracted tillering of the wild diploids has been reduced in the course of evolution of the crop, as also has the extent of investment of assimilates in root growth, and in modern wheats a higher proportion of assimilates is translocated to the grain. These changes imply, on the one hand, a decline in the ability of wheat to compete and survive in natural, mixed communities, but on the other a considerable increase in yield potential under the increasingly controlled conditions provided by cultivation.

The wheat plant has been comprehensively described by Percival (1921), and aspects of its physiology have been discussed by Peterson (1965), Quisenberry and Reitz (1967) and Wardlaw (1974).

Before considering the various steps in the life cycle of a wheat crop, it is important to emphasize the wide range and different sequences of conditions encountered by wheat around the world. Much wheat is grown under seasonal conditions and sequences comparable to those in which it evolved, namely in relatively low latitudes (around 30°) with the crop undergoing its vegetative development in cool, short winter days (10–11 h) of relatively low light intensity. Flower initiation occurs, and the inflorescences differentiate, under increasing day-length, incident radiation and temperature, while grain filling takes place under still brighter light, with day-lengths of 13–14 h. However, rapidly rising temperatures (with daily maxima of 30 °C or more) and increasing water stress frequently terminate grain filling. Thus, rapid reproductive development, once the risk of frosts is over, is essential for effective adaptation to these conditions, except for crops under irrigation. Wheat sown in Mediterranean climates once the autumn rains begin, or at the end of the monsoon, as in India (Asana, 1966), follows this sequence.

Under more continental climates, at latitudes up to 60° or more, spring-sown crops may go through their whole life cycle under long days, with a sharp rise in temperature, radiation, and often water stress, during grain filling, which may be completed within a month of midsummer, as in the Canadian and Russian prairies.

In maritime climates at moderately high latitudes, as in western Europe, on the other hand, inflorescence differentiation takes place under long and increasing day-lengths, but grain filling occurs mostly after midsummer, in relatively cool day temperatures (about 20 °C), and being thereby more prolonged, ripening occurs under declining light.

The characteristics conferring high yields are thus likely to differ substantially from one environment to another, and even within one environment there may be several strategies for success.

VEGETATIVE GROWTH

Seed germination and establishment

Germination will occur between 4 and 37 °C, with 20–25 °C optimal. The minimum moisture content for germination is 35–45 % of grain dry weight, and germination is more rapid as moisture increases above this level. Light is not of great importance in controlling germination of wheat (Grahl, 1965).

A degree of dormancy is of value, particularly since the advent of mechanical harvesting, in preventing germination in the head before harvest under moist field conditions (Belderok, 1961; Jensen, 1967). Germination can occur at a relative humidity of 97.7 %, which is below the permanent wilting point for the growing plant (Owen, 1952), and as a seedling develops it becomes more susceptible to water deficiency (Milthorpe, 1950). It is normal practice to increase sowing depth to overcome the problem of premature germination following a light initial rainfall, but this in turn reduces seedling vigour.

Following germination the seminal roots extend into the soil and the coleoptile penetrates to the surface. The growing point is subsequently raised to the surface by expansion of the rhizome, the internodes above the coleoptile. Emergence rate in wheat shows genetic variation and is positively correlated with coleoptile length and plant height, and was therefore a problem initially with dwarf wheats. Selection for long coleoptiles has been only partially effective in improving their emergence (Allan *et al.* 1962).

From the start of germination until exposure of the first green leaf to light, growth is dependent on reserve carbohydrates in the endosperm, more than half of which are utilized by the seminal roots (Williams, R. 1960). The larger the seed, the greater the reserves, the faster is seedling establishment. In weedy or sparse stands, the sowing of larger seeds often results in greater grain yield, but not always so in pure stands (Percival, 1921; Pinthus and Osher, 1966; Roy, 1973). Protein content rather than seed size may be the main factor influencing seedling development (Schlehuber and Tucker, 1967; Lowe and Ries, 1972). Crops from high-protein seed, induced for example by simazine treatment in the preceding generation, may also develop higher protein and dry weight yields (Lowe *et al.* 1972).

The drought resistance of a crop can be increased by allowing the grains to take up water to about 30% of their dry weight for 24 hours, and then air drying them, repeating the sequence several times. Following such 'hardening' treatment, the fall in relative leaf water content during stress is slowed (Woodruff, 1969). Whether or not this results in enhanced grain yield depends on the stage in the life cycle at which drought stress occurs.

Root growth and function

The growth of roots is restricted to a zone about 10 mm behind the root tip (Eliasson, 1955). The rate of extension, after starting high during dependence on grain reserves (May *et al.* 1967), varies from 0.5 to 3.0 cm day^{-1} in the main primary or adventitious roots (Barley, 1970), and is relatively constant over long periods in a uniform medium (Brouwer, 1966). There is little or no penetration of roots into dry soil (Salim *et al.* 1965).

Root growth may exceed shoot growth at low temperatures (Welbank, 1971), but as temperature rises the growth of shoots increases more than that of roots (Brouwer, 1966). Shoot growth thus appears to have a higher optimum temperature than root growth, but this difference may result from increased competition for assimilates between root and shoot at higher temperatures (Friend, 1966; Wardlaw, 1968). Similarly, the root:shoot ratio rises as light intensity increases (Nelson, 1963). Roots are poor competitors with other organs when there is a limited supply of available carbohydrate, the lower leaves of a culm being their main source of assimilates (Wardlaw, 1967; Rawson and Hofstra, 1969). With moderate water stress, on the other hand, shoot growth may be reduced more than is photosynthesis, but some root growth may remain active, leading to increase in the root:shoot ratio. Similarly, limited nitrogen supply may reduce shoot growth but increase root extension and the root:shoot ratio (Brouwer, 1966).

Two distinct root systems develop in wheat, the seminal roots arising directly from or below the seed, and the adventitious roots that arise from the nodes of the stem above the seed. Following the emergence of the primary root through the coleorhiza, the first pair of lateral roots follows and then a second pair of laterals in the same plane as the first, but a little above them. The laterals may grow initially at about 60° to the vertical, but frequently turn to the vertical when 5 to 30 cm long (Passioura, 1972; Barnard, 1974). The vertical roots generally reach a depth of between one and two metres depending on soil conditions (Barley, 1970). Adventitious roots appear somewhat later at the coleoptile node, and in turn from the nodes above it, varying in number from none at all under dry conditions to more than 100 (Locke and Clark, 1924). Each tiller normally develops its own system of adventitious roots and each root system normally supplies the shoot to which it belongs (Krassovsky, 1926; Boatwright and Ferguson, 1967).

Functionally there seems to be little difference between the two parts of the root system although they show quite distinct distribution patterns. In the field growth of the root system continues until heading, when it may cease, and roots may even degenerate during grain development (Brenchley and Jackson, 1921; Asana and Singh, 1967). Given adequate water and nutrient, however, root growth and nutrient uptake continue well into the period of grain development (Carpenter *et al.* 1952).

Winter cereals tend to produce a larger weight of roots than spring sown ones (Troughton, 1962), presumably due to the longer period of growth at low temperatures. Cultivars differ in the distribution and size of their root systems (Pinthus and Eshel, 1962; Asana and Singh, 1967; Derera *et al.* 1969). Although there is no consistent difference between tall and dwarf wheats in this respect, root weight is sometimes greater in dwarf wheat cultivars (Fischer, unpublished), as it is in plants dwarfed by applications of growth retardants (Humphries, 1968).

Optimum rooting strategies are dependent on both climate and soil type. In Mediterranean climates, for example, there may be little penetration of water to depth, whereas under monsoonal conditions there may be little rain during crop growth, which largely depends on water previously stored at depth. In the latter situation, continuous growth and penetration of roots throughout development will be beneficial in maintaining water uptake (Salim *et al.* 1965; Hurd, 1968).

Increased efficiency of water use could result from restriction on the ability of the plants to extract water from the soil during early crop growth, leaving more to support the grain-filling stage. This may be approached by selection of plants with small, erect leaves and reduced tillering. An alternative approach, suggested by Passioura (1972), would be selection for a higher resistance to water transfer, particularly in the seminal roots in which water movement is dominated by a single, large metaxylem vessel.

The mineral nutrient requirements of wheat have been reviewed by Kostic *et al.* (1957).

Shoot growth is dependent on root function and the role of roots is not confined to nutrient and water uptake. Cereal roots can reduce nitrate (Miflin, 1967), but most nitrate reduction probably occurs in leaves in the light (Stoy, 1955; Minotti and Jackson, 1970). Roots may also synthesize amino acids, and act as a source of growth substances such as cytokinins for the shoots, but the significance of their role in this respect is not clear.

Leaf growth

The rate at which leaves form on the apical meristem, emerge and unfold, as well as the shape and size of the mature lamina, depend on the temperature, light intensity, day-length and nutritional status under which the plant is grown. With constant light and temperature conditions Friend *et al.* (1962) obtained a maximum area per leaf at 10000–19000 lx and 20 °C. The temperature for maximum width is lower than that for maximum blade length (Chonan, 1971).

Leaf arrangement is an important aspect of canopy structure. Leaves formed prior to floral induction originate close to the crown, but following induction the stem internodes elongate, and the leaves are separated further in the vertical plane, giving more effective light distribution within the canopy, except in extreme dwarf types. Maximum leaf area per culm is attained before heading when the flag leaf has fully emerged (Watson *et al.* 1963; Fischer and Kohn, 1966; Puckridge, 1971).

Tillering

Following germination, lateral bud primordia are initiated at the shoot apex 2–3 plastochrones after the primordia of the subtending leaves. The first bud to grow depends on planting depth (Percival, 1921) and temperature (Taylor and McCall, 1936). At a depth of about 5 cm the third axillary bud is often the first to grow, but in shallow plantings Rawson (1971) found that up to 90 % of the coleoptile buds may produce tillers in some varieties of wheat, although a reduced number of these survive to maturity.

Alterations in environmental conditions appear to have comparatively little effect on the initiation of tiller buds at the apex, but have a marked effect on the subsequent growth of these buds. During its initial growth the tiller is enclosed in the sheath of the subtending leaf and is entirely dependent on the shoot for its supply of carbohydrate and nutrients; tillers do not become independent of their parent shoots until they have developed about three mature leaves, when adventitious roots may form at their base. Tillering is favoured by high light intensity (Khalil, 1956; Friend 1965*b*, 1966) and nutrition (Gericke, 1922; Asana *et al.* 1966). The rate of tillering is maximal at 25 °C according to Friend (1966), but Rawson (1971) found greater numbers of tillers at lower temperatures, the slower rate being more than compensated for by the greater duration of tillering.

The onset of stem elongation following flower initiation usually causes a temporary cessation of tillering (Jewiss, 1972), the maximum tiller density reached at that time being very dependent on the level of solar radiation up to that time (Figure 5.1), as well as on cultivar (Rawson, 1971). However, differences between cultivars may be lost as tiller numbers fall towards anthesis to a level largely determined by environmental conditions (Bingham, 1969; Figure 5.2). The tillers which fail to reach maturity may be considered as lost capital. However, it is clear that although tillers may become independent they are still capable of integrated activity, and under adverse conditions both nutrients and carbohydrates may be induced to move between adjacent

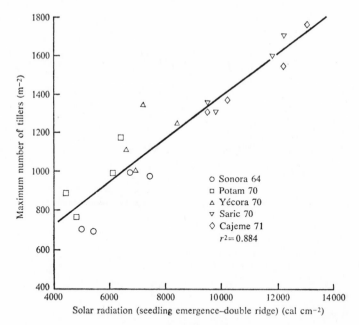

Figure 5.1. The relation between maximum tiller number, measured 6 to 9 weeks after sowing, and total solar radiation in the period from seedling emergence to floral initiation. Five semi-dwarf spring wheats of differing maturity sown at 4 dates in the field under irrigation and at high fertilizer rates at CIANO research station, Ciudad Obregon, Mexico (R. A. Fischer, unpublished data).

shoots (Smith, 1933; Wardlaw *et al.* 1965). Late tillers, although not completing their development, may therefore be of some value in ensuring a continuing supply of nutrients to the ear-bearing shoots during the later stages of development (Palfi and Deszi, 1960; Rawson and Donald, 1969; Lupton and Pinthus, 1969). Moreover, tillering confers flexibility to cope with adverse conditions, such as poor germination and frost or hail damage, and to take advantage of favourable conditions later in the season.

Stem growth and characteristics

Plant height can range from 0.3 m in extreme dwarf varieties to 1.5 m in some long-strawed European varieties. The traditional tall-growing varieties of wheat were selected over the centuries for their rapid emergence and ability to compete with weeds by shading them, to give an adequate yield with minimum care under conditions of low soil fertility (Athwal, 1971). With increased use of fertilizers and modern methods of cultivation these characteristics are no longer an asset, and may be detrimental under fertile conditions because of increased lodging. Although short-strawed forms have

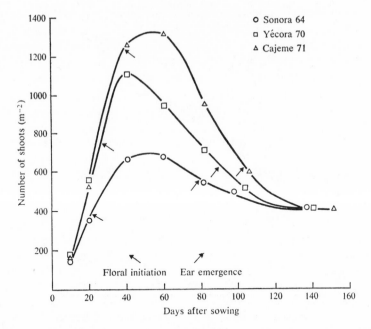

Figure 5.2. Change in shoot number with time in early (Sonora 64), mid (Yécora 70) and late (Cajeme 71) varieties of wheat grown in the field under irrigation and heavy fertilizer application at CIANO research station, Ciudad Obregon, Mexico (R. A. Fischer, unpublished data).

long been known, the first intensive breeding programme for semi-dwarf wheats began in Japan 50 years ago and it was not until the 1960s that short-strawed varieties were released extensively in both the USA and Mexico. 'Norin 10', from which most of today's semi-dwarf wheats are derived, resulted from crosses between a native Japanese variety, 'Daruma', and two American varieties (Reitz and Salmon, 1968). The success of 'Norin 10' and its crosses may have been due as much to associated increases in tillering and in the number of fertile florets per spikelet as to the reduction in stem length, however.

Early internodes are short and marked stem elongation does not occur until after floral initiation (Chinoy and Nanda, 1951). Later growth of the stem, like that of the leaf, is from an intercalary meristem, the internode extending only when the leaf inserted above it has finished expanding. The top internode (peduncle) may continue to extend until after anthesis. The encircling leaf sheaths aid in keeping the stem straight while the developing stem is still pliable. The internodes become progressively longer from the base to the top of the stem, and the uppermost internode may account for as much as half of the total shoot height (Percival, 1921; Rawson and Evans, 1971).

Stem growth occurs concurrently with that of leaves, roots and ear, rapid ear growth coinciding with that of the internode below the flag leaf (Wardlaw, 1974). Similarly, early grain growth may coincide with that of the peduncle (Carr and Wardlaw, 1965; Wardlaw, 1970). Consequently, growth of the stem may compete with that of the ear under limiting substrate conditions (Rawson and Hofstra, 1969; Patrick, 1972a; Friend, 1965a). Shortening of the upper internodes may therefore free assimilate for additional floret differentiation or grain filling, or for more tillering as suggested by Simpson (1968). There may be little advantage in the latter, however (cf. Figure 5.2).

Mature stem tissue can store carbohydrates in the form of sucrose and more complex oligosaccharides, although starch cannot usually be detected (Barnell, 1938; Lopatecki *et al.* 1962). Storage in the stem is most active at the time of anthesis before grain growth has begun, when leaf area is maximal and stem and root growth minimal. Analysis of dry weight changes suggest that some grain carbohydrate is derived from photosynthate formed prior to anthesis and stored temporarily in the stem (Asana and Mani, 1950; Asana and Saini, 1962) and this has been confirmed by the use of ^{14}C-labelled assimilates (Stoy, 1963). However, such stored carbohydrate supplies less than 10 % of the final grain yield, except under drought or other stress conditions (Asana and Saini, 1958; Wardlaw and Porter, 1967; Rawson and Evans, 1971). Long-strawed wheats might therefore have an advantage in being able to store larger amounts of assimilates to support grain growth under stress conditions. However, Rawson and Evans (1971) found that tall cultivars of wheat were no more dependent on stem reserves than short ones, and no more able to draw on reserves under limiting conditions for photosynthesis. Like the sheath, the exposed green parts of the stem are capable of photosynthesis and this may be of some value under stress conditions (Wardlaw, 1971).

Leaf photosynthesis

Photosynthesis in wheat is mediated by the Calvin cycle (Graham *et al.* 1970) and, consequently, the carbon dioxide compensation point is high, around 50 vpm in all species and cultivars of wheat (Moss *et al.* 1969; Dvorak and Nátr, 1971), while the light-compensation point increases markedly with rise in temperature (Meidner, 1970). The action spectrum for photosynthesis by wheat leaves closely resembles that of many other crop plants (McCree, 1972). The dependence of net photosynthetic rate on temperature shows a rather broad optimum between 10° and 25 °C with high rates maintained at low temperatures but a sharp fall at higher ones (Murata and Iyama, 1963; Stoy, 1965; Friend, 1966; Sawada, 1970). Dark respiration increases with rising temperature (e.g. from 0.3 to 2.5 mg CO_2 dm^{-2} h^{-1} between 15° and 35 °C; Stoy, 1965), as does photorespiration rate between 14° and 35 °C (Jolliffe and Tregunna, 1968). Photorespiration tends to be inhibited at higher carbon dioxide levels, which may account for the finding of MacDowall (1972) that a rise in carbon dioxide level stimulates growth more at low than at high light intensity.

At atmospheric carbon dioxide levels, leaf photosynthesis in modern wheats approaches light saturation in about 1/3–1/2 full sunlight, with rates of 30–35 mg CO_2 dm^{-2} h^{-1}. In the wild progenitors, however, the rates are higher and less readily light saturated, increasing to beyond 70 mg CO_2 dm^{-2} h^{-1} with increase in incident radiation up to full sunlight (Evans and Dunstone, 1970; Khan and Tsunoda, 1970), if the plants are grown under high intensity light. In the modern wheats the photosynthetic rate is much less influenced by light conditions during growth than in the diploid wheats. The pronounced differences between species are associated with parallel differences in both gas phase and residual resistances to carbon dioxide exchange (Dunstone et al. 1973).

Leaf photosynthetic rate under given environmental conditions can vary over a more than two-fold range depending on the demand for assimilates within the plant (Birecka and Dakic-Wlodkowska, 1963; King et al. 1967). The flag leaf rate may fall sharply within several hours of ear removal, and rise again when alternative sinks, such as young tillers, increase the demand for assimilate from the flag leaves. Such feedback effects may account for some of the differences in photosynthetic rate between species when grown under high light intensities. They may also account for changes with time in the rate of flag leaf photosynthesis under given conditions. In many wheats this falls from a high rate at ear emergence to a lower rate for about one week after anthesis, when stem elongation and tillering has slowed but rapid grain growth has not begun. When it does so, the flag leaf photosynthetic rate rises again (Birecka and Dakic-Wlodkowska, 1966; Evans and Rawson, 1970; Rawson and Evans, 1971). The fall in rate following anthesis is not always evident, however, presumably because alternative sinks for assimilate are available at that time (Sawada, 1970; Osman and Milthorpe, 1971).

Changes in the demand for assimilates may account also for some of the differences between early and later formed leaves in their photosynthetic rates (Dunstone et al. 1973), but anatomical factors may also be involved. For example, Chonan (1965) has shown that the number of protuberances on mesophyll cells, which increase their surface:volume ratio, increases in the later formed leaves, especially in plants grown under high light intensity (Chonan, 1966). Smaller cells may likewise be associated with higher photosynthetic rates, as in the diploid wheats (Dunstone and Evans, 1974). To some extent this may account for the tendency of species or cultivars with larger leaves to have lower photosynthetic rates (e.g. Evans and Dunstone, 1970). Since high yielding ability is strongly associated with larger leaves, it too may display a negative relation with leaf photosynthetic rate (e.g.

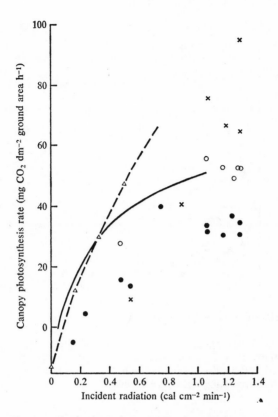

Figure 5.3. Photosynthesis by wheat crops in relation to incident radiation: (a) in field (October 14 = ○ ; 15 = × ; 16 = ●) (Denmead, 1970 and unpublished), (b) in field enclosure – solid line (from Puckridge and Ratkowsky, 1971), (c) in artificially lit cabinet – broken line (from King and Evans, 1967).

Planchon, 1969; Stoy, 1965). Nátr (1966) found a positive relation between photosynthetic rate and yield, but only late in the season when rates were falling. His results, therefore, are more an expression of the fact that leaves in the higher yielding cultivars tend to senesce later.

Canopy photosynthesis

Crops of wheat in the field have often been the subject of growth analysis for the determination of their net assimilation rate (e.g. Watson *et al.* 1963), and of studies in micrometeorology (e.g. Huber, 1952; Penman and Long, 1960; Denmead, 1969, 1970). Photosynthesis by wheat crops has also been investigated in the field by the use of enclosures (e.g. Puckridge, 1971; Puckridge and Ratkowsky, 1971), and in artificially lit cabinets (Wang and Wei, 1964; King and Evans, 1967).

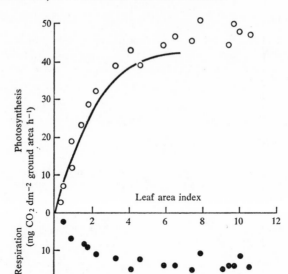

Figure 5.4. The relation between photosynthesis or respiration and leaf area index in wheat crops.

Solid line for a field crop under incident radiation of 0.6 cal cm^{-2} min^{-1} (from Puckridge and Ratkowsky, 1971); open circles for a crop in an artificially lit cabinet with a visible radiation flux of 0.27 cal cm^{-2} min^{-1}; solid circles for the same crop in darkness at 20 °C (from King and Evans, 1967).

The relation between incident radiation and net photosynthesis by crops depends on the leaf area index and canopy structure, as discussed below, but some results for well watered and developed canopies, obtained in the field by both aerodynamic and enclosure measurements, and also in growth cabinets, are given in Figure 5.3. At low radiation levels (< 0.4 cal cm^{-2} min^{-1}) crop photosynthesis increases almost linearly with increase in radiation. The extent of light saturation evident at high radiation levels varies considerably, even from day to day in the field. Denmead's (1970) measurements on 15 October, a day of high atmospheric carbon dioxide, show little evidence of light saturation, and higher maximum rates of photosynthesis than on 16 October, a day of lower carbon dioxide levels and wind speeds.

The relation between net carbon dioxide exchange and leaf area index (LAI) of wheat crops is illustrated in Figure 5.4. Net crop photosynthesis increases asymptotically with increase in LAI to about 6, but beyond that net carbon dioxide exchange rate was independent of LAI, with no clear optimum (cf. also Wang and Wei, 1964). Growth analysis experiments by Stoy (1965) also showed no fall in crop growth rate at LAI values up to 9, although net assimilation rate per unit leaf area declines with increase in

LAI. The more brightly lit uppermost leaves contribute most to net crop photosynthesis (Denmead, 1969, 1970; Puckridge, 1969), daytime carbon dioxide profiles within the crop showing marked depletion among the upper leaves. In the crop studied by Denmead the upward flux of carbon dioxide from the soil was still apparent among the lower leaves, and was probably sufficient to account for about one third of the total daytime uptake by the crop. Since net crop photosynthesis shows no decline at high LAI values there is unlikely to be a net carbon dioxide loss by respiration of the lower leaves in daylight.

Respiration and photosynthesis by ears will be considered later, but it should be noted here that the leaf sheaths and culm can contribute substantially to net photosynthesis by the crop (Puckridge, 1969; Evans and Rawson, 1970; Rawson and Evans, 1971; but cf. Sawada, 1970).

The photosynthesis–light response function of leaves is usually determined with perpendicularly-incident light. With oblique light the equivalent intensity is proportional to the cosine of the angle of incidence of the radiation (Kriedemann *et al.* 1964). This, together with a knowledge of canopy structure, has permitted crop photosynthesis under non-stress conditions to be effectively modelled (Connor and Cartledge, 1971; Osman, 1971*a*; Lupton, 1972).

Respiration

Whereas crop photosynthesis can now be modelled with reasonable confidence, at least under non-stress conditions, neither data nor concepts are adequate for satisfactory modelling of respiration by wheat crops.

Dark respiration by leaves, as influenced by age, temperature, season and cultivar, has been measured frequently (e.g. Stoy, 1965; Sawada, 1970). Rates of carbon dioxide output per unit dry weight for leaves at various ages and temperatures are comparable with those for roots, but notably lower than those for young stems (Sawada, 1970). However, the rate of stem respiration falls to much lower levels once elongation is completed (0.2–0.7 mg CO_2 g^{-1} DW h^{-1} (Stoy, 1965; Rawson and Evans, 1971), cf. 7–10 mg CO_2 g^{-1} DW h^{-1} (Sawada, 1970)), and remains fairly stable through the period of grain filling. At that stage the respiration rate per unit stem dry weight is inversely proportional to stem height, being almost four times as high in dwarf as in medium height cultivars (Rawson and Evans, 1971). Consequently, there are considerable differences between cultivars in the loss of stem weight by respiration during grain filling; in general this accounts for about one third of the total loss in stem weight, the remainder presumably being due to translocation.

Root respiration in wheat varies with time of day in relation to shoot photosynthesis, falling progressively during the night and rising rapidly once photosynthesis resumes in the light (Neales and Davies, 1966; Osman, 1971*b*).

Dark respiration by wheat crops is not proportional to accumulated crop dry weight or LAI, but reaches a plateau when LAI exceeds six (cf. Figure 5.4), the instantaneous rate at 20 °C being about one third of the net photosynthetic rate at each stage of growth (King and Evans, 1967). Dark respiration per unit dry weight therefore decreases progressively with growth of the crop, as found by Puckridge and Ratkowsky (1971). With seedlings, Sawada (1970) also found a linear relation between plant respiration and photosynthetic rate, but this showed considerable seasonal variation.

An important assumption in some methods of estimating crop respiration is that respiration is tightly coupled at all stages of crop development. However, with wheat plants in which the ears were removed during grain filling, the proportional loss of ^{14}C-activity fixed in photosynthesis was much higher than in intact plants (Birecka and Dakic-Wlodkowska, 1963; Birecka, 1968), and in subsequent experiments, Birecka *et al.* (1969) have found both oxygen uptake and carbon dioxide release by fully grown internodes of wheat stems to increase about two-fold following ear removal. Whether this apparent uncoupling of respiration is due to assimilate accumulation or to injury remains to be established.

REPRODUCTIVE DEVELOPMENT

The timing of the reproductive cycle in wheat is an important determinant of yield. If inflorescence development begins too soon, there may be extensive frost injury to young ears. On the other hand, if it is too late, or too slow, grain filling may be cut short by high temperatures and water stress, or harvesting problems may develop. The wide range of environments in which wheat is grown is matched by the equally wide range among cultivars in their response to environmental conditions during both the initiation and the development of inflorescences. Control of the reproductive cycle is mostly gained by varietal responses to vernalization and long days before inflorescence initiation.

Many of the wild progenitors and primitive wheats display marked responses to vernalization and long days which are clearly of importance in their adaptation to Mediterranean climates. Spring wheats from higher latitudes, such as northern Europe or Canada, show pronounced responses to long days, but little to vernalization. Winter wheats, on the other hand, show a strong response to, or even an obligate requirement for, vernalization. In lower latitudes, as in Australia or India, most cultivars are less delayed in their flowering by short days, but even 'spring' wheats may display some response to vernalization (Cooper, 1956, 1957; Gott, 1961; Misra, 1954; Razumov and Limar, 1971; Syme, 1968). Where winters are less cold, a need for vernalization may serve to delay flowering in the spring, but in areas with severe winters its adaptive role may be to prevent inflorescence initiation in the autumn, while the need for long days may delay it in spring until the risk of frosts has passed.

Some control of flowering time can be gained from variations in the minimum age at which plants respond to day-length. Such variations are known to occur among rice cultivars, but have scarcely been examined in wheat. Some wheats can respond to long days when only their first leaf has appeared (Cooper, 1956), whereas Gott (1961) found one Canadian cultivar to be insensitive to day-length for at least 75 days: such differences could be important in determining local adaptation and agronomic practice.

Vernalization

The vernalization response is believed to take place in the shoot apex, even in excised shoot apices of wheat (Ishihara, 1961). Thus, both imbibed grains in the soil and young green plants (Gott, 1957) can be vernalized. In fact, even developing grains exposed to low temperatures can have their vernalization response satisfied while still in the ear (Weibel, 1958; Riddell and Gries, 1958 b). According to Chujo (1966 a) the effectiveness of vernalization decreases as plants age, with complete loss of effect after three months.

Vernalization at temperatures below 0 °C appears to be very slow (Schmalz, 1958; Ahrens and Loomis, 1963). Junges (1959) suggested that 3 °C was the most effective temperature for vernalization of winter wheat, but 10 °C for spring wheats (cf. also Chujo, 1966 a). 11 °C is the upper limit for effective vernalization, but day temperatures up to 30 °C do not reduce the vernalizing effect of cold nights provided they are not prolonged beyond 8 h each day (Chujo, 1966 b). Extended periods at high temperatures during vernalization can destroy its effectiveness (Chujo, 1967).

Short day vernalization

The response to vernalization and day-length interact as they do in many long day plants, vernalization reducing the need for subsequent long days. On the other hand, exposure of unvernalized plants to short days during early growth can accelerate inflorescence initiation in plants subsequently transferred to long days, as first noted in wheat by McKinney and Sando (1935). This 'short day vernalization' can replace the need for low temperature vernalization, and Cooper (1960) found short days to be as effective as cold days in hastening the flowering of several wheats. Gott (1961), on the other hand, found no evidence of short day vernalization, and Krekule's (1964) results suggest that short days replace cold vernalization only in those cultivars in which they inhibit growth. In cultivars which do not become dormant in short days, such as most of those from low latitudes, short days cannot replace low temperatures.

Response to long days

Most wheat cultivars are quantitative long day plants, flowering sooner the longer the day-length, but without a minimum day-length below which they will not initiate inflorescences at all, although Samygin (1946) lists a number of cultivars which appeared to remain vegetative indefinitely in short days.

Cultivars differ mainly in the extent to which short days delay their flowering (e.g. Cooper, 1956; Gries *et al.* 1956), and also in the extent to which the response to day-length is modified by temperature (Riddell and Gries, 1958*a*). High temperatures have little effect when combined with long days, but may greatly increase the delay in short days for some cultivars, e.g. 'Chinese'. In others, e.g. 'White Federation', the delay is more accentuated at low temperatures. The response to long days is a true photoperiodic response, being little affected by the light intensity during the latter part of the day (Riddell *et al.* 1958).

Inflorescence development

The morphogenesis of the wheat inflorescence has been described by Bonnett (1967) and its histogenesis by Barnard (1955), while Williams, R. (1966) presents a detailed quantitative description of inflorescence development.

The rate of inflorescence development is usually faster the higher the light intensity (Friend, 1965*a*), the longer the day-length (Williams, R. and Williams, C. 1968; Riddell and Gries, 1958*a*), and the higher the temperature (Friend, 1965*a*; Riddell and Gries, 1958*a*; Halse and Weir, 1970; Rawson, 1970). In some cultivars the requirements for long days for development of inflorescences is more pronounced than that for their initiation (Halse and Weir, 1970; but cf. Gott, 1961), and short days may either prevent ear development completely, or cause abnormal differentiation (Rawson, 1971).

The number of spikelets formed is higher when light intensity is higher (Friend, 1965*a*); at high planting densities, therefore, spikelet numbers may be reduced (Puckridge, 1968), as also in densely tillered stands, or following partial defoliation (Davidson, 1965). High nitrogen levels increase spikelet number (Single, 1964; Beveridge *et al.* 1965), but only when applied before inflorescence initiation (Langer and Liew, 1973).

Conditions accelerating floral induction tend to reduce spikelet number by hastening the formation of the terminal spikelet. To some extent, therefore, the observation of Thorne *et al.* (1968) that spikelet number is determined by conditions prior to inflorescence initiation is pertinent, especially in winter wheats, but Rawson's (1970) experiments clearly establish the importance of post-inductive conditions also. For example, exposure to additional long days beyond the number required for early induction progressively reduced spikelet number, grain number and yield per ear in proportion. In fact, there was an inverse relation between day-length and spikelet number (Rawson, 1971).

Pinthus (1967) found winter wheats grown in Israel to have higher spikelet numbers and grain yields than spring cultivars. From Rawson's (1970) work, the basis of this response appears to be mainly the accumulation of more potential spikelet primordia at the apex of plants because inflorescence initiation is delayed by an unsatisfied requirement for vernalization. Selection for yield potential may therefore, unwittingly, result in the retention of a vernalization response, as in several of the Mexican dwarf wheats. Once initiation has occurred, there is little difference between vernalized and un-vernalized plants in their rate of inflorescence development (cf. also Cooper, 1956; Riddell and Gries, 1958*b*).

After the terminal spikelet has been formed, environmental conditions no longer influence spikelet number, but they may affect the number of florets differentiated within each spikelet, although such effects have scarcely been explored. The number of grains formed in an ear is dependent on light levels between initiation and anthesis (Willey and Holliday, 1971), and it appears that pollen development is particularly sensitive to water stress and high temperatures at the stage of meiosis (Bingham, 1966; Fischer, 1973). Changes with time in floret volume (Williams, R. 1966) and stage of differentiation (Langer and Hanif, 1973) for the various florets have been followed, but to what extent the development of upper florets competes with stem, awn or leaf development, as influenced by environmental conditions, is unknown. Long days hasten floret development, whereas high nitrogen levels delay it (Langer and Hanif, 1973) and increase the number of grains per spikelet (Single, 1964; Langer and Liew, 1973).

The period from inflorescence initiation to anthesis may vary from two weeks to several months in duration, depending on cultivar and environment. Because the time of initiation must not be too early, if the developing ear is to escape frost injury, whereas grain filling should begin as soon as possible to coincide with high crop photosynthesis and to avoid environmental stress at the end, one objective of breeding programmes might be to increase the rate of inflorescence development and so minimize its duration, since there appear to be heritable differences in this rate (Pinthus, 1963). To do so, however, could be counter-productive, in that it could adversely affect the number of spikelets per ear, and perhaps also fertile florets per spikelet, which are frequently powerful yield determinants in wheat. Potential ear number per unit area is largely determined by the extent of tillering before inflorescence initiation, while the other major yield component, grain size, is determined mainly by conditions after anthesis. But to take full advantage of favourable conditions during grain filling requires the formation of many spikelets, and relatively slow *initial* development of inflorescences, an explicit objective of some European wheat breeders (e.g. Bingham, 1967). Rapid late elongation of inflorescences may likewise conflict with the development of many fertile florets per spikelet. The optimum rate of inflorescence development is, therefore, a compromise between the need to develop sufficient storage capacity and the need to fill it, a compromise whose solution varies with the sequence of conditions in each environment.

Anthesis, fertilization and grain set

Not all florets reaching anthesis set grains (e.g. Evans, Bingham and Roskams, 1972), and it is important to consider what reduces grain setting below its potential level.

The biology of flowering in wheat has been reviewed by de Vries (1971), with special reference to hybrid seed production. Florets usually open in the early hours of daylight for 8 to 30 minutes, or more. Under field conditions pollen is viable for only a few hours after it is shed (Goss, 1968), and pollen germination occurs 1 to 1.5 hours after pollination (Hoshikawa, 1959), with fertilization 3 to 9 hours later depending on the temperature (Morrison, 1955; Hoshikawa, 1960, 1961b, but cf. Percival, 1921). If fertilization does not occur at anthesis, the carpel continues to grow slowly, retaining the ability to be fertilized for a further three to five days (Hoshikawa, 1961c; Evans, Bingham and Roskams, 1972), and the floret may reopen. High temperatures at anthesis may cause sterility, the optimum temperature for fertilization being 18–24 °C, the minimum 10 °C and the maximum 32 °C according to Hoshikawa (1959). Seed set is promoted by high light intensity during fertilization (Wardlaw, 1970), and is very susceptible to water stress (Asana, 1961; Wardlaw, 1971), but less so than meiosis (Fischer, 1973).

Rawson and Evans (1970) sterilized basal florets of the central spikelets, the first to reach anthesis, and found compensatory grain set not only in the more distal florets of those spikelets, but also in spikelets at the top and bottom of the ears of 'Triple Dirk' wheat. Consequently, grains per ear increased and since the replacement grains were as heavy as those they replaced, grain yield per ear was increased. This happened also in further experiments (Evans, Bingham and Roskams, 1972) in which the basal florets were not sterilized but emasculated, and their pollination delayed for up to six days. Apparently, the most advanced florets can inhibit grain set not only in distal florets of the same spikelets, but also in more distal spikelets, even when their fertilization is delayed. Both ovaries and anthers have this effect, which is presumably hormonal in nature. Cultivars differ considerably in the extent of this inhibition, which has presumably been reduced in the course of evolution of wheat from *monococcum* to *dicoccum* to modern cultivars which may set four or more grains per spikelet.

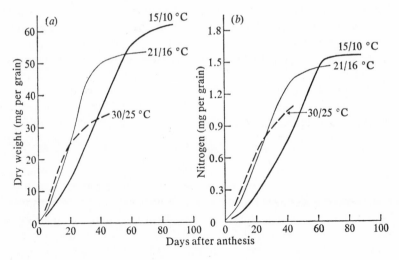

Figure 5.5. The effect of temperature on the accumulation of (a) dry weight and (b) nitrogen in grains in the basal florets of the middle spikelets of 'Late Mexico 120' wheat (from Sofield *et al.* 1974).

Grain growth

Growth of the grain, from initiation to maturity, follows a complex course, of several phases. Growth of the carpel prior to anthesis is biphasic, while that of the ovule occurs at a progressively falling rate (Williams, R. 1966). Growth is slow just prior to and following anthesis, at which stage there may be clear differences in ovary size between floret positions (Rawson and Evans, 1970) and between cultivars (Bremner, 1972). The initial growth following fertilization is predominantly in the maternal outer pericarp and the rest of the grain coat (Rijven and Cohen, 1961; Jennings and Morton, 1963a). Free endosperm nuclei are formed initially after fertilization (Morrison, 1955), at a rate depending on temperature (Hoshikawa, 1961a). Following cell wall formation the endosperm increases rapidly in cell number and then in size as starch storage begins one to two weeks after anthesis. Then follows a period of two to four weeks, depending on temperature and the incidence of water stress, of almost linear increase in grain weight (e.g. Asana and Bagga, 1966; Bremner, 1972; Evans and Rawson, 1970, and Figure 5.5), followed by an asymptotic approach to the mature grain weight. In this last phase the water content of the grain falls, at first slowly and then, once it reaches about 40 %, quite suddenly to 5–14 % (Jennings and Morton, 1963a; Rijven and Cohen, 1961; Asana and Bagga, 1966). Growth of the embryo shows an initial lag behind that of the endosperm, but continues throughout grain development.

Hoshikawa (1961 *a*) found that increase in the number of endosperm nuclei can begin within one day of pollination at 30 °C, the primary starch granules beginning to accumulate within five days and secondary granules within 10 days. Cell division was completed by 12 days, cell growth by 19 days, and maturity reached in 23 days. At 20 °C the corresponding intervals were 3, 7, 16, 19, 37 and 43 days. High temperatures increase the rate of endosperm cell formation (Wardlaw, 1970), but final cell number may be unaffected by temperature (Hoshikawa, 1961 *a*). Increase in endosperm DNA was found to cease about 14 days after anthesis in the experiments of Asana and Bagga (1966) and 19 days after anthesis in those of Jennings and Morton (1963 *b*). Low light intensity for 7–10 days after anthesis reduced endosperm cell numbers, and also final grain weight, especially at high temperatures (Wardlaw, 1970). Water stress over the same period, however, increased the rate of cell division and the initial rate of grain growth (Wardlaw, 1971; cf. also Konovalov, 1959; Asana and Joseph, 1964), but caused a reduction in final grain weight. The initial increase in cell number and grain weight was associated with a reduction in seed set by water stress (Wardlaw, 1971).

Temperature has a pronounced effect on the duration of grain filling, which continues for longer at lower temperatures, resulting in greater final grain weight (Wattal, 1965, Figure 5.5). In the experiments of Asana and Williams (1965) the main effect was due to day temperature, grain size being reduced 16 % with a rise from 25° to 31 °C, but Peters *et al.* (1971) found that a rise in night temperature from 9° to 26 °C almost halved grain yield by reducing the period of grain filling. The results presented in Figure 5.6 indicate a marked reduction in the duration of grain filling with increase in incident radiation for crops in the field. Phytotron experiments have shown, however, that it is temperature not radiation that has the predominant effect on duration of grain filling (cf. Figure 5.5, Marcellos and Single, 1972; Sofield *et al.* 1974), the results in Figure 5.6 presumably being due to covariation in radiation and temperature.

Growth rates per grain vary widely, up to a rate of 2.09 mg day^{-1} estimated from the results of Asana and Bagga (1966). Grains in different positions within an ear grow at different rates. Grains in second florets, for example, begin smaller but may grow faster and end up larger than those in basal florets (Rawson and Evans, 1970; Bremner, 1972; Rawson and Ruwali, 1972 *b*). Thus, there is no simple relation of the kind suggested by Bonnett (1967) between time of flowering and grain size. Grains in the upper spikelets grow more slowly than those below them, and receive much less assimilate from both the flag and penultimate leaves (Rawson and Evans, 1970; Evans, Bingham, Jackson and Sutherland, 1972). When the supply of assimilate from the leaves is reduced, by defoliation or shading, grain growth is most severely reduced in the upper spikelets (Bremner, 1972).

Varietal differences in grain size are frequently associated with differences

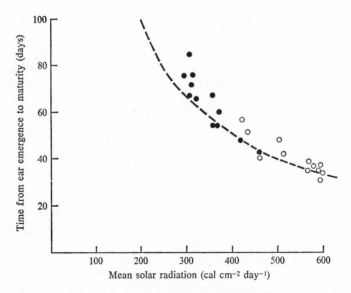

Figure 5.6. The relation between incident radiation and the length of the period from ear emergence to maturity in wheat. Solid circles, data from Welbank *et al.* (1968) for crops in England; open circles, unpublished data of Fischer for Mexican crops. The broken line represents a total radiation of 20000 cal cm⁻².

in yield, but whether they are associated with differences in rate or in duration of grain filling is not known. The latter seems more likely; Asana and Joseph (1964) found the grains of 'PBC 281' to increase at the same rate as those of 'NP 720' for the first 26 days after anthesis, but to continue growing for longer and thereby form larger grains, even when the plants were in darkness after the twenty-sixth day.

Final grain size depends to some extent on the number of grains per ear. For example, with ears emasculated in a variety of patterns, Bingham (1967) found the weight of grains in specific positions to increase as grain number per ear was reduced. This suggests that grain size may have been limited to some extent by the supply of assimilates. However, grain yield per ear fell substantially as the grain number was reduced, indicating a strong limitation on the capacity of the remaining grains to grow either faster or longer. To some extent this capacity appears to be determined by early interactions among the various florets and young grains, presumably of a hormonal nature, rather than by the supply of assimilate, which is unlikely to be limiting in the early stages of grain growth (Evans and Rawson, 1970; Wardlaw, 1971).

The processes involved in the maturation of the grain require further investigation. Grain volume may shrink from rapid loss of water by more than 50 % within a week (e.g. Asana and Bagga, 1966). It could be argued that the reduction in volume represents unused potential storage capacity, but the experiments discussed above suggest this is not so, and it is not clear why storage ceases in situations where assimilate is still available for storage. Estimations from the data of Jennings and Morton (1963 *a*) based on a final grain water content of only 5 % suggest that the concentration of sucrose in the grain does not rise above about 8 % as the water content falls, which is unlikely to inhibit storage enzymes (e.g. in Turner's (1969) analysis the sugar concentration was 9 % at the start of storage). Unfortunately, Rijven and Cohen's (1961) enzyme study stopped short of the period when grain fresh-weight suddenly falls, but enzymes involved in starch synthesis remained active until the end of the linear period of starch deposition, i.e. beyond the time when water content starts to fall. Alexandrov and Alexandrova have suggested that disintegration of the endosperm nuclei initiates grain ripening (in Frazier and Appalanaidu, 1965), but Jennings and Morton (1963 *a*, *b*) found no evidence of a fall in endosperm DNA or RNA content during ripening. It has been suggested that the rise in phytic acid content of the grain towards maturation may so reduce ATP levels that metabolism is inhibited, but Williams, S. (1970) found no evidence of a fall in ATP level, and suggested instead that the phytic acid may sequester the divalent cations in the grain.

Ear photosynthesis and respiration

The wheat ear often has a low net assimilation rate, and this has led to the conclusion that it is only of minor importance in supplying the grain with photosynthate (Gabrielsen, 1942). However, grain respiration may contribute more than 60 % of ear respiration, resulting in low apparent rates of ear photosynthesis (Carr and Wardlaw, 1965; Evans and Rawson, 1970).

Ear respiration increases rapidly with increase in the rate of grain filling, grain respiration being equivalent to about one quarter of grain storage (Evans and Rawson, 1970). During the first week after anthesis, ear photosynthesis may suffice not only for ear and grain respiration but also for up to three-quarters of the grain growth. As this latter increases, however, the contribution by ear photosynthesis may fall to less than 20 % of grain needs.

The overall contribution to grain filling by ear photosynthesis has been the subject of many investigations, reviewed by Thorne (1966). Shading of ears or leaves, interkernel competition, $^{14}CO_2$ feeding and gas analysis techniques have all been used, and have given rather different results, with the consequence that ear photosynthesis has been estimated to contribute from 10 % (Lupton, 1969) to 60 % (Saghir *et al.* 1968) of the grain dry weight.

Grain respiration generates much carbon dioxide within the ear, which is reassimilated to a considerable extent by day (Kriedeman, 1966). In fact, the grains themselves may reassimilate much of their respiratory carbon dioxide, as sufficient light to saturate their photosynthesis is transmitted through the glumes (Evans and Rawson, 1970), which may restrict gas exchange by the grains to a considerable extent (Abdul-Baki and Baker, 1970). Investing glumes could well be advantageous in this respect, as well as by reducing water loss from grains and possibly delaying maturation. Both sterile and fertile glumes are photosynthetic, but only the exposed parts of each lemma develop stomata (Teare, Law and Simmons, 1972). Nevertheless, the lemmas are photosynthetically the most active of the glumes, assimilate from them moving mostly to the enclosed grain (Bremner and Rawson, 1972).

The presence of awns can double the rate of net photosynthesis by ears (McDonough and Gauch, 1959; Evans and Rawson, 1970; Teare, Sij, Waldren and Goltz, 1972). In durum wheats, the surface area of awns may equal that of the ground surface and exceed that of flag leaves (McDonough and Gauch, 1959), but in hexaploid wheat their development is not usually so extensive (Teare and Peterson, 1971; Teare, Sij, Waldren and Goltz, 1972). Awn removal can reduce grain yield by 11–21 % (Miller *et al* 1944; Saghir *et al.* 1968), but early deawning can affect grain number as well as grain size, and the presence of awns may influence many processes besides photosynthesis, such as cytokinin movement to the grains (Michael and Seiler-Kelbitsch, 1972). Increase in the capacity of the ear for photosynthesis would appear to be a valid general objective in wheat breeding, since the ears are most favourably placed to intercept light and atmospheric carbon dioxide, the ear assimilates are mostly retained in the ear, and they provide a major contribution especially to grains in the more distal spikelets and florets (Evans *et al.* 1972). Miller *et al.* (1944), Vervelde (1953) and Grundbacher (1963) have reviewed the many field trials comparing awned with awnless or deawned wheats. Under dry conditions, awned varieties often outyield unawned ones, but in wetter climates they show little or no advantage, and may even reduce yields (McKenzie, 1972). The adverse effects of awns under wet conditions have not been identified. They do not appear to reduce grain number by competing with florets during their development (Evans, Bingham, Jackson and Sutherland, 1972), but may increase susceptibility to disease or lodging.

Pattern of carbohydrate supply to the grain

The use of $^{14}CO_2$ has shown that the flag leaf and parts above it constitute the main source of assimilates for the developing grain (Stoy, 1965; Wardlaw, 1968; Rawson and Hofstra, 1969), with photosynthate from the penultimate leaf and those below being utilized mainly in the basal parts of the plant. Overlap in the movement of assimilates from the flag and penultimate leaves does occur, with some movement up to the ear from the penultimate leaf and some movement down from the flag leaf, probably due to the pattern of insertion and anastomosis of the leaf traces in the stem of wheat (Patrick, 1972*b*). The pattern of assimilate movement is not fixed and changes with the stage of development, as the distribution of sources and sinks changes (Buttrose, 1962*a*; Birecka and Skupinska, 1963; Rawson and Hofstra, 1969).

There are genetic differences in the assimilate supply and demand patterns in wheat. Thus in the primitive wheats, which are low yielding, there is almost full reliance on ear photosynthate with little demand for flag leaf assimilates by the grain (Evans and Dunstone, 1970), while the presence of awns on some modern wheats decreases the grain requirement for assimilates from the lower leaves (Birecka *et al.* 1968). Lupton (1966) has also observed minor varietal differences in the utilization of photosynthate from penultimate leaves, and Asana and Mani (1950) obtained evidence for varietal differences in the utilization of stem reserves.

Seasonal variations in the contribution of assimilates to the grain from different organs have been observed by Asana and Mani (1955), based on the response to shading and defoliation treatments, and also by Buttrose and May (1965) using a modification of these techniques designed to overcome the problem of compensatory changes in assimilate distribution in response to the treatment. More specifically it has been shown that increased temperature increases the rate of grain growth and the demand for leaf assimilates (Hsia *et al.* 1963; Wardlaw, 1971) while nitrogen deficiency reduces photosynthesis and the movement of assimilates out of leaves (Anisimov, 1962). Water stress, which reduces photosynthesis prior to any direct effect on grain growth, results in a greater utilization of both stored carbohydrates (Asana, 1961), and photosynthate from leaves lower on the culm (Wardlaw, 1967). Many of the estimates of photosynthate contribution to the grain from different organs have been made on well spaced plants with adequate moisture and nutrition, and Kravtsova (1957) observed that with close plantings there was an increased reliance by the grains on the lower leaves for assimilates.

Starch storage

Although amylase is present in leaves of wheat, leaves store no starch except when floated on solutions of sucrose (Gates and Simpson, 1968; Wolf, 1967). Wheat stems likewise store no starch (Barnell, 1938; Lopatecki *et al.* 1962).

Starch in the grains is stored in two kinds of granules. The primary granules, which eventually became large and lenticular in shape, appear within six days of anthesis, but only within plastids with a well developed membrane system. About two weeks after anthesis, small spherical granules appear between the large granules and their enclosing membrane. These are subsequently extruded, to give efficient packing of the spaces between the large granules (Buttrose, 1963). In mature grain, the large granules comprise about 90 % of the starch.

In plants grown in a constant environment, no rings or shells are evident in the large granules (Bakhuyzen, 1937), but they are pronounced when plants are grown under diurnal periods of darkness (Buttrose, 1962*b*). This suggests there is a diurnal cycle of deposition, one shell being formed each day. However, Jenner (1968) found starch formation in detached ears to proceed for almost 24 h at the same rate as in intact plants, which suggests that deposition is not limited by diurnal changes in supply of assimilates. With such ears, higher temperatures increased the rate of starch deposition, and the rings may be caused by slower deposition during cool nights.

With plants growing in the field Jenner and Rathjen (1972*a*) found a pronounced diurnal fluctuation in sucrose levels in the leaves, but only slight changes in sucrose concentration within the grain (usually 1.9–2.3 %). Defoliation treatments had little effect on this. When sucrose supply to the ear was increased, it accumulated in parts of the ear other than the grain. They conclude, therefore, that starch deposition was limited by the capacity for sucrose transport into the grain rather than by the supply of assimilate.

Protein storage

The many proteins in the grain can be divided into four groups. The water-soluble albumins and salt-soluble globulins are enzymatic and structural proteins distributed throughout the cytoplasm of the grain. Among the globulins alone there are at least six major fractions (Coates and Simmonds, 1961). The soluble proteins constitute most of the protein present during the first two weeks after anthesis, and may increase in amount for a further two weeks, to 0.5–1.0 mg protein per grain (Rijven and Cohen, 1961; Jennings and Morton, 1963*a*). However, they are soon exceeded in amount by the prolamins (gliadins) and glutenins (Graham and Morton, 1963), stored in protein bodies in the endosperm which appear 10–16 days after anthesis (Graham *et al.* 1963; Evers, 1970). Storage of gliadins may begin later than that of glutenins (Bilinski and McConnell, 1958). It is the characteristics of the glutens (cf. Boyd *et al.* 1969; Johnson and Hall, 1965) that are so important in bread making. The glutens and prolamins (gliadins) have amino acid compositions quite different from those of the cytoplasmic proteins, being very high in glutamic acid and proline and relatively low in several essential amino acids. Thus, as protein storage proceeds the percentage of the essential amino acids falls progressively, e.g. from 8 % lysine 12 days after anthesis to 2.5 % at maturity (Jennings and Morton, 1963*a*). Proteins of the aleurone layer are much higher in the basic amino acids such as arginine (Fulcher *et al.* 1972).

Under conditions where nitrogen uptake can continue throughout grain filling, as in many modern crops with greater and later applications of nitrogen fertilizers, both protein and starch content of the grains may increase linearly until near maturity (Rijven and Cohen, 1961; Jennings and Morton, 1963a; Turner, J. 1969; Skarsaune *et al.* 1970; Bremner, 1972; Figure 5.5), and more than half of the grain protein may derive from nitrogen taken up during grain filling (Pavlov, 1969). In these circumstances the percentage nitrogen in the grain may remain high or even rise as grain filling proceeds (e.g. Asana and Sahay, 1965; Johnson, V. *et al.* 1967).

On the other hand, under lower fertility conditions where little nitrogen fertilizer is applied, soil nitrogen may be severely depleted by heading, with little further uptake during grain filling, and virtually all nitrogen in the grain is derived by remobilization from leaves and stems (e.g. Williams, R. 1955; Puckridge and Donald, 1967; Rawson and Donald, 1969), 66–75 % of total plant nitrogen ending up in the grain (McNeal *et al.* 1968). In varieties or conditions where senescence of leaves and mobilization of nitrogen from them is slow, high grain yields may be associated with low percentage nitrogen in the grain; where leaf senescence is rapid, on the other hand, starch storage may be more adversely affected than protein storage, and low yields may be associated with high percentage nitrogen (Terman *et al.* 1969; McNeal *et al.* 1972).

Thus, there is no simple relation between yield and percentage nitrogen in the grain, even for a single variety (e.g. Fernandez and Laird, 1959).

The effects of environmental conditions on protein storage have not been extensively examined. High temperatures seem to reduce starch storage more than that of protein (Campbell and Read, 1968; Figure 5.5), as does drought stress (Lipsett, 1963; Petinov and Pavlov, 1955). Reduced light intensity lowers the grain nitrogen content to about the same extent as grain weight, with the result that percentage nitrogen is little affected (Campbell *et al.* 1969; Bremner, 1972). Late applications of nitrogen fertilizers can increase grain protein substantially. For example, Abrol *et al.* (1971) found applications of 100 kg nitrogen per hectare to increase the yield of protein by 250 kg per hectare, almost entirely through increase in the storage proteins, both prolamins and glutelins, as found also by Michael (1963) and Asana and Sahay (1965). Consequently, the increase in percentage protein from use of nitrogenous fertilizers tends to result in lower percentage contents of amino acids such as lysine, valine, threonine, isoleucine and tyrosine (Abrol *et al.* 1971).

Where high grain protein results from more extensive mobilization of leaf nitrogen – whether due to water stress, high temperatures, shading of leaves, low soil nitrogen (Neales *et al.* 1963), or varietal characteristics – high nitrogen levels in grain will tend to be associated with low nitrogen levels in leaves during grain filling, as found by Johnson, V. *et al.* (1967).

High grain nitrogen may also be associated with greater uptake rather than more mobilization, in which case leaf and grain nitrogen contents may be positively correlated (e.g. Boldyrev, 1959). A longer period of development before flowering, for example, is usually associated with greater uptake and content of nitrogen in the grain (e.g. Harris *et al.* 1943; Skarsaune *et al.* 1970; Bremner, 1972). Mikesell and Paulsen (1971) found high protein lines to have a higher nitrogen content in their lower leaves at anthesis, though not at maturity. This difference was not apparent in the flag leaves, but the nitrogen content of the lower leaves was much greater, and their removal of more consequence to grain protein content. Removal of the flag leaves actually reduced grain nitrogen by more than the amount of nitrogen in the flags, as found earlier by Wardlaw *et al.* (1965), presumably because the flag leaves play an important role in maintaining uptake and reduction of nitrate during grain growth, at least in high protein lines. These lines often have higher levels of nitrate reductase in their leaves during grain growth (e.g. Croy and Hageman, 1970; Duffield *et al.* 1972). Cultivars in which greater redistribution of nitrogen is associated with high grain protein contents may have higher protease levels in their leaves, but to date this has been demonstrated only in seedlings (Rao and Croy, 1971).

The percentage of nitrogen in grains varies considerably depending on grain position in the ear. Grains from the uppermost spikelets and from the more distal florets tend to be lower in nitrogen (McNeal and Davis, 1954; Ali *et al.* 1969; Bremner, 1972). Within each grain, the density of stains for protein is highest in the outermost cells, towards the aleurone layer, but this may reflect their reduced starch content and Evers (1970) has estimated that the content of protein *per cell* is fairly uniform throughout the endosperm.

YIELD DETERMINATION

Further increases in the yield potential of wheat may not depend on, but would undoubtedly be aided by, increased recognition of those processes which limit yield most. Any of the processes discussed in the earlier sections may, on occasion, limit yield, but here we will concentrate on limitations by photosynthesis, translocation, and storage. The relative impact of these limitations will, of course, vary with cultivar and growing conditions, and especially with the sequence of conditions during inflorescence development on the one hand and grain filling on the other. In adapted high yielding cultivars, in the environment of their selection, the source, transport and sink capacities for assimilates may well be in balance, and must be raised in a co-ordinated way for yield to be increased. Increase in storage capacity, without increase in assimilate supply, would simply lead to more unfilled and wrinkled grains. More assimilate without more storage capacity would result in little gain in yield. But from such little steps, reciprocating between source and sink, come the progressive increases in potential yield, of the order of 1 % per annum (Bingham, 1971).

Photosynthesis as a determinant of yield

Under most circumstances, 90–95 % of the carbohydrate in grain is derived from carbon dioxide fixation after anthesis. Grain yield may, therefore, bear a close relation to the duration and rate of photosynthesis after anthesis, but photosynthesis before anthesis, and particularly during ear development, may profoundly influence yield through effects on the components of storage capacity.

The integral of leaf area index with respect to time from earing to maturity, Watson's leaf area duration (LAD), takes account of both the duration and extent of photosynthetic tissue, but not of the rate of photosynthesis per unit leaf area. Figure 5.7 shows the relation between LAD and yield in wheat over a wide range of environments. In spite of the very great differences in conditions during grain filling, about half of the variation in yield due to climate, agronomic practice, and variety is related to variation in LAD. Simpson (1968), Puckridge (1971) and Spiertz *et al.* (1971) have also found a close relation between yield and LAD in wheat. Nevertheless, it is evident from Figure 5.7 that this relation is much closer in some experiments than in others (cf. data of Fischer and Kohn (1966) with those of Welbank *et al.* (1968)). Some reasons for this may be deduced from Figure 5.8.

In the crops studied by Fischer and Kohn (1966) (Figure 5.8 (*a*)) and by Puckridge (1971), LAI reached its peak well before anthesis and fell progressively as water stress increased. Consequently, the LAI was less than 4, and light interception less than complete (cf. Figure 5.4), throughout most of the grain filling period. Under these circumstances, which probably apply to most wheat crops in more arid environments, yield is likely to bear a close relation to LAD, as shown in Figure 5.7. Under conditions where the increase in water stress during grain filling is less severe, the LAI may be high at anthesis and remain above 4 for most of the grain filling period (Figure 5.8 (*b*), cf. also Stoy, 1965; Spiertz *et al.* 1971). Initial values of LAI above 4 would not be associated with additional photosynthesis (cf. Figure 5.4), yet increase the LAD values considerably, which could explain why yield and LAD are not so closely related in these conditions. Yield would be much more closely related to total crop photosynthesis during this period than to LAD, as found by Puckridge (1971). On the other hand, differences in LAI towards the end of grain filling may be closely associated with yield differences between cultivars, as in Figure 5.8 (*b*) and in the work of Stoy (1965), and the rate of decline in LAI may be more relevant than any initial differences (Watson, Thorne and French, 1963). That this is not always the case is clear from the results of Asana and Williams (1965), where cv. 'Ridley' showed a greater yield than cv. 'Diadem', although 'Ridley' yellowed sooner. Also, Syme (1969) noted that the ratio of grain yield to post-anthesis LAD was much higher in semi-dwarf Mexican wheats than in standard Australian tall

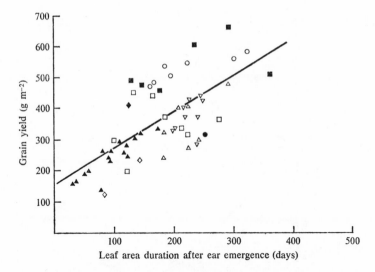

Figure 5. 7. The relation between grain yield and leaf area duration after ear emergence for wheat crops in a range of environments.

(△) UK, Watson *et al.* (1963); (◇) UK, Thorne (1966); (▽) UK, Welbank *et al.* (1966); (□) UK, Welbank *et al.* (1968); (○) UK, Thorne *et al.* (1969); (◆) Aust., Davidson (1965); (●) Aust., Turner (1966); (▲) Aust., Fischer and Kohn (1966); (■) Mexico, Fischer (unpublished).

varieties. It is not clear whether differences in the duration of photosynthetic activity cause, or are caused by, differences in the duration of grain filling. Where water or nitrogen supply is rapidly failing, as in the experiments of Fischer and Kohn (1966), storage probably depends on the duration of photosynthetic activity, but with high nutrient and water levels the reverse may be the case, the photosynthetic activity of flag leaves late in grain filling being very dependent on demand (King *et al.* 1967).

Of the variation in yield not accounted for by differences in LAD, most can be ascribed to differences in incoming radiation during grain filling. In their survey of many experiments, Welbank *et al.* (1968) found that the ratio of grain yield to LAD increased linearly with increase in daily radiation during grain filling. Grain yield itself did not do so, however, tending to plateau at high radiation levels. The reason for this is presumably that the duration of grain filling is shorter at higher radiation levels (Figure 5.6), cancelling any effect of greater photosynthesis.

When incident radiation is reduced by shading, without marked change in temperature, grain yield is usually reduced (e.g. Pendleton and Weibel, 1965; Willey and Holliday, 1971). In the experiments of Campbell *et al.* (1969), however, this occurred only in wet years, when light was limiting and water not: in dry years shading had little effect on yield. Such interactions

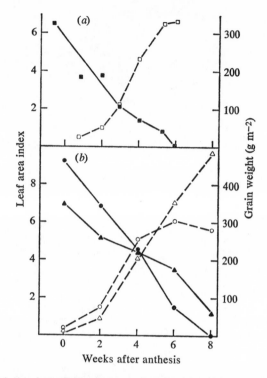

Figure 5.8. Variation in leaf area index and grain dry weight with time from anthesis in two contrasting environments. (*a*) Australia (data of Fischer and Kohn, 1966). (*b*) England (data of Watson *et al.* (1963) for two cultivars; circles, 'Squareheads Master', triangles, 'Jufy'). Solid lines, leaf area index; broken lines, grain weight.

between radiation, temperature and water supply probably account for the absence of any clear relation between radiation and grain yield in wheat in the experiments of Sibma (1970). With shading treatments applied during grain filling, Asana *et al.* (1969) found grain yield to be reduced progressively more the greater the reduction in light intensity, but far less than the reductions in total plant growth.

The effects of partial defoliation of wheat crops on their grain yield have been as variable as those of shading. Yield is often reduced (e.g. Nátr, 1967; Stoy, 1965; Womack and Thurman, 1962; Lucas and Asana, 1968). With severe defoliation, to maintain LAI at 1 or 3 throughout the growth of the crop, Davidson (1965) found grain yield to be reduced to 20% and 44% respectively, but removal of half of the leaves at heading, when the LAI = 7, did not reduce yield at all. Under comparable conditions of high radiation with irrigated crops, Fischer has found grain yield to be increased by about 20% when ambient carbon dioxide levels were doubled during grain filling. This clearly implies that increased photosynthesis can result in increased grain yield, and one might therefore expect to find higher photosynthetic rates per unit leaf area among higher yielding wheat cultivars. Only occasionally is this found, however, and some comparisons have revealed the opposite situation, with photosynthetic rate being lower in the higher yielding cultivars (e.g. Evans and Dunstone, 1970; Khan and Tsunoda, 1970; Planchon, 1969). This could be taken to imply that grain yield is not limited by photosynthetic rate, although it may be by photosynthetic *duration*. Alternatively, high photosynthetic rates are associated with other characteristics which are counter productive. They are, for example, usually associated with smaller leaves (Evans and Dunstone, 1970; Lyapshina, 1966; Planchon, 1969).

Under at least some well-defined conditions grain growth rates are not limited by photosynthetic rate, since carbohydrate balance sheets reveal that more assimilate is available for grain filling than is used (Evans and Rawson, 1970; Rawson and Evans, 1971; Wardlaw, 1971). This is evident also from the findings that grain growth at a given temperature tends to proceed linearly regardless of day to day variations in radiation (Rawson and Evans, 1970; Sofield *et al.* 1974), and that reserves built up at the time of anthesis, especially in the lower internodes of the stem, are not drawn on as fully as they are when the plants are under stress (Asana and Basu, 1963; Asana and Joseph, 1964; Yu *et al.* 1964; Rawson and Evans, 1971).

Duration of grain growth is a more powerful determinant of yield in wheat than is rate of grain growth. The latter seems more likely to be limited by translocation or storage processes than by photosynthesis.

Estimates of the photosynthetic limits to yield in wheat have been made by Stoy (1966) and Evans (1973), but depend particularly on assumptions as to the duration of grain filling at various levels of incident radiation. The latter estimate suggests that the current world record yield of 14 100 kg grain weight per hectare (209 bushels per acre) could be increased by at least one third at high radiation levels if translocation and storage capacities were not limiting (cf. Figure 1.2).

Translocation as a limitation on yield

The area of phloem tissue in the peduncle of wheats from all stages of evolution of the crop varies over a ten-fold range from wild diploid *Aegilops* to modern hexaploid (Evans *et al.* 1970). The phloem area was found to be directly proportional to the calculated maximum rate of assimilate import by the ears for the 22 lines examined, and expressed as rate per cm^2 of phloem area was similar to rates calculated for import into other rapidly growing organs. This could be taken to imply that the vascular capacity was saturated and hence limited the maximum rate of grain growth in all cases, but uncertainty about the mechanism of translocation precludes this conclusion. Moreover, feedback effects on phloem development were evident in that increasing vernalization of 'Late Mexico 120' wheat not only halved the number of spikelets and grains, and grain yield in proportion, but also halved the area of phloem in the culm. Patrick (1972a) concluded that phloem transport did not limit vegetative growth in wheat.

Within the ear, profiles of ^{14}C assimilated by flag leaves show a progressive and sharp decline in ^{14}C per spikelet above the middle of the ear (Rawson and Evans, 1970). This could imply that there is insufficient phloem to carry assimilates from the leaves to the upper spikelets. However, sterilization of basal grains in the central florets, which allowed additional grains to set in both the upper spikelets and the distal florets of the central spikelets, resulted in additional import of leaf assimilate to the top of the ear even though assimilate movement to the centre of the ear was not reduced. In this case, therefore, it was not vascular capacity but demand that was limiting.

The situation within each spikelet is no clearer. Hanif and Langer (1972) state that the three basal florets are supplied by separate groups of vascular bundles, and are therefore not in direct competition, as Rawson and Evans (1970) concluded was the case for the first and third grains. More distal grains, however, appeared to be all linked to one series of subvascular elements arising from the bundles to the third floret, in which case the third and higher grains would compete for assimilates from one poorly developed series of bundles, and the growth of the upper grains may well be limited in rate by the vascular system, if this vascular pattern applies to other cultivars. Rawson and Ruwali (1972a) have suggested that high yields with uniform grains can best be obtained by increasing spikelet number, using branched ears, rather than by increasing seed set per spikelet.

In the vascular tissue at the base of each grain Zee and O'Brien (1971) have identified two types of transfer cell. They propose that one of these extracts solutes, especially nitrogenous compounds, from the transpiration streams that supply the glumes, lemma and palea, and directs these to the grain. The other may ensure efficient transfer of assimilates from the glumes to the sieve tubes that supply the pericarp bundle of the grain. This extends from the base to the tip of the grain, within the crease. Assimilates then move from this one bundle through the chalaza and nucellar projection, and throughout the endosperm, by diffusion (Frazier and Appalanaidu, 1965). Considering the length of the grain, and its shape in cross-section, this last step in the movement of assimilates might well limit the rate of storage, particularly as starch deposition near the crease increases rapidly. Jenner and Rathjen (1972b) present evidence for the view that processes transporting carbohydrate in the final stages of its passage to the grain strongly limit starch storage. Since there are a number of conditions in which the rate of assimilate production exceeds the rate of storage, the limitations to carbohydrate transport to the ear, within the ear, within spikelets, and within grains, merits further analysis.

Storage capacity as a limitation on yield

The storage capacity of a wheat crop depends on the number of ears per unit area, the number of spikelets per ear and of grains per spikelet, and on individual grain size. The relative magnitude of these yield components varies substantially with the sequence of growing conditions, with features of agronomic practice such as sowing density and fertilizer application, and with the cultivar used.

The yield components listed above are determined sequentially during the development of the crop, both ear and spikelet number well before anthesis, grain number around anthesis, and grain size between anthesis and maturity. Thus, the storage capacity of a wheat crop can respond to environmental conditions almost until maturity. Radiation level has a pronounced effect on maximum tiller number (Figure 5.1), as do nutrient level and cultivar, but many tillers fail to produce ears (Figure 5.2), and differences in maximum tiller number become masked as sowing density is increased and as ear emergence is approached. Radiation and nutrient level during inflorescence growth, as well as temperature and day-length, influence spikelet number, while grain set is particularly influenced by light intensity and water supply just before and at anthesis. In the following weeks these factors also exert a marked effect on ultimate grain size, as does temperature throughout the grain filling period.

Consequently, the relation between grain yield and particular yield components will vary greatly depending on the sequence of environmental conditions at the various stages in the development of the crop. In some cases ear number may be the dominant component, in others grain number per ear (e.g. Apel and Lehmann, 1967; Thorne *et al.* 1968; Syme, 1969, 1970), and grain size in yet others (e.g. Simpson, 1968). There are many ways of attaining a high storage capacity, and some cultivars may make up with ear and grain number what they lack in grain size. Where similar grain yields have been obtained in a particular environment by varying combinations of the yield components, it may be suggested that this reflects yield limitation by the supply of assimilates. Inverse correlations between the various yield components have often been found (e.g. Frankel, 1935; Knott and Talukdar, 1971), and have led many plant breeders to the view that there is little point in raising one component in that there may be a compensating decrease in another. But, as discussed in various contexts above (e.g. for seed setting), these compensating mechanisms may often be due to hormonal interactions rather than to limited supply of assimilates.

Hormonal effects

The various endogenous growth substances probably play an important role in the determination of yield components and their interactions, but our knowledge of their occurrence and action in wheat is rudimentary in the extreme.

As in other plants, auxins play a major role in the control of tillering in wheat (Suge and Yamada, 1965a), interacting with the level of assimilate supply. The tendency for tillering to cease during ear development could be due to increased auxin production by young ears, but reduced assimilate supply may also be involved (Birecka, 1968). To what extent auxin production by spikelet and floret primordia, and by young grains, controls tillering, stem growth (cf. Bakhuyzen, 1947) and the pattern of grain set and growth is unknown. Auxin levels in the grain are low until late in the grain filling stage (Wheeler, 1972).

Gibberellins influence many processes in the growth and development of wheat. Maximum gibberellin levels are found in young leaves and under high levels of nutrition. Grains, seedlings and stems of dwarf cultivars contain more, not less, endogenous gibberellin than tall forms do (Radley, 1970), and the latter are more responsive to applied gibberellin (Allan *et al.* 1959). Thus, the dwarf forms appear to have a block to the utilization of gibberellins, which may affect other characters besides stem growth. For example, gibberellins influence vernalization in wheat (Suge and Hirano, 1962; Suge and Yamada, 1965b), and inhibition of gibberellin synthesis by growth retardants can retard vernalization (Suge and Osada, 1966). Gibberellins are also involved in filament growth and anther dehiscence. In the developing grains, gibberellins reach a peak early in the period of rapid grain filling (Wheeler, 1972) and again towards the end of this period according to Rejowski (1964). Since they influence not only cell expansion but also many regulatory processes, it is not surprising that the application of growth retardants which inhibit gibberellin biosynthesis has been found to have a variety of effects on wheat yield (see review by Humphries, 1968). Applications of chlorocholine chloride (CCC), for example, may increase yields by reducing the extent of lodging, with cultivars and conditions where this is serious. Along with shorter and thicker stems in the treated plants, the root system may be increased, and leaves may be shorter, broader and more upright, with a reduced photosynthetic rate (Birecka, 1967). Grain number per ear is often increased by CCC, whereas grain size is often reduced. The number of tillers and ears per unit area may also be increased by CCC.

Abscisic acid (ABA) is another endogenous growth substance likely to influence many of the yield-determining processes in wheat. Studies on its effects have so far been limited to demonstrating its rapid increase in leaves under water stress (Wright, 1969; Wright and Hiron, 1969), and its action in causing stomatal closure and reducing photosynthesis (Mittelheuser and van Steveninck, 1971). ABA is likely to mediate the effects of water stress on grain set, and may well play an important role in yield component interactions, but this has not been studied.

Similarly, little is known of the role of cytokinins in wheat, which reach their peak in grains at anthesis, falling rapidly thereafter (Wheeler, 1972). This lack of work on the roles and interactions of plant hormones in yield determining processes constitutes a major limitation to our understanding of yield development in wheat.

REFERENCES

Abdul-Baki, A. and Baker, J. E. (1970). Changes in respiration and cyanide sensitivity of the barley floret during development and maturation. *Plant Physiol.* **45**, 698–702.

Abrol, Y. P., Uprety, D. C., Ahuta, V. P. and Naik, M. S. (1971). Soil fertilizer levels and protein quality of wheat grains. *Aust. J. agric. Res.* **22**, 195–200.

Ahrens, J. F. and Loomis, W. E. (1963). Floral induction and development in winter wheat. *Crop Sci.* **3**, 463–466.

Ali, A., Atkins, I. M., Rooney, L. W. and Porter, K. B. (1969). Kernel dimensions, weight, protein content and milling yield of grain from portions of the wheat spike. *Crop Sci.* **9**, 329–330.

Allan, R. E., Vogel, O. A. and Craddock, J. C. (1959). Comparative response to gibberellic acid of dwarf, semi-dwarf, and standard short and tall winter wheat varieties. *Agron. J.* **51**, 737–740.

Allan, R. E., Vogel, O. A. and Peterson, C. J. (1962). Seedling emergence rate of fall sown wheat and its association with plant height and coleoptile length. *Agron J.* **54**, 347–350.

Anisimov, A. A. (1962). (Translocation of assimilates in wheat during paniculation stage relative to nitrogen and phosphorus nutrition.) *Doklady Akad. Nauk SSSR* **139**, 742–743.

Apel, P. and Lehmann, C. O. (1967). Photosynthese intensität von Winterweizen-Hybriden (F_1) und ihren Eltern. *Zuchter* **37**, 377–378.

Asana, R. D. (1961). Analysis of drought resistance in wheat. In *Plant–Water Relationships in Arid and Semi-Arid Conditions. Arid Zone Research* **16**, 183–190. UNESCO, Paris.

Asana, R. D. (1966). Physiological analysis of yield in wheat in relation to water stress and temperature. *J. Postgrad. School* **4** (1,2), 17–31.

Asana, R. D. and Bagga, A. K. (1966). Studies in physiological analysis of yield. VIII. Comparison of development of upper and basal grains of spikelets of two varieties of wheat. *Indian J. Plant Physiol.* **9**, 1–21.

Asana, R. D. and Basu, R. N. (1963). *Ibid.* VI. Analysis of the effect of water stress on grain development in wheat. *Indian J. Plant Physiol.* **6**, 1–13.

Asana, R. D. and Joseph, C. M. (1964). *Ibid.* VII. Effect of temperature and light on the development of the grain of two varieties of wheat. *Indian J. Plant Physiol.* **7**, 86–101.

Asana, R. D. and Mani, V. S. (1950). *Ibid.* I. Varietal differences in photosynthesis in the leaf, stem and ear of wheat. *Physiol. Plantar.* **3**, 22–39.

Asana, R. D. and Mani, V. S. (1955). *Ibid.* II. Further observations on varietal differences in photosynthesis in the leaf, stem and ear of wheat. *Physiol. Plant.* **8**, 8–19.

Asana, R. D., Parvatikar, S. R. and Saxena, N. P. (1969). *Ibid.* IX. Effect of light intensity on the development of the wheat grain. *Physiol. Plantar.* **22**, 915–924.

Asana, R. D., Ramaiah, P. K. and Rao, M. V. K. (1966). The uptake of nitrogen, phosphorus, and potassium by three cultivars of wheat in relation to growth and development. *Indian J. Plant Physiol.* **9**, 95–107.

Asana, R. D. and Sahay, R. K. (1965). A physiological analysis of the causes of mottling in two varieties of wheat. *Indian J. Plant Physiol.* **8**, 86–102.

Asana, R. D. and Saini, A. D. (1958). Studies in physiological analysis of yield. IV. The influence of soil drought on grain development, photosynthetic surface and water content of wheat. *Physiol. Plantar.* **11**, 666–674.

Asana, R. D. and Saini, A. D. (1962). *Ibid.* V. Grain development in wheat in relation to temperature, soil moisture and changes with age in the sugar content of the stem and in the photosynthetic surface. *Indian J. Plant Physiol.* **5**, 128–171.

Asana, R. D. and Singh, D. N. (1967). On the relation between flowering time, root growth and soil moisture extraction in wheat under non-irrigated cultivation. *Indian J. Plant Physiol.* **10**, 154–169.

Asana, R. D. and Williams, R. F. (1965). The effect of temperature stress on grain development in wheat. *Aust. J. agric. Res.* **16**, 1–13.

Athwal, D. S. (1971). Semi-dwarf rice and wheat in global food needs. *Quart. Rev. Biol.* **46**, 1–34.

Bakhuyzen, H. L. van der Sande (1937). *Studies on wheat grown under constant conditions.* Stanford Univ. Press, California, 400 pp.

Bakhuyzen, H. L. van der Sande (1947). Bloei en bloeihormonen in het bijzonder bij tarwe I. *Minist. Landbouw. Verslag. Onderz.* **53**, 145–212.

Barley, K. P. (1970). The configuration of the root system in relation to nutrient uptake. *Adv. Agron.* **22**, 159–201.

Barnard, C. (1955). Histogenesis of the inflorescence and flower of *Triticum aestivum* L. *Aust. J. Bot.* **3**, 1–20.

Barnard, C. (1974). The form and structure of cereals. In *Cereals in Australia*, eds A. Lazenby and E. M. Matheson. Angus and Robertson, Sydney (in press).

Barnell, H. R. (1938). Distribution of carbohydrates between component parts of the wheat plant at various times during the season. *New Phytologist* **37**, 85–112.

Belderok, B. (1961). Studies on dormancy in wheat. *Proc. Internat. Seed Testing Ass.* **26**, 697–760.

Beveridge, J. L., Jarvis, R. M. and Ridgman, W. J. (1965). Studies on the nitrogenous manuring of winter wheat. *J. agric. Sci., Camb.* **65**, 379–387.

Bilinski, E. and McConnell, W. G. (1958). Studies on wheat plants using ^{14}C-compounds. VI. Some observations on protein biosynthesis. *Cereal Chem.* **34**, 1.

Bingham, J. (1966). Varietal response in wheat to water supply in the field, and male sterility caused by a period of drought in a glasshouse experiment. *Ann. appl. Biol.* **47**, 365–377.

Bingham, J. (1967). Investigations on the physiology of yield in winter wheat by comparisons of varieties and by artificial variation in grain number. *J. agric. Sci., Camb.* **68**, 411–422.

Bingham, J. (1969). The physiological determinants of grain yield in cereals. *Agric. Progr.* **44**, 30–42.

Bingham, J. (1971). Plant Breeding. Arable Crops. In *Potential Crop Production*, eds P. F. Wareing and J. P. Cooper. Heinemann, London, pp. 273–294.

Birecka, H. (1967). Influence of 2-chloroethyl trimethylammonium chloride (CCC) on photosynthetic activity and assimilate distribution in wheat. In *Isotopes in Plant Nutrition and Physiology*. Intl. Atomic Energy Agency, Vienna, pp. 189–199.

Birecka, H. (1968). Translocation and redistribution of ^{14}C-assimilates in cereal plants deprived of the ear. I. Spring wheat. *Bull. Acad. Pol. Sci. Cl. V* **16**, 455–460.

Birecka, H. and Dakic-Wlodkowska, L. (1963). Photosynthesis, translocation and accumulation of assimilates in cereals during grain development. III. Spring wheat – photosynthesis and the daily accumulation of photosynthates in the grain. *Acta Soc. Bot. Polon.* **32**, 631–650.

Birecka, H. and Dakic-Wlodkowska, L. (1966). Photosynthetic activity before and after ear emergence in spring wheat. *Acta Soc. Bot. Polon.* **35**, 637–662.

Birecka, H., Skiba, T. and Kozlowska, Z. (1969). Translocation and redistribution of ^{14}C-assimilates in cereal plants deprived of the ear. III. Assimilate distribution in root and respiration of culm in wheat plants. *Bull. Acad. Polon. Sci. Cl. V* **17**, 121–127.

Birecka, H. and Skupinska, J. (1963). Photosynthesis, translocation and accumulation of assimilates in cereals during grain development. II. Spring barley – photosynthesis and the daily accumulation of photosynthates in the grain. *Acta Soc. Botan. Polon.* **32**, 531–552.

Birecka, H., Wojcieska, V. and Glazewski, S. (1968). Ear contribution to photosynthetic activity in winter cereals. I. Winter wheat. *Bull. Acad. Polon. Sci. Cl. V*, **16**, 191–196.

Boatwright, G. O. and Ferguson, H. (1967). Influence of primary and/or adventitious root systems on wheat production and nutrient uptake. *Agron. J.* **59**, 299–302.

Boldyrev, N. K. (1959). Relation between the chemical composition of leaves, yield of grain and quality of spring wheat as dependent on the fertilizers applied. *Akad. Nauk SSSR Dokl.* **126**, 886–889.

Bonnett, O. T. (1967). Inflorescences of maize, wheat, rye, barley and oats: their initiation and development. *Bull. Univ. Ill. Agric. Exp. Sta.* **721**, 105.

Boyd, W. J. R., Lee, J. W. and Wrigley, C. W. (1969). The D genome and the control of wheat gluten synthesis. *Experientia* **25**, 317.

Bremner, P. M. (1972). The accumulation of dry matter and nitrogen by grains in different positions of the wheat ear as influenced by shading and defoliation. *Aust. J. biol. Sci.* **25**, 657–681.

Bremner, P. M. and Rawson, H. M. (1972). Fixation of ^{14}CO$_2$ by flowering and non-flowering glumes of the wheat ear, and the pattern of transport of label to individual grains. *Aust. J. biol. Sci.* **25**, 921–930.

Brenchley, W. E. and Jackson, V. G. (1921). Root development in barley and wheat under different conditions of growth. *Ann. Bot.* **35**, 533–556.

Brouwer, R. (1966). Root growth of grasses and cereals. In *The Growth of Cereals and Grasses*, eds F. L. Milthorpe and J. D. Ivins. Butterworths, London, pp. 153–166.

Buttrose, M. S. (1962a). Physiology of cereal grain. III. Photosynthesis in the wheat ear during grain development. *Aust. J. biol. Sci.* **15**, 611–618.

Buttrose, M. S. (1962b). The influence of environment on the shell structure of starch granules. *J. Cell Biol.* **14**, 159–167.

Buttrose, M. S. (1963). Ultrastructure of the developing wheat endosperm. *Aust. J. biol. Sci.* **16**, 305–317.

Buttrose, M. S. and May, L. H. (1965). Seasonal variation in estimates of cereal-ear photosynthesis. *Ann. Bot. N.S.* **29**, 79–81.

Campbell, C. A., Pelton, W. L. and Neilson, K. F. (1969). Influence of solar radiation and soil moisture on growth and yield of Chinook wheat. *Can. J. Plant Sci.* **49**, 685–699.

Campbell, C. A. and Read, D. W. L. (1968). Influence of air temperature, light intensity and soil moisture on the growth, yield and some growth analysis characteristics of Chinook wheat grown in the growth chamber. *Can. J. Plant Sci.* **48**, 299–311.

Carpenter, R. W., Haas, H. J. and Miles, E. F. (1952). Nitrogen uptake by wheat in relation to nitrogen content of soil. *Agron. J.* **44**, 420–423.

Carr, D. J. and Wardlaw, I. F. (1965). The supply of photosynthetic assimilates to the grain from the flag leaf and ear of wheat. *Aust. J. biol. Sci.* **18**, 711–719.

Chinoy, J. J. and Nanda, K. K. (1951). Effect of vernalization and photoperiodic treatments on growth and development of crop plants. II. Varietal differences in stem elongation and tillering of wheat and their correlation with flowering under varying photoinductive and post-photoinduction treatments. *Physiol. Plantar.* **4**, 427–436.

Chonan, N. (1965). Studies on the photosynthetic tissues in the leaves of cereal crops. I. The mesophyll structure of wheat leaves inserted at different levels of the shoot. *Proc. Crop Sci. Soc. Japan* **33**, 388–393.

Chonan, N. (1966). *Ibid.* II. Effect of shading on the mesophyll structure of the wheat leaves. *Proc. Crop Sci. Soc. Japan* **35**, 78–82.

Chonan, N. (1971). *Ibid.* VII. Effect of temperature on the mesophyll structure of leaves in wheat and rice. *Proc. Crop Sci. Soc. Japan* **40**, 425–430.

Chujo, H. (1966*a*). Difference in vernalization effect in wheat under various temperatures. *Proc. Crop Sci. Soc. Japan* **35**, 177–186.

Chujo, H. (1966*b*). The effect of diurnal variation of temperature on vernalization in wheat. *Proc. Crop Sci. Soc. Japan* **35**, 187–194.

Chujo, H. (1967). The effect of thermoperiod on vernalization of wheat. *Proc. Crop Sci. Soc. Japan* **36**, 224–231.

Coates, J. H. and Simmonds, D. H. (1961). Proteins of wheat and flour, extraction, fractionation, and chromatography of the buffer-soluble proteins of flour. *Cereal Chem.* **38**, 256–272.

Connor, D. J. and Cartledge, O. (1971). Structure and photosynthesis of wheat communities. *J. appl. Ecol.* **8**, 469–475.

Cooper, J. P. (1956). Developmental analysis of populations in the cereals and herbage grasses. I. Methods and techniques. *J. Agric. Sci., Camb.* **47**, 262–279.

Cooper, J. P. (1957). *Ibid.* II. Response to low temperature vernalization. *J. agric. Sci., Camb.* **49**, 361–383.

Cooper, J. P. (1960). Short day and low temperature induction in *Lolium. Ann. Bot. N.S.* **24**, 232–246.

Croy, L. I. and Hageman, R. H. (1970). Relationship of nitrate reductase activity to grain protein production in wheat. *Crop Sci.* **10**, 280–285.

Davidson, J. L. (1965). Some effects of leaf area control on the yield of wheat. *Aust. J. agric. Res.* **16**, 721–731.

Denmead, O. T. (1969). Comparative micrometeorology of a wheat field and a forest of *Pinus radiata. Agr. Meteorol.* **6**, 357–371.

Denmead, O. T. (1970). Transfer processes between vegetation and air: measurement, interpretation and modelling. In *Prediction and Measurement of Photosynthetic Productivity.* Pudoc, Wageningen, pp. 149–164.

Derera, N. F., Marshall, D. R. and Balaam, L. M. (1969). Genetic variability in root development in relation to drought tolerance in spring wheats. *Exp. Agr.* **5**, 327–337.

Duffield, R. D., Croy, L. I. and Smith, E. L. (1972). Inheritance of nitrate reductase activity, grain protein, and straw protein in a hard red winter wheat cross. *Agron. J.* **64**, 249–251.

Dunstone, R. L. and Evans, L. T. (1974). The role of changes in cell size in the evolution of wheat. *Aust. J. Plant Physiol.* **1**, 157–165.

Dunstone, R. L., Gifford, R. M. and Evans, L. T. (1973). Photosynthetic characteristics of modern and primitive wheat species in relation to ontogeny and adaptation to light. *Aust. J. biol. Sci.* **26**, 295–307.

Dvorak, J. and Nátr, L. (1971). Carbon dioxide compensation points of *Triticum* and *Aegilops* species. *Photosynthetica* **5**, 1–5.

Eliasson, L. (1955). The connection between the respiratory gradient and the growth rate in wheat roots. *Physiol. Plantar.* **8**, 374–388.

Evans, L. T. (1973). The effect of light on plant growth, development and yield. In *Plant Response to Climatic Factors.* UNESCO, Paris, pp. 21–35.

Evans, L. T., Bingham, J., Jackson, P. and Sutherland, J. (1972). Effect of awns and drought on the supply of photosynthate and its distribution within wheat ears. *Ann. appl. Biol.* **70**, 67–76.

Evans, L. T., Bingham, J. and Roskams, M. A. (1972). The pattern of grain set within ears of wheat. *Aust. J. biol. Sci.* **25**, 1–8.

Evans, L. T. and Dunstone, R. L. (1970). Some physiological aspects of evolution in wheat. *Aust. J. biol. Sci.* **23**, 725–741.

Evans, L. T., Dunstone, R. L., Rawson, H. M. and Williams, R. F. (1970). The phloem of the wheat stem in relation to requirements for assimilate by the ear. *Aust. J. biol. Sci.* **23**, 743–752.

Evans, L. T. and Rawson, H. M. (1970). Photosynthesis and respiration by the flag leaf and components of the ear during grain development in wheat. *Aust. J. biol. Sci.* **23**, 245–254.

Evers, A. D. (1970). Development of the endosperm of wheat. *Ann. Bot.* **34**, 547–555.

Fernandez, R. and Laird, R. T. (1959). Yield and protein content of wheat in central Mexico as affected by available soil moisture and nitrogen fertilization. *Agron. J.* **41**, 33–36.

Fischer, R. A. (1973). The effect of water stress at various stages of development on yield processes in wheat. In *Plant Response to Climatic Factors.* UNESCO, Paris, pp. 233–241.

Fischer, R. A. and Kohn, G. D. (1966). The relationship of grain yield to vegetative growth and post flowering leaf area in the wheat crop under conditions of limited soil moisture. *Aust. J. agric. Res.* **17**, 281–295.

Frankel, O. H. (1935). Analytical yield investigations on New Zealand wheat. II. Five years' analytical variety trials. *J. agric. Sci., Camb.* **25**, 466–509.

Frazier, J. C. and Appalanaidu, B. (1965). The wheat grain during development with reference to nature, location, and role of its translocatory tissues. *Amer. J. Bot.* **52**, 193–198.

Friend, D. J. C. (1965a). Ear length and spikelet number of wheat grown at different temperatures and light intensities. *Canad. J. Bot.* **43**, 345–353.

Friend, D. J. C. (1965b). Tillering and leaf production in wheat as affected by temperature and light intensity. *Canad. J. Bot.* **43**, 1063–1076.

Friend, D. J. C. (1966). The effects of light and temperature on the growth of cereals. In *The Growth of Cereals and Grasses*, eds F. L. Milthorpe and J. D. Ivins. Butterworths, London, pp. 181–199.

Friend, D. J. C., Helson, V. A. and Fisher, J. E. (1962). Leaf growth in Marquis wheat, as regulated by temperature, light intensity, and day length. *Canad. J. Bot.* **40**, 299–311.

Fulcher, R. G., O'Brien, T. P. and Simmonds, D. H. (1972). Localization of arginine rich proteins in mature seeds of some members of the Gramineae. *Aust. J. biol. Sci.* **25**, 487–497.

Gabrielsen, E. K. (1942). Carbon dioxide assimilation and dry matter production in ears of wheat. *Yearbook Roy. Vet. Agric. Coll. 1942*, 28–54.

Gates, J. W. and Simpson, G. M. (1968). The presence of starch and α-amylase in the leaves of plants. *Canad. J. Bot.* **46**, 1459–1462.

Gericke, W. F. (1922). Certain relations between root development and tillering in wheat; significance in the production of high protein wheat. *Amer. J. Bot.* **9**, 366–369.

Goss, J. A. (1968). Development, physiology, and biochemistry of corn and wheat pollen. *Bot. Rev.* **34**, 333–358.

Gott, M. B. (1957). Vernalization of green plants of a winter wheat. *Nature, Lond.* **180**, 714–715.

Gott, M. B. (1961). Flowering of Australian wheats and its relation to frost injury. *Aust. J. agric. Res.* **12**, 547–565.

Graham, D., Hatch, M. D., Slack, C. R. and Smillie, R. M. (1970). Light induced formation of enzymes of the C_4-dicarboxylic acid pathway of photosynthesis in detached leaves. *Phytochem.* **9**, 521–532.

Graham, J. S. D. and Morton, R. K. (1963). Studies of proteins of developing wheat endosperm: separation by starch-gel electrophoresis and incorporation of ^{35}S sulphate. *Aust. J. biol. Sci.* **16**, 357–365.

Graham, J. S. D., Morton, R. K. and Raison, J. K. (1963). Isolation and characterization of protein bodies from developing wheat endosperm. *Aust. J. biol. Sci.* **16**, 375–394.

Grahl, A. (1965). Lichteinfluss auf die Keimung des Getreides in Abhängigkeit von der Keimruhe. *Landbouforsch.* **15**, 97–106.

Gries, G. A., Stearns, F. W. and Caldwell, R. M. (1956). Responses of spring wheat varieties to day-length at different temperatures. *Agron. J.* **48**, 29–32.

Grundbacher, F. J. (1963). The physiological function of the cereal awn. *Bot. Rev.* **29**, 366–381.

Halse, N. J. and Weir, R. N. (1970). Effects of vernalization, photoperiod, and temperature on phenological development and spikelet number of Australian wheat. *Aust. J. agric. Res.* **21**, 383–393.

Hanif, M. and Langer, R. H. M. (1972). The vascular system of the spikelet in wheat (*Triticum aestivum*). *Ann. Bot.* **36**, 721–727.

Harlan, J. R. and Zohary, D. (1966). Distribution of wild wheats and barley. *Science* **153**, 1074–1080.

Harris, R. H., Helgeson, E. A. and Sibbitt, L. D. (1943). The effect of maturity upon the quality of hard red spring and durum wheats. *Cereal Chem.* **20**, 447–463.

Helbaek, H. (1966). 1966 – Commentary on the phylogenesis of *Triticum* and *Hordeum*. *Econ. Bot.* **20**, 350–360.

Hoshikawa, K. (1959). Influence of temperature upon the fertilization of wheat grown in various levels of nitrogen. *Proc. Crop Sci. Soc. Japan* **28**, 291–295.

Hoshikawa, K. (1960). Studies on the pollen germination and pollen tube growth in relation to the fertilization of wheat. *Proc. Crop Sci. Soc. Japan* **28**, 333–336.

Hoshikawa, K. (1961*a*). Studies on the ripening of wheat. 4. The influence of temperature on endosperm formation. *Proc. Crop Sci. Soc. Japan* **30**, 228–231.

Hoshikawa, K. (1961*b*). *Ibid.* 1. Embryological observations of the early development of the endosperm. *Proc. Crop Sci. Soc. Japan* **29**, 253–257.

Hoshikawa, K. (1961*c*). Studies on the reopen floret in wheat. *Proc. Crop Sci. Soc. Japan* **29**, 103–106.

Hsia, C., Waon, S. and Wong, F. (1963). The effect of temperature on the physiological changes of wheat during grain development. *Acta bot. Sin.* **11**, 338–349.

Huber, B. (1952). Der Einfluss der Vegetation auf die Schwankungen des CO_2-Gehaltes der Atmosphäre. *Archiv. für Meteorol. Geophys und Bioklimatol.* **4**, 154–167.

Humphries, E. C. (1968). CCC and cereals. *Field Crop Abs.* **21**, 91–99.

Hurd, E. A. (1968). Growth of roots of seven varieties of spring wheat at high and low moisture levels. *Agron. J.* **60**, 201–205.

Ishihara, A. (1961). Physiological studies on the vernalization of wheat plants. III. Direct and indirect induction by low temperature in apical and lateral buds. *Proc. Crop Sci. Soc. Japan* **30**, 88–92.

Jenner, C. F. (1968). Synthesis of starch in detached ears of wheat. *Aust. J. biol. Sci.* **21**, 597–608.

Jenner, C. F. and Rathjen, A. J. (1972*a*). Factors limiting the supply of sucrose to the developing wheat grain. *Ann. Bot.* **36**, 729–741.

Jenner, C. F. and Rathjen, A. J. (1972*b*). Limitations to the accumulation of starch in the developing wheat grain. *Ann. Bot.* **36**, 743–754.

Jennings, A. C. and Morton, R. K. (1963*a*). Changes in carbohydrate, protein and non-protein nitrogenous compounds of developing wheat grain. *Aust. J. biol. Sci.* **16**, 318–331.

Jennings, A. C. and Morton, R. K. (1963*b*). Changes in nucleic acids and other phosphorus-containing compounds of developing wheat grains. *Aust. J. biol. Sci.* **16**, 332–341.

Jensen, N. F. (1967). Agrobiology: specialization or systems analysis. *Science* **157**, 1405–1409.

Jewiss, O. R. (1972). Tillering in grasses – its significance and control. *J. Brit. Grassl. Soc.* **27**, 65–82.

Johnson, B. L. and Hall, O. (1965). Analysis of phylogenetic affinities in the Triticinae by protein electrophoresis. *Amer. J. Bot.* **52**, 506–573.

Johnson, V. A., Mattern, P. J. and Schmidt, J. W. (1967). Nitrogen relations during spring growth in varieties of *Triticum aestivum* L. differing in grain protein content. *Crop Sci.* **7**, 664–667.

Jolliffe, P. A. and Tregunna, E. F. (1968). Effect of temperature, CO_2-concentration and light intensity on oxygen inhibition of photosynthesis in wheat leaves. *Plant Physiol.* **43**, 902–906.

Junges, W. (1959). Beeinflussung des Blühbeginns annueller landwirtschaftlicher und gartnerischer Kulturpflanzen durch Jarowisation bei Konstanten Temperaturen zwischen $-10°$ und $+35$ °C. *Z. für Pflanzenz.* **41**, 103–122.

Khalil, M. S. H. (1956). The interrelation between growth and development of wheat as influenced by temperature, light and nitrogen. *Meded. Landbouwhogesch. Wageningen* **56**(7), 1–73.

Khan, M. A. and Tsunoda, S. (1970). Evolutionary trends in leaf photosynthesis and related leaf characters among cultivated wheat species and its wild relatives. *Jap. J. Breeding* **20**, 133–140.

142   L. T. Evans, I. F. Wardlaw and R. A. Fischer

King, R. W. and Evans, L. T. (1967). Photosynthesis in artificial communities of wheat, lucerne, and subterranean clover plants. *Aust. J. biol. Sci.* **20**, 623–635.

King, R. W., Wardlaw, I. F. and Evans, L. T. (1967). Effect of assimilate utilization on photosynthetic rate in wheat. *Planta Berl.* **77**, 261–276.

Knott, D. R. and Talukdar, B. (1971). Increasing seed weight in wheat and its effect on yield, yield components, and quality. *Crop Sci.* **11**, 280–283.

Konovalov, J. B. (1959). The effect of a deficiency in soil moisture on grain filling of spring wheat. *Fiziol. Rast.* **6**, 183–189.

Kostic, M., Dijkshoorn, W. and de Wit, C. T. (1957). Evaluation of the nutrient status of wheat plants. *Neth. J. Agric. Sci.* **15**, 267–280.

Krassovsky, I. V. (1926). Physiological activity of the seminal and nodal roots of crop plants. *Soil Sci.* **21**, 307–325.

Kravtsova, B. Y. (1957). (Study of the role of the leaves of individual tiers in the formation of the fruit-bearing organs of spring wheat.) *Dokl. Akad. Nauk SSSR* **115**, 822–825.

Krekule, J. (1964). (Varietal differences on replacing vernalization by a short day in winter wheat.) *Biol. Plantar.* **6**, 299–305.

Kriedeman, P. (1966). The photosynthetic activity of the wheat ear. *Ann. Bot.* **30**, 349–363.

Kriedeman, P. E., Neales, T. F. and Ashton, D. H. (1964). Photosynthesis in relation to leaf orientation and light interception. *Aust. J. biol. Sci.* **17**, 591–600.

Langer, R. H. M. and Hanif, M. (1973). A study of floret development in wheat (*Triticum aestivum* L.). *Ann. Bot.* **37**, 743–751.

Langer, R. H. M. and Liew, F. K. Y. (1973). Effects of varying nitrogen supply at different stages of development on spikelet and grain production and on grain nitrogen in wheat. *Aust. J. agric. Res.* **24**, 647–656.

Lipsett, J. (1963). Factors affecting the occurrence of mottling in wheat. *Aust. J. agric. Res.* **14**, 303–314.

Locke, L. F. and Clark, J. A. (1924). Development of wheat plants from seminal roots. *J. Amer. Soc. Agron.* **16**, 261–268.

Lopatecki, L. E., Longair, E. L. and Kasting, R. (1962). Quantitative changes of soluble carbohydrates in stems of solid- and hollow-stemmed wheats during growth. *Canad. J. Bot.* **40**, 1223–1228.

Lowe, L. B., Ayers, G. S. and Ries, S. K. (1972). Relationship of seed protein and amino acid composition to seedling vigor and yield in wheat. *Agron. J.* **64**, 608–611.

Lowe, L. B. and Ries, S. K. (1972). Effects of environment on the relation between seed protein and seedling vigor in wheat. *Canad. J. Pl. Sci.* **52**, 157–164.

Lucas, D. and Asana, R. D. (1968). Effect of defoliation on the growth and yield of wheat. *Physiol. Plantar.* **21**, 1217–1223.

Lupton, F. G. H. (1966). Translocation of photosynthetic assimilates in wheat. *Ann. appl. Biol.* **57**, 335–364.

Lupton, F. G. H. (1969). Estimation of yield in wheat from measurement of photosynthesis and translocation in the field. *Ann. Appl. Biol.* **64**, 363–374.

Lupton, F. G. H. (1972). Further experiments on photosynthesis and translocation in wheat. *Ann. appl. Biol.* **71**, 69–79.

Lupton, F. G. H. and Pinthus, M. J. (1969). Carbohydrate translocation from small tillers to spike producing shoots in wheat. *Nature, Lond.* **221**, 483–484.

Lyapshina, Z. F. (1966). The effect of temperature during the period of grain formation on the productivity of spring wheat. *Soviet Plant Physiol.* **13**, 288–291.

McCree, K. J. (1972). The action spectrum, absorptance and quantum yield of photosynthesis in crop plants. *Agric. Meteorol.* **9**, 191–216.

McDonough, W. T. and Gauch, H. G. (1959). The contribution of the awns to the development of the kernels of bearded wheat. *Maryland Ag. Exp. Sta. Bull. A* **103**, 1–16.

MacDowall, F. D. H. (1972). Growth kinetics of Marquis wheat. II. Carbon dioxide dependence. *Canad. J. Bot.* **50**, 883–889.

McKenzie, H. (1972). Adverse influence of awns on yield of wheat. *Can. J. Plant Sci.* **52**, 81–87.

McKinney, H. H. and Sando, W. J. (1935). Earliness of sexual reproduction in wheat as influenced by temperature and light in relation to growth phases. *J. agric. Res.* **51**, 621–641.

McNeal, F. H., Berg, M. A., McGuire, C. F., Stewart, V. R. and Boldridge, D. E. (1972). Grain and plant nitrogen relationships in eight spring wheat crosses, *Triticum aestivum* L. *Crop Sci.* **12**, 599–602.

McNeal, F. H., Boatwright, G. O., Berg, M. A. and Watson, C. A. (1968). Nitrogen in plant parts of seven spring wheat varieties at successive stages of development. *Crop Sci.* **8**, 535–537.

McNeal, F. H. and Davis, D. J. (1954). Effect of nitrogen fertilization on yield, culm number and protein content of certain spring wheat varieties. *Agron. J.* **46**, 375–378.

Marcellos, H. and Single, W. V. (1972). The influence of cultivar, temperature and photoperiod on post-flowering development of wheat. *Aust. J. agric. Res.* **23**, 533–540.

May, L. H., Randles, F. H., Aspinall, D. and Paleg, L. G. (1967). Quantitative studies of root development. II. Growth in the early stages of development. *Aust. J. biol. Sci.* **20**, 273–283.

Meidner, H. (1970). Light compensation points and photorespiration. *Nature, Lond.* **228**, 1349.

Michael, G. (1963). Einfluss der Düngung auf Eiweissqualität und Eiweissfraktionen der Nahrungspflanzen. *Qualitas Plant. Mat. Veg.* **10**, 248–265.

Michael, G. and Seiler-Kelbitsch, H. (1972). Cytokinin content and kernel size of barley grain as affected by environmental and genetic factors. *Crop Sci.* **12**, 162–165.

Miflin, B. J. (1967). Distribution of nitrate and nitrite reductase in barley. *Nature, Lond.* **214**, 1133–1134.

Mikesell, M. E. and Paulsen, G. M. (1971). Nitrogen translocation and the role of individual leaves in protein accumulation in wheat grain. *Crop Sci.* **11**, 919–922.

Miller, E. C., Gauch, H. G. and Gries, G. A. (1944). A study of the morphological nature and physiological function of the awns in winter wheat. *Kan. Agr. Exp. Sta. Tech. Bull.* **47**, 1–82.

Milthorpe, F. L. (1950). Changes in the drought resistance of wheat seedlings during germination. *Ann. Bot. N.S.* **14**, 79–89.

Minotti, P. L. and Jackson, W. A. (1970). Nitrate reduction in the roots and shoots of wheat seedlings. *Planta* **94**, 36–44.

Misra, G. (1954). Vernalization in some varieties of Indian wheat. *J. Ind. Bot. Soc.* **33**, 239–246.

Mittelheuser, C. J. and van Steveninck, R. F. M. (1971). Rapid action of abscisic acid on photosynthesis and stomatal resistance. *Planta* **97**, 83–86.

Morrison, J. W. (1955). Fertilization and post fertilization development in wheat. *Can. J. Bot.* **33**, 168–176.

Moss, D. N., Krenzer, E. G. and Brun, W. A. (1969). Carbon dioxide compensation points in related plant species. *Science* **164**, 187–188.

Murata, Y. and Iyama, J. (1963). Studies on the photosynthesis of forage crops. II. Influence of air temperature upon the photosynthesis of some forage and grain crops. *Proc. Crop. Sci. Soc. Japan* **31**, 315–321.

Nátr, L. (1966). (Varietal differences in the intensity of photosynthesis.) *Rostl. vyroba* **12**, 163–178.

Nátr, L. (1967). (Study on grain yield formation in cereals. 3. Effect of leaf and awn removal on reduction in yield components of spring barley and winter wheat.) *Rostl. vyroba* **13**, 797–818.

Neales, T. F., Anderson, M. J. and Wardlaw, I. F. (1963). The role of the leaves in the accumulation of nitrogen by wheat during ear development. *Aust. J. agric. Res.* **14**, 725–736.

Neales, T. F. and Davies, J. A. (1966). The effect of photoperiod duration upon the respiratory activity of the roots of wheat seedlings. *Aust. J. biol. Sci.* **19**, 471–480.

Nelson, C. D. (1963). Effect of climate on the distribution and translocation of assimilates. In *Environmental Control of Plant Growth*, ed. L. T. Evans. Acad. Press, New York, pp. 149–172.

Osman, A. M. (1971*a*). Dry matter production of a wheat crop in relation to light interception and photosynthetic capacity of leaves. *Ann. Bot. N.S.* **35**, 1017–1035.

Osman, A. M. (1971*b*). Root respiration of wheat plants as influenced by age, temperature, and irradiation of shoots. *Photosynthetica* **5**, 107–112.

Osman, A. M. and Milthorpe, F. L. (1971). Photosynthesis of wheat leaves in relation to age, illuminance, and nutrient supply. II. Results. *Photosynthetica* **5**, 61–70.

Owen, P. C. J. (1952). The relation of germination of wheat to water potential. *J. exp. Bot.* **3**, 188–203.

Palfi, G. and Dezsi, L. (1960). The translocation of nutrients between fertile and sterile shoots of wheat. *Acta Bot. Acad. Sci. Hung.* **6**, 65–74.

Passioura, J. B. (1972). The effect of root geometry on the yield of wheat growing on stored water. *Aust. J. agric. Res.* **23**, 745–752.

Patrick, J. W. (1972*a*). Distribution of assimilate during stem elongation in wheat. *Aust. J. biol. Sci.* **25**, 455–467.

Patrick, J. W. (1972*b*). Vascular system of the stem of the wheat plant. I. Mature state. *Aust. J. Bot.* **20**, 49–63.

Pavlov, A. N. (1969). Sel'Khoz. *Biologia* **4**, 230–235.

Pendleton, J. W. and Weibel, R. O. (1965). Shading studies on winter wheat. *Agron. J.* **57**, 292–293.

Penman, H. L. and Long, I. F. (1960). Weather in wheat: an essay in micrometeorology. *Quart. J. Roy. Met. Soc.* **86**, 16–50.

Percival, J. (1921). *The Wheat Plant.* Duckworth, London, 463 pp.

Peters, D. B., Pendleton, J. W., Hageman, R. H. and Brown, C. M. (1971). Effect of night air temperature on grain yield of corn, wheat and soybeans. *Agron. J.* **63**, 809.

Peterson, R. F. (1965). *Wheat: Botany, Cultivation and Utilization.* Leonard Hill, London, 448 pp.

Petinov, N. S. and Pavlov, A. N. (1955). (Increase of protein content of spring wheat grain (grown under irrigation) by means of spraying with nitrogenous supplements.) *Fiziol. Rast.* **2**, 113–122.

Pinthus, M. J. (1963). Inheritance of heading date in some spring wheat varieties. *Crop Sci.* **3**, 301–304.

Pinthus, M. J. (1967). Evaluation of winter wheat as a source of high yield potential for the breeding of spring wheat. *Euphytica* **16**, 231–251.

Pinthus, M. J. and Eshel, Y. (1962). Observations on the development of the root system of some wheat varieties. *Israel J. agric. Res.* **12**, 13–20.

Pinthus, M. J. and Osher, R. (1966). The effect of seed size on plant growth and grain yield components in various wheat and barley varieties. *Israel J. agric. Res.* **16**, 53–58.

Planchon, C. (1969). (Photosynthetic activity and yield in soft wheat (*Triticum aestivum*)). *Genet. agr.* **23**, 485–490.

Puckridge, D. W. (1968). Competition for light and its effect on leaf and spikelet development of wheat plants. *Aust. J. agric. Res.* **19**, 191–201.

Puckridge, D. W. (1969). Photosynthesis of wheat under field conditions. II. Effect of defoliation on the carbon dioxide uptake of the community. *Aust. J. agric. Res.* **20**, 623–634.

Puckridge, D. W. (1971). *Ibid.* III. Seasonal trends in carbon dioxide uptake of crop communities. *Aust. J. agric. Res.* **22**, 1–9.

Puckridge, D. W. and Donald, C. M. (1967). Competition among wheat plants sown at a wide range of densities. *Aust. J. agric. Res.* **18**, 193–211.

Puckridge, D. W. and Ratkowsky, D. A. (1971). Photosynthesis of wheat under field conditions. IV. The influence of density and leaf area index on the responses to radiation. *Aust. J. agric. Res.* **22**, 11–20.

Quisenberry, K. S. and Reitz, L. P. (1967). *Wheat and wheat improvement.* Amer. Soc. Agron., Madison, 560 pp.

Radley, M. (1970). Comparison of endogenous gibberellins and response to applied gibberellin of some dwarf and tall wheat cultivars. *Planta, Berl.* **92**, 292–320.

Rao, S. C. and Croy, L. I. (1971). Protease levels in 'high' versus 'low' grain protein wheats and their association with the production of amino acids, tryptophane and IAA during early growth. *Crop Sci.* **11**, 790–791.

Rawson, H. M. (1970). Spikelet number, its control and relation to yield per ear in wheat. *Aust. J. biol. Sci.* **23**, 1–15.

Rawson, H. M. (1971). Tillering patterns in wheat with special reference to the shoot at the coleoptile node. *Aust. J. biol. Sci.* **24**, 829–841.

Rawson, H. M. and Donald, C. M. (1969). The absorption and distribution of nitrogen, after floret initiation in wheat. *Aust. J. agric. Res.* **20**, 799–808.

Rawson, H. M. and Evans, L. T. (1970). The pattern of grain growth within the ear of wheat. *Aust. J. biol. Sci.* **23**, 753–764.

Rawson, H. M. and Evans, L. T. (1971). The contribution of stem reserves to grain development in a range of wheat cultivars of different height. *Aust. J. agric. Res.* **22**, 851–863.

Rawson, H. M. and Hofstra, G. (1969). Translocation and remobilization of ^{14}C assimilated at different stages by each leaf of the wheat plant. *Aust. J. biol. Sci.* **22**, 321–331.

Rawson, H. M. and Ruwali, K. N. (1972*a*). Branched ears in wheat and yield determination. *Aust. J. agric. Res.* **23**, 541–549.

Rawson, H. M. and Ruwali, K. N. (1972*b*). Ear branching as a means of increasing grain uniformity in wheat. *Aust. J. agric. Res.* **23**, 551–559.

Razumov, V. I. and Limar, R. S. (1971). The role of temperature in the photoperiodic reaction of plant flowering. *Soviet Plant Physiol.* **18**, 783–788.

Reitz, L. P. and Salmon, S. C. (1968). Origin, history and use of Norin 10 wheat. *Crop Sci.* **8**, 686–689.

Rejowski, A. (1964). Gibberellins in maturing wheat seeds. *Bull. Acad. Polon. Sci. Cl. V* **12**, 233–236.

Riddell, J. A. and Gries, G. A. (1958*a*). Development of spring wheat. II. The effect of temperature on response to photoperiod. *Agron. J.* **50**, 739–742.

Riddell, J. A. and Gries, G. A. (1958*b*). *Ibid.* III. Temperature of maturation and age of seeds as factors influencing their response to vernalization. *Agron. J.* **50**, 743–746.

Riddell, J. A., Gries, G. A. and Stearns, F. W. (1958). Development of spring wheat. I. The effect of photoperiod. *Agron. J.* **50**, 735–738.

Rijven, A. H. G. C. and Cohen, R. (1961). Distribution of growth and enzyme activity in the developing grain of wheat. *Aust. J. biol. Sci.* **14**, 552–566.

Riley, R. (1965). Cytogenetics and the evolution of wheat. In *Essays on Crop Plant Evolution*, ed. J. B. Hutchinson. Cambridge University Press, pp. 103–122.

Roy, N. N. (1973). Effect of seed size differences in wheat breeding by single seed descent. *J. Aust. Inst. Agric. Sci.* **39**, 70–72.

Saghir, A. R., Khan, A. R. and Worzella, W. (1968). Effects of plant parts on grain yield, kernel weight, and plant height of wheat and barley. *Agron. J.* **60**, 95–97.

Salim, M. H., Todd, G. W. and Schlehuber, A. M. (1965). Root development of wheat, oats and barley under conditions of soil moisture stress. *Agron. J.* **57**, 603–607.

Samygin, G. A. (1946). (Photoperiodism in plants.) *K.A. Timiriazeva Trudy, Akad. Nauk. SSSR* **3**, 129–262.

Sawada, S. (1970). An ecophysiological analysis of the difference between the growth rates of young wheat seedlings grown in various seasons. *J. Fac. Sci. Univ. Tokyo Sect III Bot.* **10**, 233–263.

Schlehuber, A. M. and Tucker, B. B. (1967). Culture of wheat. In *Wheat and Wheat Improvement*, eds K. S. Quisenberry and L. P. Reitz. *Agron. Series* No. 13. Amer. Soc. Agron, Madison.

Schmalz, H. (1958). Die generative Entwicklung von Winterweizensorten mit unterschiedlicher Winterfestigkeit bei Fruhjahrsaussaat nach Vernalisation mit Temperaturen unter – und oberhalb des Gefrierpunktes. *Der Zuchter* **28**, 193–203.

Sibma, L. (1970). Relation between total radiation and yield of some field crops in the Netherlands. *Neth. J. agric. Sci.* **18**, 125–131.

Simpson, G. M. (1968). Association between grain yield per plant and photosynthetic area above the flag leaf node in wheat. *Can. J. Plant Sci.* **48**, 253–260.

Single, W. V. (1964). The influence of nitrogen supply on the fertility of the wheat ear. *Aust. J. exp. Agric. Anim. Husb.* **4**, 165–168.

Skarsaune, S. K., Youngs, V. L. and Gilles, K. A. (1970). Changes in wheat lipids during seed maturation. I. Physical and chemical changes in kernel. *Cereal Chem.* **47**, 522–532.

Smith, H. F. (1933). The physiological relations between tillers of a wheat plant. *J. Counc. Sci. Ind. Res. Aust.* **6**, 32–42.

Sofield, I., Evans, L. T. and Wardlaw, I. F. (1974). The effects of temperature and light on grain filling in wheat. In *Mechanisms of Regulation of Plant Growth*. Roy. Soc. New Zealand, Wellington (in press).

Spiertz, J. H. J., ten Hag, B. A. and Kupers, L. J. P. (1971). Relation between green area duration and grain yield in some varieties of spring wheat. *Neth. J. agric. Sci.* **19**, 211–222.

Stoy, V. (1955). Action of different light qualities on simultaneous photosynthesis and nitrate assimilation in wheat leaves. *Physiol. Plantar.* **8**, 963–986.

Stoy, V. (1963). The translocation of C^{14}-labelled photosynthetic products from the leaf to the ear in wheat. *Physiol. Plantar.* **16**, 851–866.

Stoy, V. (1965). Photosynthesis, respiration, and carbohydrate accumulation in spring wheat in relation to yield. *Physiol. Plantar. Suppl.* **4**, 1–125.

Stoy, V. (1966). Photosynthetic production after ear emergence as a yield limiting factor in the culture of cereals. *Acta Agric. Scand. Suppl.* **16**, 178–182.

Suge, H. and Hirano, J. (1962). Effect of gibberellin on the vernalization of winter wheat and barley. *Proc. Crop Sci. Soc. Japan* **31**, 129–134.

Suge, H. and Osada, A. (1966). Inhibitory effect of growth retardants on the induction of flowering in winter wheat. *Plant and Cell Physiol.* **7**, 617–630.

Suge, H. and Yamada, N. (1965*a*). Effect of auxin and anti-auxin on the tillering of wheat. *Proc. Crop Sci. Soc. Japan* **33**, 330–334.

Suge, H. and Yamada, N. (1965*b*). Flower-promoting effect of gibberellin in winter wheat and barley. *Plant and Cell Physiol.* **6**, 147–160.

Syme, J. R. (1968). Ear emergence of Australian, Mexican and European wheats in relation to time of sowing and their response to vernalization and daylength. *Aust. J. exp. Agric. Anim. Husb.* **8**, 578–581.

Syme, J. R. (1969). A comparison of semi-dwarf and standard height wheat varieties at two levels of water supply. *Aust. J. exp. Agric. Anim. Husb.* **9**, 528–531.

Syme, J. R. (1970). A high yielding Mexican semi dwarf wheat and the relationship of yield to harvest index and other varietal characteristics. *Aust. J. exp. Agric. Anim. Husb.* **10**, 350–353.

Taylor, J. W. and McCall, M. A. (1936). Influence of temperature and other factors on the morphology of the wheat seedling. *J. agric. Res.* **52**, 557–568.

Teare, I. D., Law, A. G. and Simmons, G. F. (1972). Stomatal frequency and distribution on the inflorescence of *Triticum aestivum*. *Can. J. Plant Sci.* **52**, 89–94.

Teare, I. D. and Peterson, C. J. (1971). Surface area of chlorophyll-containing tissue on the inflorescence of *Triticum aestivum* L. *Crop Sci.* **11**, 627–628.

Teare, I. D., Sij, J. W., Waldren, R. P. and Goltz, S. M. (1972). Comparative data on the rate of photosynthesis, respiration and transpiration of different organs in awned and awnless isogenic lines of wheat. *Can. J. Plant Sci.* **52**, 965–971.

Terman, G. L., Ramig, R. E., Dreier, A. F. and Olson, R. A. (1969). Yield-protein relationships in wheat grain, as affected by nitrogen and water. *Agron. J.* **61**, 755–759.

Thorne, G. N. (1966). Physiological aspects of grain yield in cereals. In *The Growth of Cereals and Grasses*, eds F. L. Milthorpe and J. D. Ivins. Butterworths, London, pp. 88–105.

Thorne, G. N., Ford, M. A. and Watson, D. J. (1968). Growth, development and yield of spring wheat in artificial climates. *Ann. Bot.* **32**, 425–446.

Thorne, G. N., Welbank, P. J. and Blackwood, G. C. (1969). Growth and yield of six short varieties of spring wheat derived from Norin 10 and of two European varieties. *Ann. appl. Biol.* **63**, 241–251.

Troughton, A. (1962). The roots of temperate cereals (wheat, barley, oats and rye). *Comm. Bur. Pastures and Field Crops, Hurley*, 91 pp.

Turner, J. F. (1969). Starch synthesis and changes in uridine diphosphate, glucose pyrophosphorylase and adenosine diphosphate glucose pyrophosphorylase in the developing wheat grain. *Aust. J. biol. Sci.* **22**, 1321–1327.

Turner, N. C. (1966). Grain production and water use of wheat as affected by plant density, defoliation and water status. Ph.D. Thesis, Waite Agric. Inst., Univ. Adelaide.

Vervelde, G. J. (1953). The agricultural value of awns in cereals. *Neth. J. agric. Sci.* **1**, 2–10.

Vries, A. P. de (1971). Flowering biology of wheat, particularly in view of hybrid seed production: A review. *Euphytica* **20**, 152–170.

Wang, T. D. and Wei, J. (1964). The CO_2 assimilation rate of plant communities as a function of leaf area index. *Acta Bot. Sinica* **12**, 154–158.

Wardlaw, I. F. (1967). The effect of water stress on translocation in relation to photosynthesis and growth. I. Effect during grain development in wheat. *Aust. J. biol. Sci.* **20**, 25–36.

Wardlaw, I. F. (1968). The control and pattern of movement of carbohydrates in plants. *Bot. Rev.* **34**, 79–105.

Wardlaw, I. F. (1970). The early stages of grain development in wheat: response to light and temperature in a single variety. *Aust. J. biol. Sci.* **23**, 765–774.

Wardlaw, I. F. (1971). The early stages of grain development in wheat: response to water stress in a single variety. *Aust. J. biol. Sci.* **24**, 1047–1055.

Wardlaw, I. F. (1974). The physiology and development of temperate cereals. In *Cereals in Australia*, eds A. Lazenby and E. M. Matheson. Angus and Robertson, Sydney (in press).

Wardlaw, I. F., Carr, D. J. and Anderson, M. J. (1965). The relative supply of carbohydrate and nitrogen to wheat grains, and an assessment of the shading and defoliation techniques used for these determinations. *Aust. J. agric. Res.* **16**, 893–901.

Wardlaw, I. F. and Porter, H. K. (1967). The redistribution of stem sugars in wheat during grain development. *Aust. J. biol. Sci.* **20**, 309–318.

Waterbolk, H. T. (1968). Food production in prehistoric Europe. *Science* **162**, 1093–1102.

Watson, D. J., Thorne, G. N. and French, S. A. W. (1963). Analysis of growth and yield of winter and spring wheats. *Ann. Bot. N.S.* **27**, 1–22.

Wattal, P. N. (1965). Effect of temperature on the development of the wheat grain. *Indian J. Plant Physiol.* **8**, 145–159.

Weibel, D. K. (1958). Vernalization of immature winter wheat embryos. *Agron. J.* **50**, 267–270.

Welbank, P. J. (1971). Root growth of wheat varieties. *Rothamsted Report for 1971*, Part 1, pp. 104–106.

Welbank, P. J., French, S. A. W. and Witts, K. J. (1966). Dependence of yields of wheat varieties on their leaf area durations. *Ann. Bot. N.S.* **30**, 291–299.

Welbank, P. J., Witts, K. J. and Thorne, G. N. (1968). Effect of radiation and temperature on efficiency of cereal leaves during grain growth. *Ann. Bot.* **32**, 79–95.

Wheeler, A. W. (1972). Changes in growth substance contents during growth of wheat grains. *Ann. appl. Biol.* **72**, 327–334.

Willey, R. W. and Holliday, R. (1971). Plant population, shading and thinning studies in wheat. *J. agric. Sci., Camb.* **77**, 453–461.

Williams, R. F. (1955). Redistribution of mineral elements during development. *Ann. Rev. Plant Physiol.* **6**, 25–42.

Williams, R. F. (1960). The physiology of growth in the wheat plant. I. Seedling growth and the pattern of growth at the shoot apex. *Aust. J. biol. Sci.* **13**, 401–428.

Williams, R. F. (1966). *Ibid.* III. Growth of the primary shoot and inflorescence. *Aust. J. biol. Sci.* **19**, 949–966.

Williams, R. F. and Williams, C. N. (1968). *Ibid.* IV. Effects of day length and light-energy level. *Aust. J. biol. Sci.* **21**, 835–854.

Williams, S. G. (1970). The role of phytic acid in the wheat grain. *Plant Physiol.* **45**, 374–381.

Wolf, F. T. (1967). Exogenous precursors of starch synthesis in the leaves of wheat seedlings. *Z. Pflanzenphysiol.* **57**, 128–133.

Womack, D. and Thurman, R. L. (1962). Effect of leaf removal on the grain yield of wheat and oats. *Crop Sci.* **2**, 423–426.

Woodruff, D. R. (1969). Studies on presowing drought hardening of wheat. *Aust. J. agric. Res.* **20**, 13–24.

Wright, S. T. C. (1969). An increase in the 'Inhibitor β' content of detached wheat leaves following a period of wilting. *Planta, Berl.* **86**, 10–20.

Wright, S. T. C. and Hiron, R. W. P. (1969). (+)-Abscisic acid, the growth inhibitor induced in detached wheat leaves by a period of wilting. *Nature, Lond.* **224**, 719–720.

Yu, S-W, Wang, H. C., Chien, J. and Yin, H. C. (1964). The sources of grain material, the mobilization and redistribution of material among organs and between culms of the wheat plant, and soil moisture effect. *Acta Bot. Sinica* **12**, 88–98.

Zee, S. Y. and O'Brien, T. P. (1971). Vascular transfer cells in the wheat spikelet. *Aust. J. biol. Sci.* **24**, 35–49.

Zohary, D., Harlan, J. R. and Vardi, A. (1969). The wild diploid progenitors of wheat and their breeding value. *Euphytica* **18**, 58–65.

6

Soybean

RICHARD SHIBLES, I. C. ANDERSON AND A. H. GIBSON

The soybean [*Glycine max* (L.) Merr.] probably first emerged as a domesticated species in the North China Plains around the eleventh century B.C., and later was introduced into Manchuria (Hymowitz, 1970). Eventually, it spread throughout the Orient, but the Manchurian region dominated world production for many years (Piper and Morse, 1923). Although a crop of some importance in the USA since about 1880, its primary use was as a forage prior to the late 1930s. Currently, the USA produces nearly 75 % of the world's soybeans; China is the second largest producer, with an estimated 17 %. Over 50 % of the United States' crop is exported, primarily to Europe, Japan and Canada.

In the Orient, the soybean has a long history of use as an essential dietary component, principally as foods – liquid and solid; fresh, dried, and fermented – prepared directly from the bean. In the West, however, essentially all of the crop is industrially crushed and extracted. Although products of the soybean find many uses, a high proportion of the extracted oil (90 % in the USA) is for table consumption, and almost all of the protein (98 % in the USA) goes to supplement animal diets. In recent years, similar use has developed in the more industrial areas of the Orient too.

Except in photoperiodism and symbiotic nitrogen fixation, where it has frequently been a subject for basic studies, research on the physiology of the soybean crop has lagged behind that on breeding and genetics. With the growth in world consumption of soybeans, however, interest in and financial support for all aspects of research has increased substantially over the past decade.

SEEDLING ESTABLISHMENT

The soybean is epigeous and, under favourable conditions (25–30 °C, 2.5–3.5 cm planting depth), cotyledons begin to emerge in three to four days. A day later they are fully open, revealing the expanding, unifoliolate, primary leaves, which unfold by the fifth to seventh day. The first trifoliolate leaf unfolds in nine to eleven days, and reaches full expansion by the fifteenth to seventeenth day. These and other vegetative events are substantially delayed by cooler temperatures. Secondary roots appear in three to seven days and proliferate rapidly (Mitchell and Russell, 1971; Sun, 1955).

Sugars, primarily sucrose and stachyose in the cotyledons and embryonic axis (Abrahamsen and Sudia, 1966), are the principal carbon source during germination and very early growth. Thereafter, fat utilization increases and continues for about a week after emergence. Protein utilization, which begins almost immediately, accelerates during the first week after emergence and continues up to cotyledonary senescence (McAlister and Krober, 1951).

At 26 °C, the seedling remains essentially heterotrophic for about a week after emergence, cotyledonary photosynthesis amounting to only 4–9% of cotyledonary CO_2 evolved during this time. During the first week of auto-trophic growth, however, cotyledonary photosynthesis is about 45% that of the primary leaf contribution (Abrahamsen and Mayer, 1967). Cotyledons are, therefore, important to seedling vigour for ten to fourteen days following emergence, and probably longer under unfavourable conditions. In fact, from cotyledon removal treatments Weber and Caldwell (1966) concluded 'at least one cotyledon was essential for maximal yield for about 10 days after emergence'.

Seed that develops during hot weather (> 33 °C) has poor germination and emergence (Green *et al.* 1965), but the physiological basis for this has not been elucidated. Large seed (> 220 mg) is noted for poor emergence (Edwards and Hartwig, 1972), primarily because the large cotyledons provide excessive emergence resistance in medium-to-heavy soils and viability can be lower because they are more susceptible to handling damage.

Seed viability deteriorates rapidly at warm (> 30 °C) storage temperatures. Though seedling axis dry weight and the carbohydrate and nitrogen fractions of deteriorated seeds are lower, loss of viability does not seem to be a conse-quence of depletion of food reserves. But total sugar content of seed leachate increases markedly with seed deterioriation, indicating deterioration of membranes as well as loss of immediately metabolizable substrates (Edge and Burris, 1970*a*, *b*). That this could be the important consequence of deterioration is suggested by Burris *et al.*'s (1969) finding that seed glucose level and respiration rate of imbibed seed are highly correlated with seedling vigour.

Hypocotyl elongation in certain cultivars is reduced at temperatures around 25 °C (Grabe and Metzer, 1969; Gilman *et al.* 1973). The effect evidently is a consequence of differential ethylene production, but inhibition may be magnified by soil resistance (Samimy, 1970). Fehr (1973) indicates there is a major, dominant gene imparting near insensitivity, with degrees of sensitivity being conditioned by modifying genes.

Dry weight of seven-day-old seedlings is proportional to initial embryo (and seed) weight, as is cotyledonary and primary leaf area. However, among seed of normal size range (130–220 mg), seedling performance, as indicated by percent emergence, shoot and radical length at seven days, and primary leaf photosynthesis is not substantially different; nor is stand performance, as determined by seed yield, different. Seedlings developing from very small seed (*ca* 80 mg) do exhibit substantially less vigour, and stands yield significantly lower (Burris *et al.* 1971, 1973).

ROOT GROWTH AND FUNCTION

The root system

The root system consists, principally, of branched, lateral, secondary roots. These develop in four longitudinal rows off the enlarged, upper 10–15 cm of the primary root. In row plantings, the laterals extend 34–45 cm between rows before turning sharply downward; branching profusely and growing deeply, they give rise to a system that essentially pervades the top 180 cm of a permeable, well-drained soil. Below the origin of the major laterals the primary root is of similar diameter as the major laterals, but branches less. Most root dry weight (80–90%) occurs in the top 15 cm of the soil, largely because of enlargement of the primary root and the concentration of nodule tissue there. Probably, at least 40% of root surface occurs in the top 15 cm (Mitchell and Russell, 1971; Raper and Barber, 1970*a*; Sun, 1955).

Mitchell and Russell have characterized the pattern of root development. On emerging from the seed, the primary root grows directly downward. Lateral roots first appear three to seven days after germination, and by one month, the primary root has extended downward 45–60 cm and the major laterals have extended horizontally 20–25 cm. About 60 days after planting, perhaps earlier under dry conditions, horizontal extension terminates and five to six of the major laterals turn sharply downward near mid-row, perhaps because of antagonism between roots of adjacent rows (Raper and Barber, 1970*a*). The 60–80 day period is marked by proliferation in the 0–23 cm zone and rapid downward extension of the major laterals. The final root growth phase, 80–100 days after planting and a period characterized by rapid seed filling, is when the major laterals continue elongating rapidly to depths of 120–180 cm. There is a proliferation of roots at the lower depths and a further increase in the 0–7 cm zone as well, though the latter response, undoubtedly, depends on moisture conditions. Root growth continues essentially until top growth ceases (Mitchell and Russell, 1971; Suetsugu *et al.* 1962).

In water use, depth of rooting seems of more significance than lateral development. Considerable subsoil water is utilized, but inter-row water extraction is only about one-half as efficient as extraction below the rows (Peters and Johnson, 1960). This may account for Russell *et al.*'s (1971) finding that severe pruning of laterals, as might occur in a normal intertillage operation, only decreased yield 11% when performed as late as the beginning of pod growth. Less severe treatment had no significant effect.

Lines and cultivars differ considerably in root growth, especially in pro-liferation during late development. However, yielding ability and nutrient assimilation do not seem related to known differences in rooting behaviour (Mitchell and Russell, 1971; Raper and Barber, 1970b).

Soil temperatures of 22–27 °C seem most favourable to root growth, and top:root ratio is more influenced by light than by temperature (Earley and Cartter, 1945). Root growth is exponentially and inversely related to soil bulk density (Davies and Runge, 1969).

Nodulation and nitrogen fixation

The importance of the nitrogen-fixing symbiosis between soybeans and their associated nodule-bacterium, *Rhizobium japonicum*, has been recognized in recent years by a greater research interest in all aspects of the association. Estimates of the amount of nitrogen fixed vary widely, but from a comparison with non-nodulating soybeans, Weber (1966) concluded that 160 kg nitrogen ha^{-1} could be fixed by plants yielding 2800 kg ha^{-1} of seed.

Nodules are initiated in the outer root cortex following root hair infection by *R. japonicum* (Bieberdorf, 1938). Pure cultures of the bacteria can produce indoleacetic acid (Hubbell and Elkan, 1967a) and cytokinins (Phillips and Torrey, 1972), but whether these hormones are involved in nodule initiation and development is not known. Within the developed nodules, the bacteroids containing the nitrogen-fixing enzyme system, nitrogenase, are surrounded by a solution of the red pigment, leghemoglobin (Bergersen and Goodchild, 1973) whose currently favoured role is that of facilitating oxygen diffusion to the bacteroids (Bergersen et al. 1973).

Various physical factors such as temperature (Jones and Tisdale, 1921; Maeda, 1960), oxygen concentration (Bond, 1950), light-flux density (Sampaio and Dobereiner, 1968), and day-length (Eaton, 1931; Sironval, 1958) all influence nodule development, although the critical phases affected have not been determined. Several mineral nutrients also influence nodulation, but with the exception of calcium (McCalla, 1937), nitrogenous fertilizers (see Harper and Cooper, 1971; Beard and Hoover, 1971), and saline conditions (Bernstein and Ogata, 1966), differentiation between effects on nodulation and those on other physiological processes has not been attempted. The herbicides chloramben (Olumbe and Veatch, 1969) and trifluralin (Kust and Struck-meyer, 1971) both retard nodulation in the glasshouse; in the field these initial effects may disappear and seed yield be unaffected (D. W. Staniforth and A. H. Gibson, unpublished data).

Four host genes affecting nodulation have been described (Weber et al. 1971). One prevents nodulation by most strains (Williams and Lynch, 1954) and has been incorporated into various lines for comparative studies on nitrogen nutrition (Liu and Hadley, 1971) or to investigate, so far unsuccessfully, factors likely to be responsible for nodule formation (e.g. Hubbell and Elkan, 1967 *a*; 1967 *b*).

In mixed *R. japonicum* populations, the competitive ability of strains in forming nodules is influenced by the host genotype, planting date, temperature, and their relative numerical strength (Weber *et al.* 1971).

Nodule function

Most studies of the biochemistry of nitrogen fixation by legumes have been made with soybeans (see reviews by Evans and Russell, 1970; Bergersen, 1971 *a*). Although ^{15}N was initially used in the biochemical and physiological studies, the ability of nitrogenase to reduce acetylene to ethylene has been used widely in these and agronomic studies in recent years (Hardy *et al.* 1973). Despite the wide applicability and ease of operation of the acetylene reduction technique, caution must be exercised in its use and in the interpretation of the results due to strong oxygen concentration effects, temperature influences, moisture effects, diurnal variation, and the form of the assay (e.g. Hardy *et al.* 1968; Bergersen, 1970; Sprent, 1969; Mague and Burris, 1972).

At the physiological level, most of the investigations on supply of photosynthate to nodules, and export of amino acids from them, has been done with the field pea (see Chapter 7). The applicability of these findings to soybeans remains to be ascertained, but it is likely that differences will be quantitative rather than qualitative. Asparagine is the principal nitrogenous compound transported from the nodules (Wong and Evans, 1971).

Numerous biological and physical factors influence nitrogen fixation by soybeans. Nitrate and ammonium salts (7 mM) in the nutrient solution supplied to nodulated plants rapidly retard further nodule development and the nitrogenase activity of existing nodules; however both development and activity show rapid recovery following exclusion of these salts from the nutrient solution (K. Hashimoto and A. H. Gibson, unpublished data). The decline in nitrogenase activity is paralleled by a decline in the transport of ^{14}C to the nodules, but the basic reason for the retardation of activity is not known.

Another important physical factor is oxygen concentration, with detached nodules showing increased nitrogen fixation with an increase in P_{O_2} up to about 0.50 atm (Burris *et al.* 1955; Bergersen, 1962). However, apart from Bond's (1950) conclusion that low P_{O_2} reduced nodule development and nitrogen fixation, little is known of what happens under field conditions when P_{O_2} may be less than 0.2 atm. Moisture stress has received little attention although the recent studies of Sprent (1971 *a*, 1971 *b*, 1972 *a*, 1972 *b*, 1972 *c*) indicate that the effects are very important. Slight desiccation of the nodules leads to a rapid reduction in nitrogenase activity and respiration, and this is associated with severe disruption of the cytoplasm in the vacuolate cells in the outer cortex. Osmotically applied stress has similar effects on activity.

Nodules on plants lightly water-stressed show a rapid recovery of activity on rewetting, even when water is supplied to the roots below the nodules. Weber (1966) comments that, under field conditions, nitrogen fixation is greatly reduced in a dry season.

Of the biological factors, the effectiveness of the symbiosis in fixing nitrogen is considered to be of great importance, and differences between strains of *R. japonicum* are readily demonstrated under bacteriologically-controlled conditions in the glasshouse and in the absence of naturally-occurring strains in the field (e.g. Abel and Erdman, 1964; Weber *et al.* 1971). However, relatively little is known of the reasons for such differences. Dobereiner *et al.* (1970) consider that most effective strains differ in the degree of effectiveness according to the amount of nodule tissue formed. In addition, they have described three strains of exceptional efficiency, as determined by very high rates of N_2-fixation per unit nodule weight. There will still remain the problem of achieving 'penetration' of such strains into developing nodule populations where high numbers of *R. japonicum* already exist (see Ham *et al.* 1971).

Whereas field peas show a marked decline in nitrogen fixation soon after flowering begins (Pate, 1958), the data for soybeans indicate that most of the fixation occurs after this physiological stage. Bond (1936) first showed this under glasshouse conditions, and field studies in recent years, using the acetylene reduction technique, have confirmed the observation. Cultivars differ in their 'fixation profile' (Hardy *et al.* 1971; Hardy *et al.* 1973; Mague and Burris, 1972); they differ in the time taken to initiate and terminate fixation, the form of the profile, and the total amount of activity. Elucidation of the reasons for these differences (competition for photosynthate between nodules and pods, differential longevity of symbiotic associations, differential responses to various environmental stresses or in the size of the nitrogen sink) could open the way for major increases in symbiotic nitrogen fixation by soybeans, either through the adoption of appropriate management practices or the selection of biological material with greater symbiotic potential.

TOP GROWTH AND DEVELOPMENT

The vegetative structure

The above-ground vegetative structure consists of an erect, branched stem that normally attains a height of 75–125 cm and possesses 14–26 nodes. The lowest node bears the cotyledons; the two, opposite, simple, primary leaves are next, and all other leaves on the stem and all leaves on the lateral branches are alternate, trifoliolate and borne on long petioles. Multi-foliolate lines occur, but are not common. Leaflet shape varies with genotype but most modern cultivars have ovate leaflets. Depending upon planting density, 0–6 branches per plant are usual; all nodes, however, possess the potential for both branching and floral development. With maturation the lower internodes of the stem become woody.

Two growth habit types are common, the determinate type, in which the stem and branches possess a terminal raceme with a cluster of pods, and the indeterminate type. Additionally, there are two genetic types of determinateness, one, dt_1, recessive and one, Dt_2, dominant to indeterminateness (Bernard, 1972). According to Bernard 'the primary effect of both dt_1 and Dt_2 is to hasten the termination of apical stem growth, which decreases both plant height and number of nodes, but dt_1 has a much greater effect'. Indeterminate cultivars are grown in northern United States and Canada, whereas dt_1-determinates are commonly grown in southern United States, Japan and Korea.

The vegetative canopies of indeterminate and determinate types are distinctly different. Indeterminates have a strongly tapering stem. The largest leaflets and longest petioles occur in mid-plant with gradations in size toward each end of the stem; lines vary in the degree to which leaflet size diminishes between mid-plant and top. Dt_2-types have a less tapering stem, and dt_1 stems are 'rather thick to the top' (Bernard, 1972). With determinates, leaflet size and petiole length are not smaller at the top of the plant. Because upper leaves of determinate types are large, their canopies possess poorer light distribution characteristics than do indeterminates. Determinates, especially dt_1-types, have fewer stem nodes, and they branch more. An additional difference is that indeterminate plants continue stem growth and leaf production for a long period after flowering begins, whereas stem growth of dt_1 terminates either before flowering or shortly thereafter; Dt_2 is intermediate.

Rate of leaf appearance, branching, plant height and length of the vegetative period are all strongly influenced by temperature (Table 6.1). Photoperiod strongly affects height and number of nodes produced, both increasing with length of day, but it hardly influences rate of nodal differentiation and leaf appearance (Johnson *et al.* 1960). According to Suetsugu *et al.* (1962) the number of nodes on the main stem and the number of primary branches are determined four to six days after flowering begins; but, in cultivars ranging in flowering date from 30–67 days after emergence, Johnson *et al.* (1960) found most of the nodes on the main axis (all in some cases) differentiated by 35 days after planting.

Branches do not usually emerge at the cotyledonary and primary leaf nodes (Oizumi, 1962). The period during which a branch can emerge at a given node is limited to a short time, from immediately after expansion of the leaf at the same node until the first cycle of secondary thickening has ceased in the internode below. Nodes which develop simultaneously on the main stem and branches ('co-growing nodes') tend to act similarly. They bloom at the same time and have similar flower numbers, pod numbers, and abortion percentages.

Table 6.1. *Vegetative characteristics of 'Wayne' soybeans grown under seven temperature regimes, 16-hour photoperiod*

Character	Day/Night[1] temperature, °C						
	18/13	21/16	24/19	27/22	30/25	33/28	36/31
Days between leaves[2]	6.2	4.2	3.7	3.3	2.8	2.8	2.8
Branches per plant	5.4	2.0	1.2	0.0	0.0	0.0	0.0
Height at flowering, cm	75	94	76	61	85	147	—[3]
Days to flower	83	69	53	40	43	47	—[3]

[1] Day/Night temperatures were for 8 and 16 hours, respectively.
[2] Average 1st to 11th trifoliolate.
[3] Did not flower in 55 days.
Unpublished data of R. M. Shibles.

Leaf longevity is, to some extent, a function of canopy light relations. The cotyledon, primary and the first several trifoliolate leaves usually abscise, and in high-density stands up to half the trifoliolate leaves may abscise by mid-bean-filling.

Canopy growth and dry matter accumulation

After an initial lag, leaf area production increases rapidly and nearly linearly up to the end of blooming, attaining maximal leaf area index (LAI) values of 5–8. Thereafter, LAI declines progressively, by abscission of lower leaves, during seed filling to values of 4–6 near physiological maturity, after which the remaining leaves yellow rapidly and soon abscise. At similar planting densities narrow rows reach greater LAI values than wide rows, and have more leaves abscising during seed filling (e.g. Weber *et al.* 1966; Ojima and Fukui, 1966; Buttery, 1969).

Under the contrasting production conditions of narrow rows (30–50 cm) and high populations versus wide rows (75–100 cm) and low populations, the distribution of leaves on the plant differs markedly. With the former, average leaf age is younger, because earlier closing of the canopy results in greater abscission of lower leaves. Whether or not this influences productivity is moot, because Johnston *et al.* (1969) have shown that lower leaves are photosynthetically efficient, if adequately illuminated.

Crop Growth Rate is a linear function of intercepted irradiance (Shibles and Weber, 1965, 1966), and the LAI required for 95 % light interception ranges from 3.1 to 4.5, depending upon planting density and spatial arrangement. Although Ojima and Fukui (1966) found an optimum LAI for Crop Growth Rate, others have not (Shibles and Weber, 1965 [Figure 6.1]; Buttery, 1969, 1970).

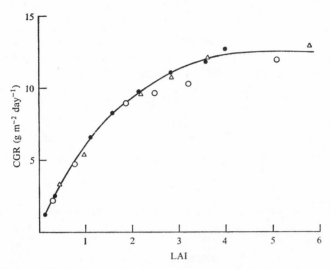

Figure 6.1. Relationship of Crop Growth Rate (CGR) and LAI for three equi-distant spacings; ●, 4.3; ○, 7.7; △, 12.9 plants m $^{-2}$. (Adapted from Shibles and Weber, 1965.)

Cultivars differ in Crop Growth Rate, Hanway and Weber (1971*a*) finding rates during the linear phase varying from 8.8 to 14.9 g m $^{-2}$ day $^{-1}$ among cultivars. Despite this, rate of dry weight increase in the bean fraction of all cultivars was similar, about 9.9 g m $^{-2}$ day $^{-1}$. Major differences in seed yield were attributable to length of the bean development period rather than rate of daily growth. The highest Crop Growth Rate reported for soybeans is 17.2 g m $^{-2}$ day $^{-1}$ (Buttery, 1970).

Seasonal dry matter accumulation exhibits essentially a linear trend between about mid-bloom and late-seed-filling (Figure 6.2*a*), with maximum vegetative dry weight occurring at about mid-bean-filling. Vegetative and pod dry weight decline during the later stages of bean-filling as tissues lose dry matter because of respiration and mobilization to the bean fraction. $^{14}CO_2$ studies (D. J. Hume and J. G. Criswell, personal communication) and carbohydrate analyses (Dunphy, 1972) at various stages of development confirm that substantial quantities of labile carbohydrates accumulate in leaves, petioles and stems prior to seed development and are later utilized in seed growth. In Hume and Criswell's study, as much as 8 % of $^{14}CO_2$ assimilated by whole plants as early as two weeks prior to flowering eventually was recovered in seed.

Transpiration

Soybeans use a lot of water. Kato (1967) estimates their water requirement to be 580 g g $^{-1}$ dry matter. Using plastic-covered, irrigated field plots,

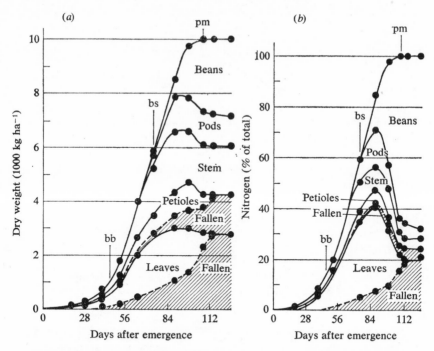

Figure 6.2. Cumulative dry weight (*a*) and relative amount of nitrogen (*b*) in above-ground plant parts during the growing season. bb, bs and pm indicate beginning bloom, beginning seed-filling and physiological maturity stages respectively. The shaded portions indicate abscised petioles and leaves. (Adapted from Hanway and Weber, 1971*b*, 1971*d*.)

Peters and Johnson (1960) assessed their transpiration potential to be about 129 mm from beginning bloom until maturity. With frequent rainfall and abundant water availability, they found total water use was 392 mm (95 % of open-pan evaporation) over the same period.

Water consumption is determined largely by leaf area and its distribution, water supply, and atmospheric evaporative demand. Prior to full ground-cover, leaf area is the most influential variable; thereafter, atmospheric demand is the principal controlling factor (Laing, 1966; Kato, 1967). Reducing row spacing from 100 cm to 50 cm and doubling population density, Peters and Johnson (1960) increased transpiration from the 129 mm mentioned above to 226 mm. It is likely that this response was mainly a consequence of increasing the amount of exposed leaf surface in the earlier part of the season; but, remembering that they also found roots to be not as efficient in extracting water from the inter-row space as directly below the row, some of the increased transpiration was probably related to root distribution too.

According to Boyer (1971) there is a very high resistance (1.6×10^6 s cm^{-1})

to water movement through soybean plants, which is probably why leaves show wilting even in wet soil on days of very high evaporative demand. The high plant resistance can be traced to an exceptionally high resistance in root radial tissue, the resistance in other tissues being similar or less than those measured in other broad-leaved species. Because of this high root resistance (compared to other plants such as *Phaseolus* and *Helianthus*) Boyer estimates soybean leaf water potential would have to decrease about twice as much to maintain water flow at the same rate. Hence, soybeans may be under water stress more frequently and more severely than many plants.

Mycorrhizally infected soybeans have lower root resistance to water transport. The effect seems unrelated to additional absorptive surface area or the pathway through root radial tissues provided by hyphae, but possibly to enhanced nutrient status (Safir *et al.* 1972). Root resistance might be reduced through breeding, because rootstocks seem to differ somewhat in their ability to supply water to tops (Sullivan, 1972). This, plus differences in stomatal resistance (Dornhoff and Shibles, 1970), may account for Hansen's (1972) finding that cultivars differ in evapotranspiration independently of differences in LAI.

Stevenson and Shaw (1971*a*) believe that under water stress there is a preferential flow of water to upper leaves, because stomatal resistance of mid-canopy leaves increases with time more rapidly and to a greater extent than resistance of upper leaves on high-demand days. The observation of Farkas and Rajhathy (1955) that water content of lower leaves decreased more than upper leaves under drought lends support to the preferential flow hypothesis. Even under high water availability and low atmospheric demand, Stevenson and Shaw found leaf resistance to be substantially less in upper than mid-canopy leaves.

Leaflets of upright display (10–15° to the vertical) have lower temperature and lower stomatal diffusion resistance than those of horizontal display, except on overcast days (Stevenson and Shaw, 1971*b*). Hence, selection for upright display could aid in moisture conservation and reduce stress. Introduction of the dense pubescence character also is worth consideration, as it seems to reduce transpiration substantially (e.g. by 26 %) while not affecting net photosynthetic rate (Ghorashy, Pendleton, Bernard and Bauer, 1971).

REPRODUCTIVE DEVELOPMENT

Regulation

The soybean is a short-day plant, and cultivars differ in critical day-length (Garner and Allard, 1930; Borthwick and Parker, 1939; Hartwig, 1954; Johnson *et al.* 1960). In general, the shorter-seasoned the cultivar the longer the day under which it will flower and mature.

However, adaptation is not only controlled by photoperiod. Temperature also influences the time required to reach floral differentiation, floral expression, and subsequent reproductive stages under inductive photoperiods. For induction, the major effect is attributed to night temperature of the leaves (Parker and Borthwick, 1943). Successive reproductive stages seem to have narrower temperature limits (e.g. Van Schaik and Probst, 1958).

At latitudes greater than 40°, where soybeans are planted during a cold spring, the length of time between emergence and flowering for adapted cultivars is mainly a function of accumulated temperature and is little influenced by natural photoperiod. Adapted cultivars are photoinduced and usually exhibit flowers by the longest day of the year (Brown, 1960; Pandey, 1972). At lower latitudes with warmer spring temperatures, planting usually is delayed until approximately a month before the longest day of the year, particularly for early cultivars, so that plants are not stunted by premature photoinduction which may decrease yields, particularly in wide rows. This suggests that the rate of reproductive development of cultivars adapted to lower latitudes is controlled mainly by their response to changes of the natural photoperiod.

Cultivars and lines are classified according to length of growing period. 'Maturity Group' designations range from 00 through X, those with the longest growing season. Many Group 00 and 0 lines are day neutral with respect to flowering, but height is affected by photoperiod – i.e. in longer than normal photoperiods they grow taller (Criswell and Hume, 1972; Polson, 1972; Guthrie, 1972). In photoperiodically responsive cultivars, successive reproductive stages require progressively shorter days. Guthrie (1972) grew 20 cultivars in the field at Ames, Iowa (42° N) with a 12-hour photoperiod, a range from natural to 21 hours, and also a 24-hour photoperiod. The cultivar 'Shinsei' (Group 00) was day neutral for all vegetative and reproductive attributes. An adapted cultivar, 'Hawkeye' (Group II), had a critical daylength of 18 hours for flowering, 17 hours for flowering to the apex, 15.5 or less hours for podding to the apex, and 14.5 or less hours for the 'green-bean' stage. In a warm glasshouse Hawkeye has a critical photoperiod of about 15.5 hours. The longer critical photoperiod in the field may be indicative of a slower rate of decay of the far-red form of phytochrome with the cooler night temperatures in the field. Short-seasoned cultivars were not delayed by small increases in day-length until later stages of reproduction were reached. Conversely, long-seasoned cultivars were influenced at earlier stages of reproduction in comparison to the adapted cultivar. Natural photoperiod has little effect in determining the time of flowering, but the interaction between natural photoperiod and photoperiodic characteristics of a particular cultivar regulates the rate at which reproduction is allowed and thereby determines its time of maturity.

The difference in time of flowering between cultivars at the higher latitudes may be controlled by their varying response to temperature, or to a quantitative effect of the difference between critical and experienced photo-period, or to a plant age factor (Fisher, 1955).

The illumination level which inhibits the dark reactions leading to flowering is approximately 5 lx, but light of only 0.1 lx has a quantitative effect on number of floral primordia (Borthwick and Parker, 1938).

Bernard (1971) described an early gene (e_2) and a late gene (E_1); they affect both time of flowering and maturity. A third photoperiod gene (e_3) has been described by Buzzell (1971) and by Kilen and Hartwig (1971). Cultivars with the recessive gene respond to fluorescent light used to extend day-length as though they were in darkness. The (e_3) gene in 'Harosoy' (Group II) length-ened the critical photoperiod from 17 to 21 hours for flowering in the field (Guthrie, 1972). A very significant observation is that isolines carrying this gene mature earlier in the field (Kilen and Hartwig, 1971), which supports the concept of a quantitative effect of photoinduction. These three (E) genes in soybeans are simply inherited, with two alleles at each locus. In addition, time of maturity is controlled by other factors more complexly inherited (Byth, 1968).

Flowering, pod and seed development

The soybean is self-pollinated, pollination occurring when the flower opens or slightly previously. The inflorescence is an axillary raceme, except that determinates also have a terminal raceme. Axillary racemes vary greatly in length, being up to 10 cm long in cultivars adapted to low latitudes and much shorter in those adapted to high latitudes.

Flowering occurs over a four to six week period, depending upon seasonal adaptation, but the sequence of flowering differs with growth habit. In indeterminate lines, flowering begins at the fourth to eighth node on the main stem and progresses upward. Branches begin flowering a few days later than the main stem. Flowering is most abundant at the middle nodes. Dt_2-determinates behave similarly, but the dt_1-determinate, in its area of adapta-tion, behaves quite differently (E. E. Hartwig, personal communication). Flowering begins at about the eighth or ninth node and proceeds rapidly, so that within a few days there is flowering at every node, including branches and the terminal raceme. It takes 10 to 18 days for full flowering on a primary raceme. Secondary racemes usually develop and flower, and if few flowers set on the primary and secondary racemes, tertiary racemes may be produced also. Hence, in dt_1-determinates, length of flowering is associated with dura-tion of flowering at a node, whereas in indeterminates and Dt_2-determinates, it depends upon sequential flowering of nodes.

Flowering is poor below 20 °C and increases up to about 32 °C (Van Schaik and Probst, 1958). Below 20 °C cleistogamy is often observed. Spraying the foliage with tri-iodobenzoic acid (TIBA) 10 days before bloom increases flower production (Ishihara, 1956).

The number of ovules per pod is genetically determined, and varies somewhat with environment. There are genetic 2-, 3-, and 4-ovule types. Most modern cultivars are 3-ovule types. They may have some 2- and 4-ovule pods, but about 85 % of mature pods will contain three ovules (R. G. Palmer, personal communication). According to Kato *et al.* (1954) the later flowers tend to be those having fewer ovules.

Pods develop slowly for the first few days following fertilization. Rapid elongation begins about the fifth day, and full length is attained by the fifteenth to twentieth day. Cell division in the cotyledon is completed two weeks after fertilization. This is followed by a 10–12 day period of rather slower development, with lipid granules, protein globules, and starch appearing during the latter part. Rapid cotyledon development, characterized by virtually linear accumulation of dry weight, protein, and fat, occurs over about the next three weeks. During this rapid growth phase, up to 80 % of the cotyledon dry weight, protein, and fat is accumulated. At full development, cotyledons contain numerous starch grains, but these presumably are mobilized during dehydration and seed maturation because none are present at seed maturity. The embryo, with the first trifoliolate leaf differentiated, reaches full size about a week before maximum seed dry weight is attained. For most full-season cultivars, the latter point is reached 45–52 days following fertilization, but in instances where maturation is accelerated it can be as soon as 30 days. (See Kamata, 1952; Kato *et al.* 1954; Ozaki *et al.* 1956; Howell and Cartter, 1958; Suetsugo *et al.* 1962; Bils and Howell, 1963.)

Seeds within a pod do not grow at the same rate (Kato *et al.* 1954). For the first 10–15 days after fertilization, the apical seed grows most rapidly. To some extent, this may be a function of its earlier fertilization. In the second phase, just after full elongation of the pod, the basal seed shows more rapid growth. In the final stages, the central seed(s) is dominant.

Soybeans exhibit substantial loss of reproductive organs. Estimates of flower and pod abscission range from 40 to 80 % (Van Schaik and Probst, 1958; W. R. Hansen and R. M. Shibles, unpublished data). Additionally, ovule and seed abortion in mature pods ranges from 9 to 22 % (Woodworth, 1930; R. G. Palmer, personal communication).

Although aborted seeds and ovules of all sizes are found, certain stages of development are more prone to abortion. Kato *et al.* (1955) examined abscised buds, flowers, and pods, and classified them according to stage of ovule or seed development (Table 6.2). By far the greatest abscission takes place after fertilization, and most of this occurs during early stages of embryo development. Abscised buds are frequently the last buds to differentiate at a node, whereas abscised pods are among the earliest to differentiate. Abscission is disproportionately greater at lower nodes and on branches (Hansen and Shibles, unpublished data). Abortion, as recorded in mature pods, is most frequent in the basal position, accounting for about 60 % of all abortions

Table 6.2. *Frequency (%) of anatomical stages of bud, flower and pod abortion (adapted from Kato* et al. *1955)*

Stage of development reached at failure	1952	1953
Flower differentiation	5.2	1.4
Reproductive cell division	16.3	13.2
Flowering and fertilization	4.3	3.6
Initial stage of proembryo*	43.3	
Later stage of proembryo*	12.8	81.8
Cotyledon stage*	17.9	

* 'Initial stage of proembryo' is 3–8 cells, 3–7 days after fertilization; 'later stage of proembryo' is about 100 cells, 10–15 days after fertilization; 'cotyledon stage' is 20–25 days after fertilization, just after cotyledon differentiation (Kato and Sakaguchi, 1954).

in three-ovule pods. Abortion at the central position is slightly greater than at the apex (R. G. Palmer, personal communication). Kato and Sakaguchi (1954) explain this on the basis of phasic development of the individual seeds. Most abortion occurs when the apical seed is dominant in growth.

Kato and Sakaguchi (1954) observed that the principal stages of abortion are coincident with stages of rapid embryo growth, and retardation of development is first manifest in cessation of embryo elongation. Because of this and the fact that water deficit has a marked effect on abortion, they conclude that the principal reason for seed or ovule abortion is inadequate water. Competition among seeds and organs, as well as inadequate external supply, is considered an important factor. Agronomists have long observed that abscission of flowers and pods is particularly high during periods of water stress. Whether the high percentage of abscission that occurs under seemingly good conditions also is related to water deficit is debatable.

Abscission is promoted by long photoperiods and also by high temperature; flowering is also promoted by high temperature. Consequently, Van Schaik and Probst (1958) found no effect of mean day temperatures between 16 °C and 32 °C on average number of pods set per node. At temperatures above 40 °C, however, Mann and Jaworski (1970) found pod set reduced by 57 to 71 %. Light stress (shading to 5–16 klx) also induced severe abscission. Cool temperatures (15 °C) two weeks prior to blooming severely reduce pod setting (Saito *et al.* 1970). At later stages or for shorter periods they have less effect. The cool temperature effect is intensified at high nitrogen levels (Hashimoto and Yamamoto, 1970).

Because abscission of reproductive organs seems to be proportional to flowering, except for the lower five to six nodes which have greater loss, pod and seed distribution is also related to flowering. In indeterminate cultivars there is much greater production from middle nodes (Hansen and Shibles, unpublished data), whereas in dt_1-determinates podding is heavier in the upper half of the plant, and branches and the terminal racemes make a significant contribution (E. E. Hartwig, personal communication).

Final seed size is influenced both environmentally and genetically. Each cultivar tends to have a characteristic average seed weight. Year to year and location to location variation in average seed size of a cultivar ranges up to 60 %, but can be greater under unusual circumstances. Average seed weight of cultivars ranges from 120–280 mg per seed (at 13 % moisture). Cultivars with very large seed size do not yield more. They simply produce fewer seeds.

Protein and fat

The seed of modern cultivars contains about 21 % fat and 41 % protein. Recent prices illustrate the greater need of protein rather than oil from soybeans. Breeders have lines available with over 50 % protein, but the seed yields of these are 10–20 % below those of standard cultivars. It is not known whether the lower yield is a reflection of less effort in breeding or represents a barrier of some type. In North America fat content increases 2 % but protein changes little from Canada to the Gulf. At a given latitude, variations in protein and fat contents are usually inversely related, with about two weights of protein equivalent to one of fat (Hanson *et al.* 1961).

Relatively high temperatures increase fat content appreciably without much influence on protein content. The maximum effect of high day temperatures begins during the earlier phase of seed development and continues until half maximum seed weight is reached (Howell and Cartter, 1958). Saito (1961) found a similar relationship between day temperatures and fat content and also reported that high night temperatures increase protein content. Stresses such as late planting, moisture, and mineral nutrient deficiencies (except nitrogen), decrease fat and increase protein contents (Weiss *et al.* 1952; Leffel, 1961).

The amino acid composition of soybean meal is well balanced as a source of protein for growth of monogastric animals, methionine being the most deficient amino acid. The proteins of the seed are heterogeneous and even the major globulin, glycinin, is not homogeneous (Altschul *et al.* 1966).

Much progress has been made in elucidating the path of fatty acid synthesis in developing seeds (Rinne and Canvin, 1971*a*, 1971*b*; Inkpen and Quakenbush, 1969). An oxidative desaturase seems to be able to dehydrogenate the acids – i.e. stearic→oleic→linoleic→linolenic, the content of linolenic acid decreasing with increasing temperature during seed development (Howell and Collins, 1957).

The fat of soybean contains 7–8 % linolenic acid which causes poor flavour stability in comparison to corn oil with 1 % linolenic. The major fatty acid of soybean is linoleic (50 %) which is diunsaturated. Another undesirable characteristic is the high level of lipoxygenase. Hammond *et al.* (1972) were able to reduce linolenic acid to 3.5 % in one cycle of selection from crosses of low linolenic lines. They also found lines with as high as 64 % oleic acid which would confer stability to the oil by reducing the linoleic content.

YIELD DETERMINING PROCESSES

Photosynthesis

The leaf

The soybean leaf has characteristic Calvin cycle photosynthesis.

As a leaflet expands, its CO_2-assimilating capacity increases, reaching a maximum a few days after full expansion. The maximum rate is maintained for about a week before the onset of a gradual decline (Ojima *et al.* 1965). Successively produced leaves, through the ninth trifoliolate at least, have higher net photosynthetic capacity (Kumura and Naniwa, 1965). But, subsequent leaves probably do not vary much – until the beginning of seed filling when upper, young leaves have photosynthetic rates much higher than any previous leaf (Dornhoff and Shibles, 1970; Ghorashy, Pendleton, Peters, Boyer and Beuerlein, 1971).

Saturating light-flux density is a function of genotype, ontogeny and environment. Leaves of moderate capacity (*ca* 33 mg CO_2 dm^{-2} h^{-1}) saturate at about 120 w m^{-2} whereas those of high capacity (*ca* 45 mg CO_2 dm^{-2} h^{-1}) saturate near 300 w m^{-2} (Dornhoff, 1969). Leaves of spaced, field-grown plants have exceptionally high rates (*ca* 60 mg CO_2 dm^{-2} h^{-1}) and are not light-saturated, even at full sunlight (Beuerlein and Pendleton, 1971).

Stems and pods exhibit a small net CO_2-uptake but pods, in particular, have high respiration (Kumura and Naniwa, 1965). Calculating gross CO_2-uptake from the net CO_2-uptake and respiration data of Kumura and Naniwa, we estimate that during rapid seed development pods might fix 14 % of the plant's daytime uptake; stems probably fix about 3 % and the leaf laminae about 82 %.

Cultivars and lines differ in leaf net photosynthetic potential (Ojima and Kawashima, 1968; Curtis *et al.* 1969; Dornhoff and Shibles, 1970). Genotypic differences are not related to differential photorespiration (Dornhoff and Shibles, 1970; Curtis *et al.* 1969). In fact, higher photosynthesis is associated with higher photorespiration (Ogren and Bowes, 1971). Rather, genotypic variation in net photosynthesis seems related both to stomatal conductivity and internal leaf factors, such as mesophyll and carboxylation resistance (Dornhoff and Shibles, 1970; Dornhoff, 1971).

Photosynthesis in the F_1 tends to be similar to the low parent, whereas in the F_2 it is normally distributed, leading Ojima *et al.* (1969) to conclude that photosynthetic potential is quantitatively inherited. Because photosynthetic rate varies considerably with environment over short intervals, some more stable character, such as specific leaf weight, leaflet thickness or leaflet nitrogen content per unit area – all of which are correlated with genetic photosynthetic potential (Ojima and Kawashima, 1968; Dornhoff and Shibles, 1970; Dornhoff, 1971) – might be a better selection index than carbon dioxide uptake itself. In this regard Ojima and Kawashima (1970) found the correlation of F_3 progeny with F_2 parents in two crosses was inconsistent for net photosynthesis (0.45, 0.66) and nitrogen content per unit area (0.47, 0.72), only moderately good for leaf thickness (0.55, 0.59), but reasonably high for specific leaf weight (0.71, 0.79). However, because the majority of their F_3 lines with high net photosynthetic ability came from F_2 plants with high net photosynthesis, Ojima and Kawashima conclude that selection for high net photosynthesis is possible. But the relationship between F_3 progeny and F_2 plants suggests that specific leaf weight may be the better selection criterion. Regardless of criterion used, however, care must be exercised in selection because all these characters vary greatly with stage of development. Thus, selection within a population segregating widely in rate of development would be difficult.

The canopy

The principal factors influencing canopy net photosynthesis are genotype, light, and LAI (Hansen, 1972). Fully-developed, non-stressed canopies do not show light-saturation, but saturation is evident at LAI of 4 or less. Maximum carbon dioxide assimilation occurs at an LAI of 5–6 under high irradiance, whereas under low irradiance it occurs at an LAI of 3–4. With low irradiance, there is occasional evidence of a decline in assimilation rate at LAI greater than 3–4.

Soybean leaflets have no preferred azimuth and canopies are only slightly planophile (Blad and Baker, 1972). Nevertheless, a high proportion of light is intercepted in the periphery of the canopy (Shaw and Weber, 1967; Singh *et al.* 1968; Kumura, 1969), which probably accounts for Kumura's (1968) finding that carbon dioxide assimilation increased by up to 30 % as the proportion of diffuse light increased from less than 25 % to more than 75 % at 90 klx. Kumura's results suggest that a less dense, more erectophile canopy could be beneficial to photosynthetic productivity. The fact that lower canopy leaves will exhibit increased photosynthesis under supplemental light (Johnston *et al.* 1969) supports this suggestion. But differing extinction coefficients, and differing photosynthesis as a consequence thereof, have yet to be demonstrated in soybeans.

Independent of change in LAI or environment, there is a declining trend in canopy photosynthetic potential with maturation. Starting about two weeks after the beginning of seed filling, the four cultivars Hansen (1972) monitored declined an average $1.4\,g\,CO_2\,m^{-2}\,h^{-1}$ over 24 days, probably due to increasing leaf age. Using growth analysis formulae, Koller *et al.* (1970) calculated a secondary peak in net assimilation rate (NAR) which they attributed to an increase in sink demand during seed filling. In neither Hansen's nor Jeffers and Shibles' (1969) data is there any evidence of a secondary peak in net carbon dioxide uptake, however. We interpret the secondary peak in NAR as being a consequence of the rapid abscission of lower, shaded leaves that occurred in Koller *et al.*'s study.

Cultivars differ in canopy photosynthesis independently of LAI and other effects. Jeffers and Shibles reported differences of $0.8\,g\,CO_2\,m^{-2}\,h^{-1}$, but Hansen (1972), in a more extensive study, found differences of up to $2\,g\,CO_2$ $m^{-2}\,h^{-1}$ among cultivars under high irradiance.

Maximum canopy net photosynthetic rates reported for soybeans range as high as $7\,g\,CO_2\,m^{-2}\,h^{-1}$ (Kumura, 1968; Hansen, 1972), and rates exceeding $5\,g\,CO_2\,m^{-2}\,h^{-1}$ under high irradiance commonly are attained.

Respiration and photorespiration

Dark respiration occurs during the day in soybeans, although it may be inhibited somewhat by light (Mulchi *et al.* 1971). Dark respiration, on a daily basis, equals approximately one-third of net photosynthesis, but may be higher or lower. The rate of dark respiration is mainly controlled by temperature and is modified by supply of available substrates, moisture, and nutrients. Tissues such as expanding leaves and pods respire at high rates whereas old leaves deep in a heavy canopy respire at low rates.

The soybean exhibits the typical photorespiratory properties of C_3 plants. It gives a dark carbon dioxide burst, has a high carbon dioxide compensation point (which is a linear function of oxygen concentration), and oxygen inhibits net photosynthesis (Forrester, Krotkov and Nelson, 1966). Recent studies using soybeans show two effects of oxygen on ribulose 1,5-diphosphate carboxylase: (1) oxygen competitively inhibits carbon dioxide fixation (Ogren and Bowes, 1971; Bowes and Ogren, 1972), and (2) oxygen reacts as a substrate to give phosphoglycollate, a photorespiratory intermediate, and presumably 3-phosphoglyceric acid (Bowes *et al.* 1971).

Evolution of carbon dioxide into carbon dioxide-free air in light (Hew *et al.* 1969; Hofstra and Hesketh, 1969; Dornhoff and Shibles, 1970) and extrapolation of net photosynthesis to zero carbon dioxide (Forrester *et al.* 1966) give similar estimates of photorespiration – approximately $3\,mg\,CO_2\,dm^{-2}\,h^{-1}$ when corrected for dark respiration. However, because of refixation and diffusive resistances these methods seem to underestimate photorespired-carbon dioxide by as much as three-fold (Samish *et al.* 1972).

Menz *et al.* (1969) described a technique of screening material for low carbon dioxide compensation points by growing mixtures of maize and soybeans in a closed, lighted chamber. Cannell *et al.* (1969) used this technique to screen much of the available soybean germplasm, but they failed to find any genotypes with low compensation points. Ogren (1971) is using mutagens to increase the probability of obtaining a type with a minimal effect of oxygen on net photosynthesis.

Translocation

Anatomy

The primary vascular system of the stem consists of a network of interconnected leaf traces (Bell, 1934; Thaine *et al.* 1959). The vascular supply to the leaf has been described by Watari (1934). Leaf traces emerge from three gaps, and at the pulvinus they are fused into a continuous ring. In the pulvinar–petiole transition region, the vascular tissue divides into five main bundles and several minor bundles which extend through the petiole. At the distal end, the five major bundles undergo subdivision and a complicated recombination, the consequence of which is that the vascular supply to each lateral leaflet emerges as elements from three adjacent major petiolar bundles, whereas the supply to the terminal leaflet is composed of elements from all five major bundles.

The leaf has a typical net venation pattern, with large interveinal dimensions, averaging about 157 to 239 μm, and with a unique layer of parenchyma (the paraveinal mesophyll) which is 'one cell thick, extending horizontally between the vascular bundles in the plane of the phloem . . . in contact with the lower layer of palisade parenchyma' (Fisher, D. 1967). These cells are large, more than 100 μm, and nearly all of them are in contact with a border (bundle-sheath) parenchyma cell. They do contain chloroplasts, but these are smaller and paler than those of other mesophyll cells. Fisher, D. (1970b) believes the paraveinal mesophyll functions in assimilate transport rather than production, but its actual role in transport is still unclear.

Assimilate distribution

The primary sugar translocated is sucrose, which accounts for 90–95 % of the translocate (Vernon and Aronoff, 1952; Fisher, D. 1970a).

In general, the destination of exported assimilate is primarily to the nearest sink. In the vegetative plant, upper leaves export principally to the apex and young, expanding leaves; lower leaves export to roots. Middle leaves supply both (Thaine *et al.* 1959; Belikov and Kostetskii, 1958). However, superimposed on this generalized directional pattern there is a strong phyllotactic influence. In vegetative plants, Thrower (1962) found the expanding leaf to import more from the second leaf below than from the one immediately below. In the fruiting plant, although export from a leaf is primarily to its own pods and seeds (Belikov, 1955; R. A. Stephenson, personal communication), the pods and seeds in the axil of the second node below (Blomquist and Kust, 1971) and the second above (Stephenson) receive more assimilate from a given leaf than do those in the axil immediately above and below. Evidently, the alternate phyllotaxis results in more vascular connections among leaves on the same side of the stem, thus facilitating translocation among them.

Lower and upper leaves differ in the proportion of assimilate translocated to other nodes. Stephenson found the lower leaves to retain a much higher proportion in fruit at their own axil, perhaps because they also fix considerably less carbon dioxide. Main stem leaves may translocate to pods and seeds on axillary branches (Belikov, 1957).

Developmental habit seems to play a role in directional movement of assimilates too. With indeterminate types, there seems very little upward movement of assimilates during fruiting (Belikov, 1957; Belikov and Pirskii, 1966; Blomquist and Kust, 1971). Thus, there is a distinct change in pattern with transition from the vegetative to the fruiting phase. With a determinate type, however, Stephenson found substantial quantities translocated upwards; in fact, more assimilates moved from an upper source leaf to the second node above than to the second node below. At first, this might seem a function of sink requirement, because the determinate plant possesses a terminal raceme with a cluster of pods, whereas the indeterminate usually has only a few pods in the axils of the upper two nodes. However, when Belikov (1957) defoliated the three nodes above a mid-canopy leaf of an indeterminate plant, he obtained no increase in upward movement to pods at those nodes, even though pods had been removed from four nodes immediately below the source leaf.

There is evidence, however, that sink strength influences assimilate distribution among sinks on an already established pathway. If lower leaves are removed on fruiting plants, downward translocation to pods and seeds at defoliated nodes is increased (Belikov and Pirskii, 1966). Undoubtedly, this is why seeds at lower nodes with shaded or abscised leaves will develop normally and, in fact, usually comprise a significant proportion of yield. In vegetative plants, Thrower (1962) defoliated below a leaf translocating in both directions, and the proportion of assimilates moving from that leaf to the roots was increased.

Nutrient uptake and nitrogen metabolism

Mineral nutrition studies with soybeans published prior to 1963 were reviewed by Ohlrogge (1963), and new information on nutrient uptake patterns has come from various sources in recent years (e.g. De Mooy and Pesek, 1970; Hanway and Weber, 1971*b*, 1971*c*, 1971*d*; Leggett and Frere, 1971; Harper, 1971). These results show that the concentration of nitrogen, phosphorus, and potassium remains relatively constant, or tends to fall, in all plant parts except the seeds as the plant develops. Although levels are increased by fertilizer application, the trends are not affected. Approximately 80 % of nutrient accumulation occurs between beginning pod growth and physiological maturity, with the plants showing a constant daily rate of nutrient uptake. Figure 6.2*b* shows nitrogen uptake and loss patterns. For plants yielding 2200–2500 kg ha^{-1}, the harvested seeds contain about 70 % of the nitrogen and phosphorus, and 62 % of the potassium, found in the above-ground parts late in the growth cycle. Hanway and Weber calculate that half of the nutrients in the seeds are taken up directly by the seeds with the remainder being stored in, and then transferred from, the leaves, stems and petioles.

In any discussion of the nitrogen nutrition of soybeans, consideration must be given to the two sources, mineral (soil and fertilizer) nitrogen and symbiotically-fixed nitrogen, and to the strong, and at times paradoxical, interaction between these sources. Higher levels of mineral nitrogen retard early nodule development, but under some conditions, marked beneficial effects of mineral nitrogen on the total amount of N_2 fixed during the growing season have been recorded. Plants completely dependent on symbiotic fixation do not receive substantial supplies of nitrogen until four weeks after germination; early supplies of mineral nitrogen enable the plant to maintain a reasonable growth rate from the outset, enable nodules to develop more rapidly once the inhibition ceases, and because of the increased plant size may increase nodule mass. Unfortunately, the analysis of conditions under which such responses are found has received little attention. During later stages of development, the two sources of nitrogen have a complementary relationship, an increase in the supply of mineral nitrogen proportionately retarding symbiotic N_2-fixation. It is this relationship which creates difficulties in accurately assessing the contribution from each source under field conditions, and in determining the potential of the symbiotic system to meet the requirements of the crop.

Using solution culture with both nodulating and non-nodulating soybeans, Hashimoto (1971) has shown that all principal yield components are affected by the rate, duration, and time of nitrogen-supply. Mineral nitrogen improved yield when applied from the early stages of development through to mid-flowering, but after this time, symbiotic N_2-fixation was adequate to achieve high yields and protein levels. This is similar to the situation under many field conditions where soil nitrogen levels are high during the early developmental stages, but as the season progresses, the plants become more dependent on symbiotic N_2-fixation (e.g. Streeter, 1972*a*; Harper and Hageman, 1972; Hardy *et al.* 1973).

Recent studies have shown that nitrate reductase in soybean leaves tends to be higher in nodulating than in non-nodulating lines (Liu and Hadley,

1971), that as nitrate reductase activity declines glutamate dehydrogenase activity increases (Streeter, 1972*b*), and that there is genetic variability in enzyme activity among cultivars (Harper *et al.* 1972). However, in studies with moderate to high levels of supplied nitrate-nitrogen, and where cultivar differences in uptake are evident, no differences in yield or protein level are obtained, presumably due to the ability of the symbiotic system to adequately compensate for lower mineral nitrogen uptake. Under such circumstances, it is not valid to consider the symbiotic system as capable of meeting only 25–50% of total nitrogen requirements; the symbiotic system has provided all that is required, and its potential contribution, in absolute or relative terms, remains to be ascertained. It is at this level, where the demands for nitrogen are high, that differences in the N_2-fixing effectiveness of various symbiotic combinations are most likely to have an impact on yield and protein levels.

Cultivars differ in their tolerance to high levels of phosphorus (Foote and Howell, 1964) and in efficiency of iron utilization. Efficiency of iron utilization is conditioned by a single gene (Weiss, 1943), and on alkaline soils inefficient lines may show marked iron-deficiency symptoms. Phosphorus-tolerant lines accumulate less phosphorus and more readily incorporate it into nucleotides than sensitive lines, as well as showing higher rates of RNA synthesis and turnover (Lee *et al.* 1970). The fact that phosphorus levels giving greatest growth at the seven-leaf stage may be inferior to lower levels with respect to final yield (De Mooy and Pesek, 1970) indicates that phosphorus exerts an important influence on, as yet undetermined, yield components. Phosphorus is known to increase protein levels and lower fat content of seeds (Jones and Lutz, 1971). Molybdenum has a vital role in the assimilation of both mineral and atmospheric nitrogen, and while responses are usually associated with low pH soils (see De Mooy, 1970), further attention to this element is required.

Grafting experiments have shown that the genotype of the shoot rather than that of the root exerts the major influence on nutrient uptake (Kleese and Smith, 1970), although this is at variance with earlier reports concerning phosphorus (Foote and Howell, 1964). Root growth is less dependent on nutrient concentration than is shoot growth (Leggett and Frere, 1971).

Recent reports that show soil micro-organisms favourably influencing nutrient uptake open up a new area for the investigation of physiological aspects of soybean ecology. Plant growth may be increased following inoculation of sterile soil with a bulk inoculant of uncharacterized soil micro-organisms (Miller and Chau, 1970), and the mycorrhiza, *Endogone*, increased nutrient uptake and yield (29%) when added to fumigated plots (Ross and Harper, 1970).

The two principal deficiencies in nutritional studies are the lack of precise information on the qualitative and quantitative effects of various elements on yield components, and the lack of information on rates of nutrient uptake required if yields are to be greatly increased. Many studies have been made with crops yielding 2000–3000 kg ha^{-1}, which is relatively low in relation to ultimate aims. For most elements, the highest rate of uptake occurs during seed-filling, and it is pertinent to consider whether previously observed rates can be, or need be, doubled to achieve 4000–5000 kg ha^{-1}. For elements such as phosphorus, potassium and nitrogen, which are readily transferred from leaves and stems to developing seeds, greater accumulation prior to flowering may be possible although toxicity effects cannot be ignored. Furthermore, the extent of storage is limited as much of the dry weight increase occurs post-flowering. In seeking genotypes best able to achieve high levels of production, consideration should be given to seeking types that are most efficient in extracting and assimilating nutrients from the soil. With increasing pressures upon limited resources of some fertilizer elements, consideration also should be given to seeking genotypes making the most efficient use of assimilated nutrients in their various metabolic processes.

Stress physiology

The most frequently encountered stresses are those imposed by water deficit and storm injury. Temporary flooding, lodging, and the various types of mechanical damage wrought by hail are the most common consequences of severe weather.

Within the context of a moderate productivity level (2000–3000 kg ha^{-1}), yielding ability shows remarkably strong resistance to stress. The soybean possesses great recoverability and this, plus the fact there is no especially critical phase in its development when yield can be said to hinge on favourable conditions, gives the crop considerable stability of performance. Its lack of a critical phase and to some extent its recoverability are a consequence of the prolonged and overlapping nature of each reproductive phase and of its ability to branch profusely, even through late vegetative development.

Lodging, as a consequence of wind-driven rain, occurs not infrequently, especially with late-maturing, heavily-vegetative lines. Susceptibility to lodging increases as the plant approaches maximum vegetative size. In instances of severe lodging, yield reductions of up to 23 % have been recorded (Cooper, 1971). The physiological consequences of lodging are not well understood. However, some of the yield reduction is probably related to poor canopy light relations (Shaw and Weber, 1967) and high diffusive resistance caused by compression of the canopy.

Defoliation, topping, and bruising of tissues in varying degrees are the usual outcome of hail. In the American midwest, crop hail insurance is common and much work evaluating the results of simulated hail damage has been conducted (Kalton *et al.* 1949; Camery and Weber, 1953; Weber and Caldwell, 1966; Burmood and Fehr, 1973).

Working with whole plants, Boyer (1970*a*, 1970*b*) has shown that leaf enlargement is the first process to respond to water deficit, followed by respiration and then photosynthesis. The photosynthetic response was attributable almost solely to stomatal behaviour. Tracing the development of water deficit in leaves, Chen *et al.* (1971) found a maximum leaf relative water content (RWC) of about 93.5 % in the dark following watering. Upon illumination, RWC dropped to 90 % and remained fairly constant as long as water uptake kept pace with transpirational demand. As RWC dropped below 90 %, photosynthesis declined linearly, reaching half maximum at *ca* 80 % RWC. As RWC declined further, net photosynthesis declined less rapidly and plateaued at about one quarter of maximum at RWC's between 70 and 60 %. In recovering from severe water deficit (RWC = *ca* 60 %), net photosynthesis lagged behind rehydration. On watering, RWC increased to 90 % within 45 minutes, but net photosynthesis only recovered to 56 % of its former maximum within 40 minutes. Full recovery of photosynthesis did not occur until the following day.

Imposing short-term, moderate water stress on plants grown outdoors in large potometers, Shaw and Laing (1966) estimated the sensitivity of yield to stress at various stages of reproductive development (Figure 6.3). Their data illustrate, in particular, the compensation among yield components that endows the soybean with some of its yielding stability. Stress during early flowering (period 1) resulted in less than 10 % yield reduction. Flower and pod drop occurred in the lower parts of the plant, but compensation in the form of more pods set on upper nodes almost negated the pod loss. During mid-flowering and early-podding (periods 2 and 3), substantial pod loss from lower nodes was partially compensated for by increased bean set on upper nodes and increased bean size at lower nodes. As development proceeds, compensatory capacity diminishes, all components eventually coming under reduction throughout the plant and major effects shifting from pod number to beans per pod and seed size (see also Laing, 1966).

YIELD

Limits and limitations to yield

The highest documented seed yield on a field-size basis is 5560 kg ha^{-1} (Anon. 1966). Assuming average composition, this translates as a fat and protein yield of about 1110 and 2220 kg ha^{-1}. Average farmer seed yields are much lower, the highest estimated average being 2420 kg ha^{-1} on 2450000 ha for Iowa, USA in 1972 (US Dept. of Agr. 1973). Average yields, however, do not reflect current producer capability. Farm yields

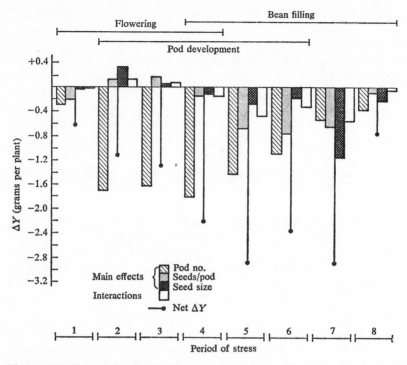

Figure 6.3. Changes in yield, ΔY, as a consequence of changes in yield components because of water stress. Stress periods were successive one-week intervals, during which RWC of upper leaves was at or below 85 % for 4 days. Non-stressed plants yielded 6.49 g of seed. (Adapted from Shaw and Laing, 1966.)

averaging 3000 kg ha^{-1} are very frequent and, occasionally, 4000 kg ha^{-1} is attained.

There is considerable evidence that seed yield is principally a function of seed number produced (e.g. Weber, C. *et al.* 1966; Greer and Anderson, 1965; Chase, 1971). Except in the case of stress, delayed planting, or where a substantial proportion of the yield is attributable to branches, which usually produce smaller seed (O. G. Carter, personal communication), yield is usually poorly related to seed size. Thus, factors limiting high yield are operating not during filling, but earlier and the most important components of yield variation are number of fruiting nodes and, assuming pods abscise because embryos abort, seed set per node.

Considerable speculation has centred around the adequacy of biological N$_2$-fixation to sustain high yields. Much of it is based on the large disparity between calculated nitrogen requirements for high yields (*ca* 320 kg ha^{-1} in seed, 150–200 kg ha^{-1} in vegetative matter, for a 5000 kg ha^{-1} crop) and estimated fixation rates under normal cultural conditions of up to 160 kg ha^{-1}

(Weber, 1966; Hardy *et al.* 1968; Hardy *et al.* 1973). Added concern derives from observations that nitrogen fixation declines sharply during seed-filling (Brun, 1972; J. W. Tanner and I. C. Anderson, personal communication). That this latter observation is not corroborated by other results (Hardy *et al.* 1971; Mague and Burris, 1972) indicates that different environmental conditions, host-bacterial associations, and requirements for nitrogen all probably influence the results obtained. Furthermore, nitrogen fertilizer trials, in general, have yielded inconsistent results, with responses running the gamut from substantial (rare) to nil (frequent). In no case has it been shown that yields can be increased by nitrogen-fertilization at a stage when N_2-fixation is declining, while none of the estimates of N_2-fixation have been made on high-yielding (greater than 4000 kg ha^{-1}) crops.

Perhaps duration of nitrogen fixation is not as important as the amount of fixation and/or uptake during earlier stages. We have already suggested that, except under conditions of stress, potential yield is determined largely by events prior to seed-filling, and various sources (Lathwell and Evans, 1951; Hashimoto and Yamamoto, 1970; Mann and Jaworski, 1970; Hashimoto, 1971) indicate the critical importance of adequate nitrogen to high yield during vegetative development, flowering, and seed set. Fairly substantial quantities of nitrogen are mobilized from vegetative tissues during seed-filling, Hanway and Weber's (1971*c*, 1971*d*) data indicating that 80 kg ha^{-1} (50% of requirement in their crop) were transferred to seed during the latter half of filling (Figure 6.2*b*). Whether a significant proportion of the 50 kg ha^{-1} lost as leaf fall could also have been used, if required, is conjectural.

Concern that symbiotic nitrogen fixation is energetically demanding, placing other yield-limiting stresses on the plant, seems unfounded (Gibson, 1966; Bergersen, 1971*b*; Minchin and Pate, 1973), though we recognize that the energetics of the soybean–*R. japonicum* association have not been worked out.

The potential for symbiotic N_2-fixation seems substantially greater than estimated fixation. Under carbon dioxide enrichment, field plots have fixed up to 450 kg ha^{-1} (Hardy and Havelka, 1973). That symbiotic fixation may be limited by carbohydrate supply is supported by Brun's (1972) finding that during seed filling, carbohydrate-stressed (defoliated, shaded) plants showed reduced fixation, whereas carbohydrate-enriched plants (supplemental light, depodded) showed enhanced fixation compared with a control. Thus, the potential for N_2-fixation seems adequate to sustain high yields, and it is probable that, where an effective association has been formed, N_2-fixation is not the primary yield-limiting factor.

There is convincing evidence, from differing experimental approaches, that increased photosynthesis increases yield. Johnston *et al.* (1969) provided supplemental irradiance to the lower, partly-shaded leaves of a canopy and increased their photosynthesis. Yield was increased because more nodes were produced on branches. Hardman and Brun (1971) grew plants in a carbon dioxide enriched atmosphere (1200 ppm) and increased yield through greater seed set. Additionally, comparison of net photosynthetic or NAR rates among cultivars and their parents suggests selection for yield has resulted in new cultivars with rates superior to their parents (Ojima *et al.* 1968; Dornhoff and Shibles, 1970; Buttery and Buzzell, 1972). Although this could be a consequence of selection for increased sink demand, this seems unlikely, for when Hardman and Brun increased seed number by temporarily enriching the atmosphere with carbon dioxide (to 1200 ppm) during flowering and early seed development and then returned the plants to a normal atmosphere during filling, they obtained no increase in yield. Many seeds aborted and others failed to fill completely. Presumably, photosynthesis failed to respond to the increased need.

With indeterminate types, the vegetative sink is strongly dominant during flowering and early seed set, up to 40 % of all top vegetative dry matter accumulating during this period (Hanway and Weber, 1971 *b*). Apical dominance seems to play a significant role in limiting seed set, and strongly influences the response to increased photosynthesis. When Hardman and Brun provided additional carbon dioxide only during flowering and early seed set, vegetative growth was increased markedly relative to reproductive growth. When CO_2-enrichment was given after decline in vegetative dominance, however, seed set and yield were increased substantially. Greer and Anderson (1965) earlier proposed that attenuating apical dominance of indeterminates might enhance seed set and yield. By removing apices or spraying the foliage with TIBA during early flowering they increased yields by 10–15 %, but whether seed set was increased is not known. Application of TIBA altered canopy morphology, allowing better light distribution, and increased dry matter production and branching.

Approaches to yield improvement

Most soybeans are grown under wide-row (75–100 cm) culture, yet many tests clearly demonstrate that narrow spacings (30–50 cm) give substantially greater yield. Whether the yielding superiority of narrow rows is primarily because of greater light interception during flowering and seed set (Shibles and Weber, 1966) or more efficient exploitation of moisture (Peters and Johnson, 1960) or nutrients has not been elucidated. Selection of specific cultivars for the narrow row environment – shorter, earlier-flowering, open-canopied types – because they are more responsive to close spacing than the presently-grown, densely-foliaged types (Moraghan, 1970), should further improve yielding ability.

In our view, research approaches to maximize yield will need to give primary consideration to increasing seed set, or reducing ovule and seed abortion. There are practical limitations to increasing the number of fruiting nodes, as increased height would lead to greater lodging, already a problem under intensive cultivation. There is some genetic variation for internode length, and shortening internodes might allow a few additional nodes. But, canopy considerations preclude much change in internode length. Branching and stand density vary inversely, so increasing either has little effect on number of nodes or yield. Hence, we are inclined toward the study of factors limiting seed set as potentially the most profitable course.

In general, we conclude that increasing net carbohydrate production holds greatest potential for increasing seed set. With indeterminate types, in particular, some attenuation of vegetative dominance in conjunction with increased carbohydrate availability should be a useful approach. Though it may be possible to select for reduced dominance within indeterminate germplasm, serious consideration also should be given to development of determinate cultivars for the high latitudes (above 38° N) of the USA and Canada where indeterminates are now grown. The dt_1-type may be too limiting to vegetative development (node number) to allow high yields at these latitudes, though in conjunction with appropriate 'maturity genes' (see Bernard, 1971), it might prove useful. The recently discovered Dt_2-type seems a more logical choice. It also reduces node number, but only by a few (Bernard, 1972). The uppermost nodes of indeterminates set few seeds, and the terminal raceme of Dt_2 with its cluster of pods may largely compensate for loss of these nodes.

In view of past experience, it seems the problem of increasing photosynthesis is probably amenable to a breeding approach, especially if, as Buttery and Buzzell (1972) suggest, there is transgressive segregation for net photosynthesis. Heritability of net photosynthesis has not been estimated, but Buttery and Buzzell have estimated heritability of NAR (55 %) and specific leaf weight (90 % for one group, 52 % for another). To what extent net photosynthesis can be advanced by breeding is speculative.

REFERENCES

Abel, G. H. and Erdman, L. W. (1964). Response of Lee soybeans to different strains of *Rhizobium japonicum. Agron. J.* **56**, 423–424.

Abrahamsen, M. and Mayer, A. M. (1967). Photosynthetic and dark fixation of $^{14}CO_2$ in detached soybean cotyledons. *Physiol. Plantarum*, **20**, 1–5.

Abrahamsen, M. and Sudia, T. W. (1966). Studies on the soluble carbohydrates and carbohydrate precursors in germinating soybean seed. *Amer. J. Bot.* **53**, 108–114.

Altschul, A. M., Yatsu, L. Y., Ory, R. R. and Engleman, E. M. (1966). Seed proteins. *Ann. Rev. Plant Physiol.* **17**, 113–136.

Anonymous (1966). Reiser's second staggering yield: a new record. *Soybean Dig.* **26** (4), 10–11.

Beard, B. H. and Hoover, R. M. (1971). Effect of nitrogen on nodulation and yield of irrigated soybeans. *Agron. J.* **63**, 815–816.

Belikov, I. F. (1955). The local utilization of photosynthetic products of soybean. *Dokl. Akad. Nauk SSSR* **102**, 379–381 (from *Field Crops Abstr.* **9**, 111).

Belikov, I. F. (1957). The distribution of the products of photosynthesis in the soya plant with partial removal of beans and leaves. *Dokl. Akad. Nauk SSSR, Bot. Sci. Sect.* **117**, 272–273.

Belikov, I. and Kostetskii, E. (1958). Distribution of the products of photosynthesis in soybean at early stages of its development. *Dokl. Akad. Nauk SSSR, Bot. Sci. Sect.* **119**, 112–115.

Belikov, I. F. and Pirskii, L. I. (1966). Violation of the local distribution of assimilates in soybean. *Soviet Plant Physiol.* **13**, 361–364.

Bell, W. H. (1934). Ontogeny of the primary axis of *Soja max*. *Bot. Gaz.* **95**, 622–635.

Bergersen, F. J. (1962). The effects of partial pressure of oxygen upon respiration and nitrogen fixation by soybean root nodules. *J. gen. Microbiol.* **29**, 113–125.

Bergersen, F. J. (1970). The quantitative relationship between nitrogen fixation and the acetylene-reduction assay. *Aust. J. biol. Sci.* **23**, 1015–1025.

Bergersen, F. J. (1971 a). Biochemistry of symbiotic nitrogen fixation in legumes. *Ann. Rev. Plant Physiol.* **22**, 121–140.

Bergersen, F. J. (1971 b). The central reactions of nitrogen fixation. In *Biological Nitrogen Fixation in Natural and Agricultural Habitats*, eds T. A. Lie and E. G. Mulder. Martinus Nijhoff, The Hague, pp. 511–524.

Bergersen, F. J. and Goodchild, D. J. (1973). Cellular location and concentration of leghaemoglobin in soybean root nodules. *Aust. J. biol. Sci.* **26**, 741–756.

Bergersen, F. J., Turner, G. L. and Appleby, C. A. (1973). Studies on the physiological role of leghaemoglobin in soybean root nodules. *Biochim. biophys. Acta* **292**, 271–282.

Bernard, R. L. (1971). Two major genes for time of flowering and maturity in soybeans. *Crop Sci.* **11**, 242–244.

Bernard, R. L. (1972). Two genes affecting stem termination in soybeans. *Crop Sci.* **12**, 235–239.

Bernstein, L. and Ogata, G. (1966). Effects of salinity on nodulation, nitrogen fixation, and growth of soybeans and alfalfa. *Agron. J.* **58**, 201–203.

Beuerlein, J. E. and Pendleton, J. W. (1971). Photosynthetic rates and light saturation curves of individual soybean leaves under field conditions. *Crop Sci.* **11**, 217–219.

Bieberdorf, F. W. (1938). The cytology and histology of the root nodules of some Leguminosae. *J. amer. Soc. Agron.* **30**, 375–389.

Bils, R. F. and Howell, R. W. (1963). Biochemical and cytological changes in developing soybean cotyledons. *Crop Sci.* **3**, 304–308.

Blad, B. L. and Baker, D. G. (1972). Orientation and distribution of leaves within soybean canopies. *Agron. J.* **64**, 26–29.

Blomquist, R. V. and Kust, C. A. (1971). Translocation pattern of soybeans as affected by growth substances and maturity. *Crop Sci.* **11**, 390–393.

Bond, G. (1936). Quantitative observations on the fixation and transfer of nitrogen in the soya bean, with especial reference to the mechanism of transfer of fixed nitrogen from bacillus to host. *Ann. Bot.* **50**, 559–578.

Bond, G. (1950). Symbiosis of leguminous plants and nodule bacteria. IV. The importance of the oxygen factor in nodule formation and function. *Ann. Bot. N.S.* **15**, 95–108.

Borthwick, H. A. and Parker, M. W. (1938). Photoperiodic perception in Biloxi soybeans. *Bot. Gaz.* **100**, 374–387.

Borthwick, H. A. and Parker, M. W. (1939). Photoperiodic responses of several varieties of soybeans. *Bot. Gaz.* **101**, 341–365.

Bowes, G. and Ogren, W. L. (1972). Oxygen inhibition and other properties of soybean ribulose 1,5-diphosphate carboxylase. *J. biol. Chem.* **247**, 2171–2176.

Bowes, G., Ogren, W. L. and Hageman, R. H. (1971). Phosphoglycolate production catalyzed by ribulose diphosphate carboxylase. *Biochem. biophys. Res. Commun.* **45**, 716–721.

Boyer, J. S. (1970a). Leaf enlargement and metabolic rates of corn, soybean, and sunflower at various leaf water potentials. *Plant Physiol.* **46**, 233–235.

Boyer, J. S. (1970b). Differing sensitivity of photosynthesis to low leaf water potentials in corn and soybean. *Plant Physiol.* **46**, 236–239.

Boyer, J. S. (1971). Resistances to water transport in soybean, bean, and sunflower. *Crop Sci.* **11**, 403–407.

Brown, D. M. (1960). Soybean ecology. I. Development-temperature relationships from controlled environment studies. *Agron. J.* **52**, 493–495.

Brun, W. A. (1972). Nodule activity of soybeans as influenced by photosynthetic source–sink manipulations. *Agron. Abstr.* p. 31.

Burmood, D. T. and Fehr, W. R. (1973). Variety and row spacing effects on recoverability of soybeans from simulated hail injury. *Agron. J.* **65**, 301–303.

Burris, J. S., Edge, O. T. and Wahab, A. H. (1969). Evaluation of various indices of seed and seedling vigor in soybeans [*Glycine max* (L.) Merr.]. *Ass. Offic. Seed Anal. Proc.* **59**, 73–81.

Burris, J. S., Edge, O. T. and Wahab, A. H. (1973). Effects of seed size on seedling performance in soybeans: II. Seedling growth and photosynthesis and field performance. *Crop Sci.* **13**, 207–210.

Burris, J. S., Wahab, A. H. and Edge, O. T. (1971). Effects of seed size on seedling performance in soybeans: I. Seedling growth and respiration in the dark. *Crop Sci.* **11**, 492–496.

Burris, R. H., Magee, W. E. and Bach, M. K. (1955). The P_{N_2} and the P_{O_2} function for nitrogen fixation by excised soybean nodules. In Biochemistry of nitrogen. *Ann. Acad. Sci. Fennicae. Ser. A. Sect.* II, **60**, 190–199.

Buttery, B. R. (1969). Analysis of the growth of soybeans as affected by plant population and fertilizer. *Can. J. Plant Sci.* **49**, 675–684.

Buttery, B. R. (1970). Effects of variation in leaf area index on growth of maize and soybeans. *Crop Sci.* **10**, 9–13.

Buttery, B. R. and Buzzell, R. I. (1972). Some differences between soybean cultivars observed by growth analysis. *Can. J. Plant Sci.* **52**, 13–20.

Buzzell, R. I. (1971). Inheritance of a soybean flowering response to fluorescent-daylength conditions. *Can. J. Genet. Cytol.* **13**, 703–707.

Byth, D. E. (1968). Comparative photoperiodic responses for several soya bean varieties of tropical and temperate origin. *Aust. J. agr. Res.* **19**, 879–890.

Camery, M. P. and Weber, C. R. (1953). Effects of certain components of simulated hail injury on soybeans and corn. *Iowa Agr. Exp. Sta. Res. Bull.* **400**, 465–504.

Cannell, R. Q., Brun, W. A. and Moss, D. N. (1969). A search for high net photosynthetic rate among soybean genotypes. *Crop Sci.* **9**, 840–841.

Chase, D. L. (1971). Effect of variety, planting date, and 2,3,5-triiodobenzoic acid on the yield of soybeans, *Glycine max* (L.) Merrill. B.S. Thesis. University of Sydney Library, Sydney.

Chen, L. H., Mederski, H. J. and Curry, R. B. (1971). Water stress effects on photosynthesis and stem diameter in soybean plants. *Crop Sci.* **11**, 428–431.

Cooper, R. L. (1971). Influence of early lodging on yield of soybean [*Glycine max* (L.) Merr.]. *Agron. J.* **63**, 449–450.

Criswell, J. G. and Hume, D. J. (1972). Variation in sensitivity to photoperiod among early maturing soybean strains. *Crop Sci.* **12**, 657–660.

Curtis, C. E., Ogren, W. L. and Hageman, R. H. (1969). Varietal effects in soybean photosynthesis and photorespiration. *Crop Sci.* **9**, 323–327.

Davies, D. B. and Runge, E. C. A. (1969). Comparison of rooting in soil parent materials using undisturbed soil cores. *Agron. J.* **61**, 518–521.

De Mooy, C. J. (1970). Molybdenum response of soybeans (*Glycine max* (L.) Merrill) in Iowa. *Agron. J.* **62**, 195–197.

De Mooy, C. J. and Pesek, J. (1970). Differential effects of P, K, and Ca salts on leaf composition, yield and seed size of soybean lines. *Crop Sci.* **10**, 72–77.

Dobereiner, J., Franco, A. A. and Guzman, I. (1970). Estirpes de *Rhizobium japonicum* de excepcional eficiencia. *Pesq. Agropec. Brasil.* **5**, 155–161.

Dornhoff, G. M. (1969). Genotypic variation in net photosynthesis of *Glycine max* (L.) Merr. leaves. M.S. Thesis. Iowa State University Library, Ames.

Dornhoff, G. M. (1971). Net photosynthesis of soybean leaves as influenced by anatomy, respiration and variety. Ph.D. Thesis. Iowa State University Library, Ames.

Dornhoff, G. M. and Shibles, R. M. (1970). Varietal differences in net photosynthesis of soybean leaves. *Crop Sci.* **10**, 42–45.

Dunphy, E. J. (1972). Soybean water-soluble-carbohydrate, nutrient, yield, and growth pattern responses to phosphorus and potassium fertility differences. Ph.D. Thesis. Iowa State University Library, Ames.

Earley, E. B. and Cartter, J. L. (1945). Effect of temperature of the root environment on growth of soybean plants. *J. amer. Soc. Agron.* **37**, 727–735.

Eaton, S. V. (1931). Effect of variation in daylength and clipping of plants on nodule development and growth of soybean. *Bot. Gaz.* **91**, 113–143.

Edge, O. T. and Burris, J. S. (1970*a*). Seedling vigor in soybeans. *Ass. Offic. Seed Anal. Proc.* **60**, 149–157.

Edge, O. T. and Burris, J. S. (1970*b*). Physiological and biochemical changes in deteriorating soybean seeds. *Ass. Offic. Seed Anal. Proc.* **60**, 158–166.

Edwards, C. J., Jr. and Hartwig, E. E. (1972). Effect of seed size upon rate of germination in soybeans. *Agron. J.* **63**, 429–430.

Evans, H. J. and Russell, S. A. (1970). Physiological chemistry of symbiotic nitrogen fixation by legumes. In *The chemistry and biochemistry of nitrogen fixation*, ed. J. R. Postgate. Plenum Press, London, pp. 191–244.

Farkas, G. L. and Rajhathy, T. (1955). Cited by Howell, R. W. 1963. Physiology of the soybean. In *The soybean*, ed. A. G. Norman. Acad. Press, New York, pp. 75–124.

Fehr, W. R. (1973). Breeding for soybean hypocotyl length at 25 C. *Crop Sci.* **13**, 600–603.

Fisher, D. B. (1967). An unusual layer of cells in the mesophyll of the soybean leaf. *Bot. Gaz.* **128**, 215–218.

Fisher, D. B. (1970*a*). Kinetics of C-14 translocation in soybean. I. Kinetics in the stem. *Plant Physiol.* **45**, 107–113.

Fisher, D. B. (1970*b*). Kinetics of C-14 translocation in soybean. III. Theoretical considerations. *Plant Physiol.* **45**, 119–125.

Fisher, J. E. (1955). Floral induction in soybeans. *Bot. Gaz.* **117**, 156–165.

Foote, B. D. and Howell, R. W. (1964). Phosphorus tolerance and sensitivity of soybeans as related to uptake and translocation. *Plant Physiol.* **39**, 610–613.

Forrester, M. L., Krotkov, G. and Nelson, C. D. (1966). Effect of oxygen on photosynthesis, photorespiration, and respiration of detached leaves. I. Soybean. *Plant Physiol.* **41**, 422–427.

Garner, W. W. and Allard, H. A. (1930). Photoperiodic responses of soybeans in relation to temperature and other environmental factors. *J. agr. Res.* **41**, 719–735.

Ghorashy, S. R., Pendleton, J. W., Bernard, R. L. and Bauer, M. E. (1971). Effect of leaf pubescence on transpiration, photosynthetic rate, and seed yield of three near-isogenic lines of soybeans. *Crop Sci.* **11**, 426–427.

Ghorashy, S. R., Pendleton, J. W., Peters, D. B., Boyer, J. F. and Beuerlein, J. E. (1971). Internal water stress and apparent photosynthesis with soybeans differing in pubescence. *Agron. J.* **63**, 674–676.

Gibson, A. H. (1966). The carbohydrate requirements for symbiotic nitrogen fixation: a 'whole-plant' growth analysis approach. *Aust. J. biol. Sci.* **19**, 499–515.

Gilman, D. F., Fehr, W. R. and Burris, J. S. (1973). Temperature effects on hypocotyl elongation of soybeans. *Crop Sci.* **13**, 246–249.

Grabe, D. F. and Metzer, R. B. (1969). Temperature-induced inhibition of soybean hypocotyl elongation and seedling emergence. *Crop Sci.* **9**, 331–333.

Green, D. E., Pinnell, E. L., Cavanah, L. E. and Williams, L. F. (1965). Effect of planting date and maturity date on soybean seed quality. *Agron. J.* **57**, 165–168.

Greer, H. A. L. and Anderson, I. C. (1965). Response of soybeans to triiodobenzoic acid under field conditions. *Crop Sci.* **5**, 229–232.

Guthrie, M. L. (1972). Soybean response to photoperiodic extension under field conditions. M.S. Thesis. Iowa State University Library, Ames.

Ham, G. E., Frederick, L. R. and Anderson, I. C. (1971). Serogroups of *Rhizobium japonicum* in soybean nodules sampled in Iowa. *Agron. J.* **63**, 69–72.

Hammond, E. G., Fehr, W. R. and Snyder, H. E. (1972). Improving soybean quality by plant breeding. *J. Amer. Oil Chem. Soc.* **49**, 33–35.

Hansen, W. R. (1972). Net photosynthesis and evapotranspiration of field-grown soybean canopies. Ph.D. Thesis. Iowa State University Library, Ames.

Hanson, W. D., Leffel, R. C. and Howell, R. (1961). Genetic analysis of energy production in the soybean. *Crop Sci.* **1**, 121–126.

Hanway, J. J. and Weber, C. R. (1971 *a*). Dry matter accumulation in eight soybean [*Glycine max* (L.) Merrill] varieties. *Agron. J.* **63**, 227–230.

Hanway, J. J. and Weber, C. R. (1971 *b*). Dry matter accumulation in soybean [*Glycine max* (L.) Merrill] plants as influenced by N, P, and K fertilization. *Agron. J.* **63**, 263–266.

Hanway, J. J. and Weber, C. R. (1971 *c*). N, P, and K percentages in soybean [*Glycine max* (L.) Merrill] plant parts. *Agron. J.* **63**, 286–290.

Hanway, J. J. and Weber, C. R. (1971 *d*). Accumulation of N, P, and K by soybean [*Glycine max* (L.) Merrill] plants. *Agron. J.* **63**, 406–408.

Hardman, L. L. and Brun, W. A. (1971). Effect of atmospheric carbon dioxide enrichment at different developmental stages on growth and yield components of soybeans. *Crop Sci.* **11**, 886–888.

Hardy, R. W. F., Burns, R. C., Hebert, R. R., Holsten, R. D. and Jackson, E. K. (1971). Biological nitrogen fixation: a key to world protein. In *Biological Nitrogen Fixation in Natural and Agricultural Habitats*, eds T. A. Lie and E. G. Mulder. Martinus Nijhoff, The Hague, pp. 561–590.

Hardy, R. W. F., Burns, R. C. and Holsten, R. D. (1973). Applications of the acetylene–ethylene assay for measurement of nitrogen fixation. *Soil Biol. Biochem.* **5**, 47–81.

Hardy, R. W. F. and Havelka, U. D. (1973). Symbiotic N_2 fixation: multifold enhancement by CO_2-enrichment of field-grown soybeans. *Plant Physiol. Suppl.* **51**, 35.

Hardy, R. W. F., Holsten, R. D., Jackson, E. K. and Burns, R. C. (1968). The acetylene–ethylene assay for N_2 fixation: laboratory and field evaluation. *Plant Physiol.* **43**, 1185–1207.

Harper, J. E. (1971). Seasonal nutrient uptake and accumulation patterns in soybeans. *Crop Sci.* **11**, 347–350.

Harper, J. E. and Cooper, R. L. (1971). Nodulation response of soybeans (*Glycine max* L. Merr.) to application rate and placement of combined nitrogen. *Crop Sci.* **11**, 438–440.

Harper, J. E. and Hageman, R. H. (1972). Canopy and seasonal profiles of nitrate reductase in soybeans (*Glycine max* L. Merr.) *Plant Physiol.* **49**, 146–154.

Harper, J. E., Nicholas, J. C. and Hageman, R. H. (1972). Seasonal and canopy variation in nitrate reductase activity of soybean (*Glycine max* L. Merr.) varieties. *Crop Sci.* **12**, 382–386.

Hartwig, E. E. (1954). Factors affecting time of planting soybeans in the southern states. *U.S. Dep. Agr. Circ.* **943**, 13 pp.

Hashimoto, K. (1971). The significance of combined and symbiotically fixed nitrogen on soybeans at successive stages of growth. *Hokkaido Nat. Agr. Exp. Sta. Res. Bull.* **99**, 17–29.

Hashimoto, K. and Yamamoto, T. (1970). Studies on the cool injury in bean plants. II. Effects of time and period of nitrogen application on the soybeans treated with low temperature before flowering. *Crop Sci. Soc. Japan Proc.* **39**, 164–170.

Hew, C., Krotkov, G. and Canvin, D. T. (1969). Effects of temperature on photosynthesis and CO_2 evolution in light and darkness of green leaves. *Plant Physiol.* **44**, 671–677.

Hofstra, G. and Hesketh, J. D. (1969). Effect of temperature on the gas exchange of leaves in light and dark. *Planta* **85**, 228–237.

Howell, R. W. and Cartter, J. L. (1958). Physiological factors affecting composition of soybeans: II. Response of oil and other constituents of soybeans to temperature under controlled conditions. *Agron. J.* **50**, 664–667.

Howell, R. W. and Collins, F. I. (1957). Factors affecting linolenic and linoleic acid content of soybean oil. *Agron. J.* **49**, 593–597.

Hubbell, D. H. and Elkan, G. H. (1967a). Correlation of physiological characteristics with nodulating ability in *Rhizobium japonicum. Can. J. Microbiol.* **13**, 235–241.

Hubbell, D. H. and Elkan, G. H. (1967b). Host physiology as related to nodulation of soybean by rhizobia. *Phytochemistry* **6**, 321–328.

Hymowitz, T. (1970). On the domestication of the soybean. *Econ. Bot.* **24**, 408–421.

Inkpen, J. A. and Quakenbush, F. W. (1969). Desaturation of palmitate and stearate by cell-free fractions from soybean cotyledons. *Lipids.* **4**, 539–543.

Ishihara, A. (1956). The effect of 2,3,5-triiodobenzoic acid on the flower initiation of soybeans. *Crop Sci. Soc. Japan Proc.* **24**, 211.

Jeffers, D. L. and Shibles, R. M. (1969). Some effects of leaf area, solar radiation, air temperature, and variety on net photosynthesis in field-grown soybeans. *Crop Sci.* **9**, 762–764.

Johnson, H. W., Borthwick, H. A. and Leffel, R. C. (1960). Effects of photoperiod and time of planting on rates of development of the soybean in various stages of the life cycle. *Bot. Gaz.* **122**, 77–95.

Johnston, T. J., Pendleton, J. W., Peters, D. B. and Hicks, D. R. (1969). Influence of supplemental light on apparent photosynthesis, yield, and yield components of soybeans (*Glycine max* L.). *Crop Sci.* **9**, 577–581.

Jones, F. R. and Tisdale, W. B. (1921). Effect of soil temperature upon the development of nodules on the roots of certain legumes. *J. agr. Res.* **22**, 17–31.

Jones, G. D. and Lutz, J. A. (1971). Yield of wheat and soybeans and oil and protein content of soybean as affected by fertility treatments and deep placement of limestone. *Agron. J.* **63**, 931–934.

Kalton, R. R., Weber, C. R. and Eldredge, J. C. (1949). The effect of injury simulating hail damage to soybeans. *Iowa Agr. Exp. Sta. Res. Bull.* **359**, 736–796.

Kamata, E. (1952). Studies on the development of fruit in soybeans. (1). Histological observations. *Crop Sci. Soc. Japan Proc.* **20**, 296–298.

Kato, I. (1967). *Studies on the transpiration and evapotranspiration amount by the chamber method.* Tokai-Kinki Nat. Agr. Exp. Sta. 14 pp.

Kato, I. and Sakaguchi, S. (1954). Studies on the mechanism of occurrence of abortive grains and their prevention in soybeans, *Glycine max* M. *Tokai-Kinki Nat. Agr. Exp. Sta. Bull.* **1**, 115–132. [From transl. by Ming-Hung Yu, Iowa State Univ. 1972]

Kato, I., Sakaguchi, S. and Naito, Y. (1954). Development of flower parts and seed in the soybean plant, *Glycine max* M. *Tokai-Kinki Nat. Agr. Exp. Sta. Bull.* **1**, 96–114. [From transl. by Soon-Woo Hong, Univ. of Neb. July 1956.]

Kato, I., Sakaguchi, S. and Naito, Y. (1955). Anatomical observations on fallen buds, flowers and pods of soya-bean, *Glycine max* M. *Tokai-Kinki Nat. Agr. Exp. Sta. Bull.* **2**, 159–168.

Kilen, T. C. and Hartwig, E. E. (1971). Inheritance of a light-quality sensitive character in soybeans. *Agron. J.* **11**, 559–561.

Kleese, R. A. and Smith, L. J. (1970). Scion control of genotypic differences in mineral salts accumulation in soyabean (*Glycine max* L. Merr.) seeds. *Ann. Bot. N.S.* **34**, 183–188.

Koller, H. R., Nyquist, W. E. and Chorush, I. S. (1970). Growth analysis of the soybean community. *Crop Sci.* **10**, 407–412.

Kumura, A. (1968). Studies on dry matter production of soybean plant. 3. Photosynthetic rate of soybean plant population as affected by proportion of diffuse light. *Crop Sci. Soc. Japan Proc.* **37**, 570–582.

Kumura, A. (1969). Studies on dry matter production in soybean plant. V. Photosynthetic system of soybean plant population. *Crop Sci. Soc. Japan Proc.* **38**, 74–90.

Kumura, A. and Naniwa, I. (1965). Studies on dry matter production of soybean plant. I. Ontogenic changes in photosynthetic and respiratory capacity of soybean plant and its parts. *Crop Sci. Soc. Japan Proc.* **33**, 467–472.

Kust, C. A. and Struckmeyer, B. E. (1971). Effects of trifluralin on growth, nodulation and anatomy of soybeans. *Weed Sci.* **19**, 147–152.

Laing, D. R. (1966). The water environment of soybeans. Ph.D. Thesis. Iowa State University Library, Ames.

Lathwell, D. J. and Evans, C. E. (1951). Nitrogen uptake from solution by soybeans at successive stages of growth. *Agron. J.* **43**, 264–270.

Lee, K. W., Clapp, C. E. and Caldwell, A. C. (1970). P^{32} distribution in phosphorus fractions of phosphorus-sensitive and -tolerant soybeans. *Plant Soil* **33**, 707–712.

Leffel, R. C. (1961). Planting date and varietal effects on agronomic and seed compositional characters in soybeans. *Maryland Agr. Exp. Sta. Bull.* A-117. 69 pp.

Leggett, J. E. and Frere, M. H. (1971). Growth and nutrient uptake by soybean plants in nutrient solutions of graded concentrations. *Plant Phsyiol.* **48**, 457–460.

Liu, M. C. and Hadley, H. H. (1971). Relationships of nitrate reductase activity to protein content in related nodulating and nonnodulating soybeans. *Crop Sci.* **11**, 467–471.

McAlister, D. F. and Krober, O. A. (1951). Translocation of food reserves from soybean cotyledons and their influence on the development of the plant. *Plant Physiol.* **26**, 525–538.

McCalla, T. M. (1937). Behaviour of legume bacteria (*Rhizobium*) in relation to exchangeable calcium and hydrogen ion concentration of the colloidal fraction of the soil. *Missouri Agr. Exp. Sta. Res. Bull.* **256**, 44 pp.

Maeda, K. (1960). Studies on the rhizosphere temperature affecting the leguminous plant–*Rhizobium* symbiotic complex. I. Early growth and nodule formation in soybean as affected by the rhizosphere temperatures. *Crop Sci. Soc. Japan Proc.* **29**, 158–160.

Mague, T. H. and Burris, R. H. (1972). Reduction of acetylene and nitrogen by field-grown soybeans. *New Phytol.* **71**, 275–286.

Mann, J. D. and Jaworski, E. G. (1970). Comparison of stresses which may limit soybean yields. *Crop Sci.* **10**, 620–624.

Menz, K. M., Moss, D. N., Cannell, R. Q. and Brun, W. A. (1969). Screening for photosynthetic efficiency. *Crop Sci.* **9**, 692–694.

Miller, R. H. and Chau, T. J. (1970). The influence of soil microorganisms on the growth and chemical composition of soybeans. *Plant Soil* **32**, 146–160.

Minchin, F. R. and Pate, J. S. (1973). The carbon balance of a legume and the functional economy of its root nodules. *J. expt. Bot.* **24**, 259–271.

Mitchell, R. L. and Russell, W. J. (1971). Root development and rooting patterns of soybeans [*Glycine max* (L.) Merrill] evaluated under field conditions. *Agron. J.* **63**, 313–316.

Moraghan, B. J. (1970). Plant characters related to yield response of unselected lines of soybeans in various row widths and plant populations. Ph.D. Thesis. Iowa State University Library, Ames.

Mulchi, C. L., Volk, R. J. and Jackson, W. A. (1971). Oxygen exchange of illuminated leaves at carbon dioxide compensation. In *Photosynthesis and Photorespiration*, eds M. D. Hatch, C. B. Osmond and R. O. Slatyer. Wiley–Interscience, New York, pp. 35–50.

Ogren, W. L. (1971). Photorespiration in soybeans. A key to increased productivity. *Soybean News* **23**, (1) 3.

Ogren, W. L. and Bowes, G. (1971). Ribulose diphosphate carboxylase regulates soybean photorespiration. *Nature New Biol.* **230**, 159–160.

Ohlrogge, A. J. (1963). Mineral nutrition of soybeans. In *The soybean*, ed. A. G. Norman. Academic Press, New York, pp. 125–160.

Oizumi, H. (1962). Studies on the mechanism of branching and its agronomic considerations in soybean plants. *Tohoku Nat. Agr. Exp. Sta. Bull.* **25**, 1–95.

Ojima, M. and Fukui, J. (1966). Studies on the seed production of soybean. 3. An analytical study of dry matter production in the soybean plant community. *Crop Sci. Soc. Japan Proc.* **34**, 448–452.

Ojima, M., Fukui, J. and Watanabe, I. (1965). Studies on the seed production of soybean. II. Effect of three major nutrient elements supply and leaf age on the photosynthetic activity and diurnal changes in photosynthesis of soybean under constant temperature and light intensity. *Crop Sci. Soc. Japan Proc.* **33**, 437–442.

Ojima, M. and Kawashima, R. (1968). Studies on the seed production of soybean. 5. Varietal differences in photosynthetic rate of soybean. *Crop Sci. Soc. Japan Proc.* **37**, 667–675.

Ojima, M. and Kawashima, R. (1970). Studies on the seed production of soybean. VIII. The ability of photosynthesis in F_3 lines having different photosynthesis in their F_2 generation. *Crop Sci. Soc. Japan Proc.* **39**, 440–445.

Ojima, M., Kawashima, R. and Mikoshiba, K. (1969). Studies on the seed production of soybean. VII. The ability of photosynthesis in F_1 and F_2 generations. *Crop Sci. Soc. Japan Proc.* **38**, 693–699.

Ojima, M., Kawashima, R. and Sakamoto, S. (1968). Studies on the seed production of soybean. 6. Relationship between activity of photosynthesis of improved varieties and that of the parent ones. *Crop Sci. Soc. Japan Proc.* **37**, 676–679.

Olumbe, J. W. K. and Veatch, C. (1969). Organic matter–Amiben interactions on nodulation and growth of soybeans. *Weed Sci.* **17**, 264–265.

Ozaki, K., Saito, M. and Nitta, K. (1956). Studies on the seed development and germination of soybean plants at various ripening stages. *Hokkaido Nat. Agr. Exp. Sta. Res. Bull.* **70**, 6–14.

Pandey, R. K. (1972). Photoperiodic and temperature effects on flower initiation of diverse genotypes of soybeans. Ph.D. Thesis. Univ. of Ill. Library, Urbana-Champaign.

Parker, M. W. and Borthwick, H. A. (1943). Influence of temperature on photoperiodic reactions in leaf blades of Biloxi soybean. *Bot. Gaz.* **104**, 612–619.

Pate, J. S. (1958). Nodulation studies in legumes. I. Synchronization of host and symbiotic development in the field pea, *Pisum arvense* L. *Aust. J. biol. Sci.* **3**, 366–381.

Peters, D. B. and Johnson, L. C. (1960). Soil moisture use by soybeans. *Agron. J.* **52**, 687–689.

Phillips, D. A. and Torrey, J. G. (1972). Studies on cytokinin production by *Rhizobium*. *Plant Physiol.* **49**, 11–15.

Piper, C. V. and Morse, W. J. (1923). *The Soybean.* McGraw-Hill, New York, 329 pp.

Polson, D. E. (1972). Day-neutrality in soybeans. *Crop Sci.* **12**, 773–776.

Raper, C. D., Jr. and Barber, S. A. (1970*a*). Rooting systems of soybeans. I. Differences in root morphology among varieties. *Agron. J.* **62**, 581–584.

Raper, C. D., Jr. and Barber, S. A. (1970*b*). Rooting systems of soybeans. II. Physiological effectiveness as nutrient absorption surfaces. *Agron. J.* **62**, 585–588.

Rinne, R. W. and Canvin, D. T. (1971*a*). Fatty acid biosynthesis from acetate and CO_2 in the developing soybean cotyledon. *Plant Cell Physiol.* **12**, 387–393.

Rinne, R. W. and Canvin, D. T. (1971*b*). Fractionation of the fatty acid synthesizing system from the developing soybean cotyledon into a particulate and soluble component. *Plant Cell Physiol.* **12**, 395–403.

Ross, J. P. and Harper, J. A. (1970). Effect of *Endogone* on soybean yields. *Phytopathology* **60**, 1552–1556.

Russell, W. J., Fehr, W. R. and Mitchell, R. L. (1971). Effects of row cultivation on growth and yield of soybeans. *Agron. J.* **63**, 772–774.

Safir, G. R., Boyer, J. S. and Gerdemann, J. W. (1972). Nutrient status and mycorrhizal enhancement of water transport in soybean. *Plant Physiol.* **49**, 700–703.

Saito, M. (1961). Studies on the influence of low temperature on soybean plants. I. Effects of day and night temperature on the growth and ripening of plants. *Hokkaido Nat. Agr. Exp. Sta. Res. Bull.* **76**, 9–14.

Saito, M., Yamamoto, T., Goto, K. and Hashimoto, K. (1970). The influence of cool temperature before and after anthesis, on pod-setting and nutrients in soybean plants. *Crop Sci. Soc. Japan Proc.* **39**, 511–519.

Samimy, C. (1970). Physiological bases for the temperature-dependent short growth of hypocotyls in some varieties of soybean. Ph.D. Thesis. Iowa State University Library, Ames.

Samish, Y. B., Pallas, J. E., Jr., Dornhoff, G. M. and Shibles, R. M. (1972). A re-evaluation of soybean leaf photorespiration. *Plant Physiol.* **50**, 28–30.

Sampaio, I. B. M. and Dobereiner, J. (1968). Effects of shading and lime treatments on nitrogen fixation rates and on the efficiency of soybean (*Glycine max*) nodules. *Pesq. Agropec. Brasil.* **3**, 255–262.

Shaw, R. H. and Laing, D. R. (1966). Moisture stress and plant response. In *Plant environment and efficient water use*, eds W. H. Pierre, D. Kirkham, J. Pesek and R. H. Shaw. Amer. Soc. Agron. and Soil Sci. Soc. Amer., Madison, Wisconsin, pp. 73–94.

Shaw, R. H. and Weber, C. R. (1967). Effects of canopy arrangements on light interception and yield of soybeans. *Agron. J.* **59**, 155–159.

Shibles, R. M. and Weber, C. R. (1965). Leaf area, solar radiation interception and dry matter production by soybeans. *Crop Sci.* **5**, 575–577.

Shibles, R. M. and Weber, C. R. (1966). Interception of solar radiation and dry matter production by various soybean planting patterns. *Crop Sci.* **6**, 55–59.

Singh, M., Peters, D. B. and Pendleton, J. W. (1968). Net and spectral radiation in soybean canopies. *Agron. J.* **60**, 542–545.

Sironval, C. (1958). Relation between chlorophyll metabolism and nodule formation in soybean. *Nature, Lond.* **181**, 1272–1273.

Sprent, J. I. (1969). Prolonged reduction of acetylene by detached soybean nodules *Planta* **88**, 372–375.

Sprent, J. I. (1971 a). Effects of water stress on nitrogen fixation in root nodules. In *Biological Nitrogen Fixation in Natural Habitats*, eds T. A. Lie and E. G. Mulder Martinus Nijhoff, The Hague, pp. 225–230.

Sprent, J. I. (1971 b). The effects of water stress on nitrogen-fixing root nodules. I. Effects on the physiology of detached soybean nodules. *New Phytol.* **70**, 9–17.

Sprent, J. I. (1972 a). *Ibid.* II. Effects on the fine structure of detached soybean nodules. *New Phytol.* **71**, 443–450.

Sprent, J. I. (1972 b). *Ibid.* III. Effects of osmotically applied stress. *New Phytol.* **71**, 451–460.

Sprent, J. I. (1972 c). *Ibid.* IV. Effects on whole plants of *Vicia faba* and *Glycine max*. New Phytol. **71**, 603–611.

Stevenson, K. R. and Shaw, R. H. (1971 a). Diurnal changes in leaf resistance to water vapor diffusion at different heights in a soybean canopy. *Agron. J.* **63**, 17–19.

Stevenson, K. R. and Shaw, R. H. (1971 b). Effects of leaf orientation on leaf resistance to water vapor diffusion in soybean (*Glycine max* L. Merr.) leaves. *Agron. J.* **63**, 327–329.

Streeter, J. G. (1972 a). Nitrogen nutrition of field-grown soybean plants. I. Soil nitrogen and nitrogen composition of stem exudate. *Agron. J.* **64**, 311–314.

Streeter, J. G. (1972 b). Nitrogen nutrition of field-grown soybean plants. II. Seasonal variations in nitrate reductase, glutamate dehydrogenase and nitrogen constituents in plant parts. *Agron. J.* **64**, 315–319.

Suetsugu, I., Anaguchi, I., Saito, K. and Kumano, S. (1962). Developmental process of the root- and top-organs in the soybean varieties. *Hokuriku Agr. Exp. Sta. Bull.* **3**, 89–96.

Sullivan, T. P. (1972). The effect of root genotype on the water relations of soybean [*Glycine max* (L.) Merrill]. Ph.D. Thesis. University of Minnesota Library, St Paul.

Sun, C. N. (1955). Growth and development of primary tissues in aerated and non-aerated roots of soybean. *Bull. Torrey Bot. Club* **82**, 491–502.

Thaine, R., Ovenden, S. L. and Turner, J. S. (1959). Translocation of labelled assimilates in the soybean. *Aust. J. biol. Sci.* **12**, 349–372.

Thrower, S. L. (1962). Translocation of labelled assimilates in the soybean. II. The pattern of translocation in intact and defoliolated plants. *Aust. J. biol. Sci.* **15**, 629–649.

U.S. Department of Agriculture. Crop Reporting Board, Statistical Research Service (1973). In *Crop production – 1972 annual summary*. Publ. CrPr 2–1 (73), 1327 pp.

Van Schaik, P. H. and Probst, A. H. (1958). Effects of some environmental factors on flower production and reproductive efficiency in soybeans. *Agron. J.* **50**, 192–197.

Vernon, L. P. and Aronoff, S. (1952). Metabolism of soybean leaves. IV. Translocation from soybean leaves. *Arch. Biochem. Biophys.* **36**, 383–398.

Watari, S. (1934). Anatomical studies on some leguminous leaves with special reference to the vascular system in petioles and rachises. *J. Fac. Sci. Imp. Univ. Tokyo. Sect. III, Bot.* **4**, 225–365.

Weber, C. R. (1966). Nodulating and non-nodulating soybean isolines. II. Response to applied nitrogen and modified soil conditions. *Agron. J.* **58**, 46–49.

Weber, C. R. and Caldwell, B. E. (1966). Effects of defoliation and stem bruising on soybeans. *Crop Sci.* **6**, 25–27.

Weber, C. R., Shibles, R. M. and Byth, D. E. (1966). Effect of plant population and row spacing on soybean development and production. *Agron. J.* **58**, 99–102.

Weber, D. F., Caldwell, B. E., Sloger, C. and Vest, H. G. (1971). Some USDA studies on the Soybean–Rhizobium symbiosis. In *Biological Nitrogen Fixation in Natural and Agricultural Habitats*, eds E. G. Mulder and T. A. Lie. Martinus Nijhoff, The Hague, pp. 293–304.

Weiss, M. G. (1943). Inheritance and physiology of efficiency in iron utilization in soybeans. *Genetics* **28**, 253–268.

Weiss, M. G., Weber, C. R., Williams, L. F. and Probst, A. H. (1952). Correlation of agronomic characters and temperature with seed compositional characters in soybeans, as influenced by variety and time of planting. *Agron. J.* **44**, 289–297.

Williams, L. F. and Lynch, D. L. (1954). Inheritance of a non-nodulating character in the soybean. *Agron. J.* **46**, 28–29.

Wong, P. P. and Evans, H. J. (1971). Poly-β-hydroxybutyrate utilization by soybean (*Glycine max.* Merr.) nodules and assessment of its role in maintenance of nitrogenase activity. *Plant Physiol.* **47**, 750–755.

Woodworth, C. M. (1930). Abortive seeds in soybeans. *J. Amer. Soc. Agron.* **22**, 37–50.

7

Pea

J. S. PATE

Pisum, one of five genera in the tribe Vicieae of the family Papilionaceae (Leguminosae) (see Heywood, 1971), is a temperate Old World genus in which the distribution of 'wild' species suggests several centres of origin and genetic diversity, including Central Asia, Asia Minor, Ethiopia and the Mediterranean basin (Yarnell, 1962; Blixt, 1970). The large-seeded forms exploited by man since Stone Age times (Hawkes and Wooley, 1963; Bland, 1971) are probably of Mediterranean origin and were possibly brought to Europe from Palestine or Egypt (Cole, 1961). The pea is unlikely to have been used widely until Greek or Roman times (Janick *et al.* 1969) but since then it has been introduced into virtually every temperate region of the globe, including high altitude locations in tropical countries such as Nigeria and Kenya.

Considerable confusion surrounds the naming of species of *Pisum*, and the status of its many hundreds of cultivated forms. Anything from four to eight species are recognized by various authors (see Hector, 1936; Bland, 1971), the agriculturally important forms being known variously as *Pisum hortense*, *P. humile*, *P. macrocarpon*, or the more commonly used *P. sativum* and *P. arvense* (see Bailey, 1949; Yarnell, 1962). A recent hybridization study between alleged species by Makasheva (1971) suggests that only *P. formosum*, *P. fulvum*, *P. syriacum* and *P. sativum* (*sensu amplissima*) deserve the rank of species, the last named embracing the white- and pink-flowered, large-seeded forms most commonly used by man.

The present day importance of the pea as an agricultural plant stems largely from its use as a protein-rich seed crop for human or animal consumption, although there are still regions where the whole plant is utilized as a fodder crop for hay or silage, often in company with a cereal such as oats. The increasing demand by affluent palates for tender, sweet green peas has caused a virtual revolution in methods of pea cultivation over the past 25 years, involving a decrease in acreages of dried peas and an increase in production of uniformly unripe peas for quick freezing, dehydration or other forms of processing (Bundy, 1971). A highly organized 'vining pea' industry has evolved to meet this demand, using special factory-simulated and contract-based techniques for sowing, managing and harvesting the crop and controlling its quality (Bland, 1971).

SEED VIABILITY AND GERMINATION: FIELD EMERGENCE AND
ESTABLISHMENT

Seeds of pea are not particularly long-lived, nor do they exhibit after-ripening or secondary dormancy. As with other species, viability decreases markedly at high storage temperatures and high seed moisture content, and on the basis of the predictability of the effects of these factors, a nomograph has been constructed for the species (Roberts and Roberts, 1972), relating percentage viability to storage conditions.

It has been known for a long time that germination tests on peas carried out at optimum laboratory temperatures are often very poorly correlated with ability to germinate and become established in the field, partly due to varying tolerance of prolonged exposure to damp cold conditions (Torfason and Nonnecke, 1959) and partly to attack by pathogens whose growth may be stimulated by solutes exuded by the seeds. Most of the leaked solutes come from the cotyledons, one suggestion being that it is the sudden inrush of water during imbibition which causes the injuries resulting in leakage (see Larson, 1968; Perry and Harrison, 1970), another being that drying out of the embryo during seed ripening causes cell membranes to lose their integrity, thus rendering cellular contents susceptible to leaching (see Simon and Harun, 1973). Losses can involve substantial fractions of the sugar, amino acid and inorganic solutes of the seed so that particularly leaky seeds may give rise to seedlings of subnormal vigour (Larson and Kyagaba, 1969).

Although the presence of the relatively impermeable seed coat reduces leakage from cotyledons during imbibition (Larson, 1968; Simon and Harun, 1973), the coat's resistance to oxygen diffusion creates problems of anaerobiosis within the seed, as witnessed by the accumulation of ethanol and lactic acid during the first few days after imbibition (Cossins, 1964; Leblova *et al.* 1969), and by the concomitant rise in alcohol dehydrogenase activity (Cameron and Cossins, 1967; Kollöffel, 1970). Eventually, on the second and third day after imbibition, mitochondria became fully organized (Bain and Mercer, 1966*b*; Nawa and Asahi, 1971), and with the penetration of the seed coat by the radicle an improvement in oxygen supply allows an aerobic pattern of respiration to be established (Kollöffel and Sluys, 1970). As this is occurring, alcohol dehydrogenase activity declines rapidly, possibly through inactivation of the enzyme by fatty acids. It is highly likely that if the transition to aerobic metabolism be delayed, permanent damage to seed metabolism will result.

Mobilization of reserves of the cotyledons starts at the seed periphery and proceeds inwards (Smith and Flinn, 1967) but does not reach significant proportions until several days after emergence of the radicle and plumule, by which time full complements of various hydrolytic enzymes may be present in the seedling (Juliano and Varner, 1969; Abbott and Matheson,

1972). Proteases and peptidases are involved in protein breakdown (Beevers and Splittstoesser, 1968; Nakano and Asahi, 1972), the debranching enzyme – amylopectin 1-6-glucosidase – and various amylases and phosphorylases in starch utilization (Shain and Mayer, 1968; Juliano and Varner, 1969; Lee and Shallenberger, 1969), ribonuclease in the release of stored RNA (Beevers and Splittstoesser, 1968), and phytase in the utilization of phytic acid, a phosphorus reserve in the seed (Guardiola and Sutcliffe, 1971b). During germination sugars and amino acids are translocated to the seedling axis, protein breakdown being associated with the synthesis of large amounts of the non-protein amino acid homoserine (Virtanen *et al*. 1953; Mitchell and Bidwell, 1970), and some arginine (Jones and Boulter, 1968) and γ-glutamyl alanine (Virtanen and Berg, 1954). These syntheses suggest that selection by man for high protein content in pea seeds may have necessitated the development of elaborate mechanisms for ridding the seedling of amino nitrogen surplus to its requirements in growth.

Part of the enzymic complement of seedling cotyledons is a legacy from seed ripening, released and activated on imbibition. A larger part, however, appears to be either synthesized, or activated, during the utilization of reserves (Shain and Mayer, 1968; Juliano and Varner, 1969; Guardiola and Sutcliffe, 1971a), but fails to be so if cotyledons are detached before imbibition is complete. This suggests that stimuli from the axis of the seedling direct the whole pattern of structural and enzymic events which occur in the cotyledons during germination (Varner *et al*. 1963; Bain and Mercer, 1966b; Guardiola and Sutcliffe, 1971a; Chin *et al*. 1972). According to Varner *et al*. (1963), substances moving to the cotyledons during the first 24–48 hours after imbibition are responsible for triggering these changes, and it is perhaps significant that, just prior to this, differentiation of conducting elements occurs in the vascular network of the cotyledons (Smith and Flinn, 1967). The precise nature of the stimuli has not been discovered, although in contrast to the situation in cereals, it is unlikely that gibberellins are involved (Sprent, 1968a).

In summary then, the epigeal pattern of germination, the large and rich cotyledon reserves, and the overall sensitivity of pea seeds to factors in the soil environment are likely to create many serious problems for the agronomist. Much has to be overcome before germination and establishment of pea crops becomes as reliable as it is with most other species.

SHOOT MORPHOLOGY, GROWTH AND BRANCHING

The pea is a scrambling annual plant bearing, usually, only one dominant shoot. The first two or three nodes bear trifid scale leaves, the next few nodes foliage leaves with a single pair of leaflets, and nodes above these have increasingly larger leaves with greater numbers of pairs of leaflets. Stipules are inserted at the petiole base to each foliage leaf; in upper leaves the terminal, and sometimes subterminal leaflets are modified as tendrils. With increase in size and complexity of leaves there is a corresponding increase in the length and diameter of successive internodes, this trend being evident at least until flowering is under way.

Reproduction is by means of axillary inflorescences bearing one or more flowers, the basic pattern of fruit maturation being therefore a sequential one. 'Early' varieties tend to be dwarf, to flower at a lower node, and to bear fewer reproductive nodes than 'late' ones, although shoot morphology and reproductive behaviour may be greatly influenced by genotype and environment. There is good evidence for at least one cultivar ('Alaska') that reproduction, and particularly seed filling, hastens senescence of the shoot apex. However even when flower buds are removed continuously the apex will die after completing a certain number of plastochron cycles. Apices of old plants cannot have their functional life span extended beyond this limit if grafted on to younger plants (Lockhart and Gottschall, 1961).

The cytogenetical basis of morphological variation in *Pisum* has been well explored (Yarnell, 1962; Blixt, 1970) and a wide variety of natural and artificially induced mutants of potential interest to plant breeders has been described.

Of equal interest to the physiologist is the morphological plasticity displayed by the individual genotype, for this above all will determine how the plant reacts and yields in various environments. In conditions adverse to growth, plants generally develop only a single slender stem, bearing short-lived leaves and few inflorescences. Low temperatures – which favour root more than shoot growth (Stanfield *et al.* 1966) – water stress (McIntyre, 1971), and localized cooling of the shoot apex (Husain and Linck, 1967) all promote outgrowth of laterals, although these may become inhibited once the first-formed fruits drain the shoot of assimilates (Husain and Linck, 1966). Decapitation of plants allows a quite specific hierarchy of sister axillary shoots to be established. If the shoot of a young seedling is removed at the level of the epicotyl, two cotyledonary shoots grow out. Dwarf varieties tend to throw unequal shoots, the less vigorous of which senesces early (Sachs, 1966; Dostal, 1967; Lovell, 1969), whereas tall varieties generate shoots of more equal size and both of these may survive to maturity (Snow, 1929, 1931, 1937). Any treatment reducing the vigour of one of a pair of equal shoots, or the dominant one of a pair of unequal shoots, allows the other shoot to grow more vigorously and even to deprive its sister of assimilates (Lovell, 1969). The reason for imbalance between cotyledonary shoots of dwarf peas is not completely understood, although the position of the seed during germination, or even its orientation during development in the parent pod may have some effect (Dostal, 1968). Removal of one cotyledon usually leads to the shoot in the axil of the removed cotyledon becoming dominant

(Sebanek, 1965), although this effect may be reversed if benzyladenine is applied to the remaining cotyledon (Sprent, 1968 *b*). Growth of a partially inhibited junior shoot may be restored by application to it of cytokinin. Auxin applications lead to internode elongation, and application of both auxin and kinin allows a junior shoot to grow almost as well as it would have done had its dominant sister shoot been removed (Sachs and Thimann, 1967).

Essentially similar correlative effects are observed between buds if older plants are decapitated higher up the stem. At first all axillary buds grow equally and receive equal amounts of nutrients (e.g. leaf-applied ^{14}C or ^{32}P), but soon one shoot, usually that closest to the top of the stem, outgrows its companions and begins to monopolize the nutrient supply (Husain, 1967). Detailed studies on buds during the first 48 hours after release of apical dominance show that a bud swells, increases in sugar content, and traps significantly more ^{14}C-labelled assimilates within 24 hours, a day or more before differentiation of new vascular tissues can be detected in the traces supplying the bud (Wardlaw and Mortimer, 1970). Thus, although only a poorly developed vascular link is present between dominant bud and stem (Nakamura, 1964; Sorokin and Thimann, 1964), it does not appear to act as an impediment to transfer of assimilates to the germinating bud. Rather, the destruction of apical dominance appears to trigger off the production of growth substances by the bud, these possibly attracting assimilates to the bud locus. Conversely, auxin applied to the apical stump simulates the dominance of the natural apex (Thimann and Skoog, 1934) and restores the apically-dominated translocation pattern typical of the intact plant (Seth and Wareing, 1964, 1967; Morris and Thomas, 1968; Bowen and Wareing, 1971).

So the branching pattern of the pea shoot and the correlations between its developing parts are likely to be shaped primarily by the interactions of endogenous growth substances. By controlling cell division, vascular development (Sachs, 1968 *a*, *b*) and assimilate transport, these growth substances provide the physiological constraints against which metabolic and nutritional influences will operate. It is possible, therefore, that growth regulators could be used on pea crops to manipulate plant morphology and leaf longevity (e.g. see Herner, 1969; Linke and Marinos, 1970).

ROOT GROWTH AND NODULATION

The early growth of the seedling root ensures that it receives the major share of cotyledon reserves and that its primary axis and main framework of laterals are well developed before the first foliage leaf expands. According to Yorke and Sagar (1970), seedling root elongation proceeds at alternate high and low rates, each period of slow growth coinciding with the emergence of a higher order of roots. Observations on older, soil-grown plants indicate that growth rate of roots reaches a maximum at about the time of initiation of flower primordia, and then declines abruptly even before flowering starts (Salter and Drew, 1965). All main roots bear a dense covering of fine 'tertiary' roots. Certain of the laterals show greater growth potential than others and penetrate downwards almost to the depth of the primary root (Jean, 1928).

Infection by *Rhizobium* takes place through root hairs (Dixon, 1964), so that initiation of nodules is inevitably linked with root extension. Nodulation of the primary root and older laterals occurs largely at the expense of cotyledon reserves, and since fixation commences promptly it is unusual to find any marked 'nitrogen hunger' stage in seedling establishment (Pate, 1958). Nevertheless, a penalty is paid for early nodulation: seedling establishment is slower, and rate of growth and final size of the root is less than in uninoculated plants supplied with combined forms of nitrogen (Virtanen and Miettinen, 1963; Oghoghorie, 1971).

Peak nodule numbers are present in mid-vegetative growth and at about this time a maximum is recorded for the ratio of nodule weight to plant weight. During later growth, increases in average size and efficiency of nodules more than offset any decline in nodule numbers and allow a maximum mass of nodules of peak activity to be attained by about the time of flowering (Pate, 1958). Large losses of nodules occur at fruiting (Pate, 1958) and whole sections of the root decay at that time (Salter and Drew, 1965).

Nodule formation in legumes is generally more sensitive to environmental stress than is host plant growth, and in peas this applies to stresses such as low light intensity (Diener, 1950), low pH (Lie, 1969), waterlogging (Minchin, unpublished) and extremes of temperature (Lie, 1971). However, because there are compensatory interactions for nodule number and size, an environmental experience restricting root infection by *Rhizobium* need not necessarily engender an equivalent reduction in nodule tissue production (see Nutman, 1956). Nodulation may also be affected by materials applied to leaves (e.g. ethrel [Drennan and Norton, 1972], and urea [Cartwright and Snow, 1962]), a fact to be borne in mind when devising spraying programmes.

Jean (1928) has shown that the characteristics of root penetration are genetically controlled, and two studies (Gelin and Blixt, 1964; Federova, 1966) suggest that the same applies to nodule number. The ratio of primary root nodule number to lateral root nodule number is also cultivar specific, and dwarf varieties tend to bear more organs (nodules + laterals) per unit length of primary root than do deeper rooted, tall varieties (Pate, 1956).

NITROGEN FIXATION BY ROOT NODULES

It is not usual practice to dress peas with nitrogen fertilizers (Reynolds, 1960; Bland, 1971) since mineral nitrogen released from the soil during growth, together with fixation by nodules, is normally sufficient to support a maximum rate of growth. However, if root growth has been affected by *Fusarium*, or nodules destroyed by larvae of *Sitonia*, plants may benefit from added nitrogen (Mulder, 1948 *b*). The reason behind the general lack of response to nitrogen is that fixation by nodules and their formation and growth are increasingly inhibited as the amount of inorganic nitrogen available to the plant is increased (Schalldach and Schilling, 1966; Oghoghorie and Pate, 1971), an uneconomically large fertilizer application, almost totally suppressing fixation, being required to achieve maximum yield and nitrogen content (Oghoghorie, 1971). A low level of inorganic nitrogen (say 30–60 ppm) is suggested as the best compromise since it achieves almost a maximum in yield whilst still permitting fixation to occur at 60–80 % of the rate achieved by plants relying solely on symbiotic fixation. This is the very situation likely to be encountered in the reasonably fertile soils in which the crop is grown.

Indigenous populations of *Rhizobium* are usually adequate to ensure fully effective nodulation of a pea crop, and, indeed, several instances are recorded where the local strains are superlatively adapted for symbiosis with the locally-grown cultivars in the home environment (Vartiovaara, 1937; Roponen *et al.* 1970; Mishustin and Shilnikova, 1971). Thus little benefit is likely to be gained from seed inoculation, although if the necessity should arise there is no shortage of effective isolates for peas.

In view of the reliance placed on symbiosis, it is desirable to be able to determine the current rate of fixation by a crop, and how much nitrogen it has gained from the atmosphere during its life cycle. Since haemoglobin pigmentation of the central bacterial tissues of a nodule is a reliable visual indicator of nodule activity (Virtanen *et al.* 1947), the weight of red nodules per plant is probably as good a simple guide as any to the crop's potential for nitrogen fixation, although it must be borne in mind that nodules formed by different *Rhizobium* strains vary considerably in their efficiency, and that age of the plant influences the efficiency of fixation (Pate, 1958).

A relatively simple gas chromatographic technique has become available recently for estimating fixation in laboratory and field (Hardy *et al.* 1968; Schwinghamer *et al.* 1970; Hardy *et al.* 1971), exploiting the finding that the nitrogen-fixing system will reduce acetylene to ethylene in preference to its natural substrate, molecular nitrogen (Dilworth, 1966; Schöllhorn and Burris, 1966). So far the technique has been used for *Rhizobium* strain effectiveness trials, for life cycle studies of fixation under field conditions, and for short-term studies of symbiotic response to environmental stress. In peas, acetylene reduction (nitrogen fixation) is greater if nodules are left attached to the parent root, is highest during the photoperiod when photosynthesis is occurring, is suppressed by waterlogging or droughting of roots, and exhibits a broad temperature optimum, implying great versatility in performance under field conditions (Halliday, Minchin and Pate, unpublished).

NUTRIENT UPTAKE AND MINERAL NUTRITION

Pea seeds contain a well balanced set of mineral elements as evidenced by the ability of seedlings to grow quite healthily and even to reach the stage of flowering in the absence of added nutrients. Guardiola and Sutcliffe (1971 a,b, 1972) show that the pattern of mobilization of nitrogen, phosphorus, sulphur, potassium and calcium from the cotyledons to seedling tissues is strictly related to the rate of breakdown and release of reserves in storage cells of the cotyledon, yet apparently controlled by forces operating within the seedling axis. Because of the large seed reserves, 'carry over' effects from one generation to another are likely, as shown by Austin (1966) for seeds high or low in phosphorus. Indeed, micronutrient reserves (e.g. boron, molybdenum) may be sufficiently large to permit several successive generations to grow quite normally in environments deficient in such elements (see Woodbridge, 1969).

There are many papers which define fertilizer requirements for pea crops under different soil conditions (see Bland, 1971), responses to phosphorus, calcium, and potassium being most commonly recorded. Nutrient removal by the crop in its growth cycle is not noticeably different from that by other species, although, as with other legumes, calcium requirements seem to be relatively high. Certainly, peas do not do well in acid soils low in calcium (Lie, 1969).

Judging from studies on nutrient accumulation by different zones of the pea root, the species behaves conventionally in exhibiting highest rates of uptake of ions such as PO_4^{3-} (Canning and Kramer, 1958) and NO_3^- and NH_4^+ (Grasmanis and Barley, 1969) in young actively growing regions close to the root apex. These regions also display most intense enzyme activity (e.g. for glutamic acid dehydrogenase (Pahlich and Joy, 1971) and nitrate reductase (Carr and Pate, 1967)). The bulk of the solute reserves of the root are located in the apical regions (Pate, 1973).

A guide to visual symptoms of various element deficiencies in peas has been compiled (Wallace, 1961). Boron appears to have special significance in nodulation (Mulder, 1948a), molybdenum in the enzyme systems for nitrate reduction and nitrogen fixation (Mulder et al. 1959), and calcium in root growth (Sorokin and Sommer, 1940). Salt tolerance (Bernstein and Hayward, 1958) and selective capacities in respect of sodium and potassium (Collander, 1941) suggest that the pea is intermediate among crop plants in its ability to withstand saline conditions. The requirements of peas for zinc have been well studied (Reed, 1942; Mulder, 1948a) as have those for boron (Woodbridge, 1969) and manganese (Reynolds, 1955). Deficiency of the latter element causes 'marsh spot' disease, readily discernible in the field as a characteristic browning of the seedling plumule.

PHOTOSYNTHESIS, RESPIRATION AND DRY MATTER ACCUMULATION

Pea leaves, in common with those of many other species, reach their maximum photosynthetic activity at the time of full expansion, losing activity thereafter at a rate somewhat faster than the loss of chlorophyll (Smillie, 1962). Respiratory activity, on the other hand, tends to decline steadily with leaf age. The rise and fall in photosynthetic activity of the first foliage leaves of the seedling is telescoped into a relatively short period (Smillie, 1962), but a near maximum rate of apparent photosynthesis may be maintained for 20 days or more in blossom leaflets committed to the nourishment of developing fruits (Flinn and Pate, 1970). These leaflets are also slower to lose their chlorophyll and reserves of protein, and seed and pod removal studies suggest that extension of functional life is promoted by the presence of the developing seeds (Flinn, unpublished).

Different pea cultivars differ significantly in their maximum rates of carbon dioxide assimilation per unit area of newly expanded leaf, but within a variety this rate does not appear to be different for leaves developing at different levels on the stem (Harvey, 1971). Photosynthetic and respiratory activity of stems and petioles does not appear to have been studied, but stipules are reported by Flinn (1969) to be as efficient in photosynthesis as sister leaflets.

Under atmospheric concentrations of carbon dioxide and saturating light intensity (17.6 klx), the rate of net photosynthesis exhibits a broad optimum spanning the range 25–35 °C (Hellmuth, 1971), but because dark respiration of the shoot rises steadily with increasing temperature over the range 18–42 °C, the shoot functions much more effectively in conserving carbon when grown at low night temperatures. The carbon dioxide compensation point is relatively high (70 ppm at 27 °C, 17.6 klx), and the rate of net photosynthesis, relatively low, from which Hellmuth concludes that the pea operates on the Calvin pathway for photosynthesis.

In a study of the carbon economy of nodulated pea plants Minchin and Pate (1973) find that during the pre-flowering stages of growth some 47 % of the net gain of carbon by the photosynthesizing shoot is regularly lost to the root environment through respiration of underground organs (Figure 7.1). Approximately one quarter of this loss is due to nodules, the rest to respiration of the supporting root. Respiration of a nodulated root fixing nitrogen (5.9 mg C respired per mg N_2 fixed) is not noticeably different from that of one utilizing nitrate (6.2 mg C respired per mg NO_3-N assimilated). As root respiration increases with increasing temperature, the carbon balance of a plant depends closely on the temperature of its root environment. For example, in two cultivars nitrogen fixation and dry matter accumulation was twice as great in plants exposed to a low night temperature (10–17 °C below the day temperature) as in constant temperature (Roponen *et al.* 1970).

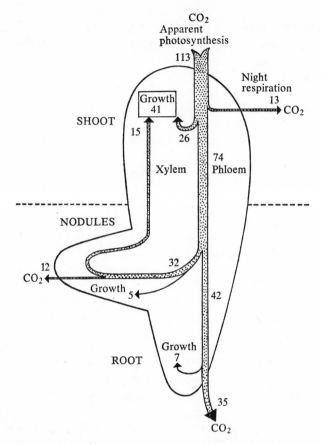

Figure 7.1. Flow sheet for carbon in effectively nodulated plants of *Pisum sativum* (cv. 'Meteor'), relying solely on the atmosphere for their nitrogen. The study period covered is 21–30 days after sowing, a time when the roots and nodules are still growing actively. Data are expressed on the basis of a net gain by the shoot of 100 units of carbon from the atmosphere (from Minchin and Pate, 1973).

Other studies in controlled environments (e.g. Brouwer and van Vliet, 1960; Stanfield *et al.* 1966) have also suggested that low night temperature regimes benefit pea growth, even when plants are not relying on nodules.

If dry matter production by growing pea plants and their composite organs is plotted against time, sigmoid growth curves are obtained (Brouwer, 1962). The proportion of total dry weight contributed by each organ, and the rate at which it changes, gives some indication of the relative importance of various organs as sources or sinks of mobilized nutrients. On this basis, leaves and stems, but not roots, would appear to contribute significantly to the dry matter required by the developing fruits. At maturity the fruits comprise a fairly constant proportion (40–60 %) of the total dry weight, and dry weight of vegetative structure attained by the onset of flowering has been shown to give a reliable prediction of seed yield (see Brouwer, 1962).

The leaf arrangement in pea is distichous (1/2 phyllotaxy). The nodal anatomy is trilacunar (Dormer, 1946), the central leaf trace going to the petiole, while the two lateral ones each serve a stipule. The primary architecture of the vascular system is essentially as described for *Trifolium* by Devadas and Beck (1972), a series of four axial bundles running from base to top of the stem, giving rise in acropetal sequence to the various traces supplying organs at the nodes. One pair of axial bundles (A_1 and A_4, Figure 7.2) contributes mid leaf traces to the nodes alternately (i.e. A_1 gives rise to L_2, L_4, L_6, etc. A_4 to L_1, L_3, L_5, etc.), so that a direct vascular link exists between vertically adjacent leaves on one side of the stem. However, since the lateral leaf traces of a node (*S* series, Figure 7.2) arise from axial traces other than that supplying the mid leaf trace (cf. Figure 7.2), at least three and sometimes all four axial strands contribute to the vasculature of a leaf. Two traces (marked B in Figure 7.2) supply each axillary bud, one arising from the same axial strand as supplies the mid trace of the subtending leaf, the other from an adjacent axial bundle. Axillary shoots or reproductive structures are therefore likely to be nourished by the leaf which subtends them and also by other leaves not necessarily of the same orthostichy.

The pea root usually shows a triarch arrangement of its xylem, and lateral roots and nodules tend to arise opposite the proto-xylem points (Oinuma, 1948; Phillips, 1971). Study of the anatomy of the root–stem transition region (Hayward, 1938) has shown that each cotyledon is connected with a specific pole of xylem and hence, eventually, with a vertical file of lateral roots and nodules. The third xylem pole, not so connected, serves the seedling shoot. The configuration of phloem in this region allows cotyledons to transfer assimilates directly to the plumule and to all sectors of the root. The transition is not complete until the fourth internode, and only at this level do the four axial strands and square stem outline become prominent (Hayward, 1938).

Short-term feeding of individual leaves with $^{14}CO_2$ (Brennan, 1966; Carr and Pate, 1967) or $^{15}NO_3^-$ (Oghoghorie and Pate, 1972) confirms that the links between leaves of the same vertical series are of greater importance in assimilate transfer than are any connections with leaves on the opposite side of the stem. Indeed, two peaks in upward export during the life of a leaf coincide with the inception and early growth of the next two leaves vertically above it on the stem (Carr and Pate, 1967). This pattern of flow between alternate nodes, though manifest throughout the life of the plant, tends to be obscured as reproductive units begin to command increasing proportions of the plant's photosynthate. Studying flow of ^{32}P and ^{14}C assimilates at blossom nodes, Linck and Swanson (1960) and Linck and Sudia (1962) record an increasing commitment on the part of leaflets to reproductive structure(s), until at the time of seed filling as much as 90 % of their current photosynthate may be translocated to the subtended fruit. Essentially similar findings are recorded for blossom leaflets, stipules, and carpels (Lovell and Lovell, 1970; Flinn and Pate, 1970; Harvey, 1971). Hence, in a plant whose lower parts have senesced only a small photosynthetic surface will be available for maintenance of the root and shoot apex. It is not surprising therefore that non-reproductive regions cease to grow and begin to senesce at this time.

Complementing the outward flow of assimilates from photosynthetic organs is the upward flow in the xylem of water and dissolved solutes from root and nodules. In this case solutes tend to be dispensed in proportion to an organ's ability to transpire. Fruits and young, unexpanded organs at the shoot apex tend to fare badly in this respect, leaflets and stipules much better. A complication is that xylem parenchyma lining the xylem conduits can trap amino acids and other solutes from the ascending transpiration stream and so allow storage parenchyma to benefit directly from the xylem stream (Pate and O'Brien, 1968). Nevertheless, mature leaves are still likely to gain more solutes from the root than they currently require, and hence are in the position of acting as donors of root-derived solutes to younger leaves or growth centres elsewhere in the plant (Pate *et al.* 1965).

Although nodules possess only a sparse, peripheral vascular network (Bond, 1948; Pate *et al.* 1969), they fix up to two or three times their soluble content of nitrogen in a day, requiring assimilate equivalent to 10.3 mg carbohydrate for every mg N fixed (Minchin and Pate, 1973). Import of sugars is believed to be phloem-mediated but the xylem bleeding activity of many leguminous root nodules, when detached from their parent root, suggests that export of fixed nitrogen can be osmotically-operated and involves secretion to the xylem of highly concentrated nitrogenous solutes (Pate *et al.* 1969; Wong and Evans, 1971). Thus, export may occur very economically in terms of water loss.

Regardless of whether the pea root is relying on atmospheric nitrogen or a combined form of nitrogen such as ammonium, nitrate or urea, four compounds dominate the spectrum of organic products released to shoots in the xylem (Wieringa and Bakhuis, 1957; Wieringa, 1958; Pate and Wallace, 1964). Two of these, asparagine and homoserine, are readily diverted into storage pools of nitrogen, the other two – aspartic acid and glutamine – functioning more effectively as substrates for protein synthesis (Pate *et al.* 1965).

Export of nitrogen to the shoot shows a marked diurnal rhythm with a maximum activity near noon. The amounts and proportions of nitrogenous solutes released to the xylem change throughout the life cycle (Pate, 1962;

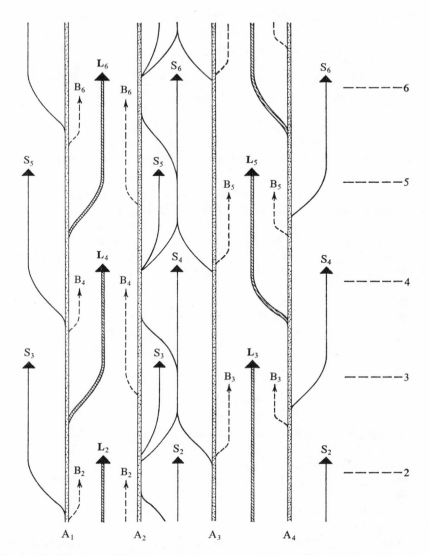

Figure 7.2. Diagram of the shoot vascular system of *Pisum*, following the schematic representation depicted by Devadas & Beck (1972) for the 3-trace nodal anatomy of the related genus *Trifolium*. A_1–A_4, axial bundles of shoot; B, traces to axillary shoots; L, mid leaf trace (serving leaflets); S, lateral traces (serving stipules). Nodes of stem are numbered from base upwards and their positions indicated on the right of the figure.

The vascular network is shown as if split along one side and spread out in one plane. Arrows mark the level of departure of traces to the various organs.

Pate and Greig, 1964; Pate *et al.* 1965). Nevertheless, the amounts of amino nitrogen leaving the root, and the amounts of compounds such as asparagine, homoserine, *o*-acetylhomoserine and glutamic acid accumulating in a shoot, serve as useful indicators of a plant's ability to generate and store soluble reserves of nitrogen (see Wieringa and Bakhuis, 1957; Grobbelaar and Steward, 1969; Oghoghorie, 1971). Analysis of xylem sap for inorganic solutes, gibberellin-like substances (Carr *et al.* 1964) and cytokinins (Carr and Burrows, 1966) provides additional information on current root activity.

The phloem sap of peas contains sucrose as its major organic constituent, but a variety of amino compounds are also present, some typical of root and nodule metabolism, and probably derived from these organs, others (e.g. serine, glycine and the aromatic amino acids) supplied to the phloem by photosynthetic activity (Lewis and Pate, 1973). Indeed, the phloem supplies growing regions such as the shoot apex with solutes collected from virtually every mature organ of the plant, and dispenses these in almost correctly balanced proportions for synthesis of new cellular material in growth (Pate, 1966; Lewis and Pate (1973)) .

SOURCE–SINK INTERRELATIONSHIPS

Once the amounts and proportions of compounds moving in the transport channels are known, and the rates of production and utilization in donor and receptor organs determined, it is possible to construct balance sheets and circulatory profiles for the whole plant and its parts. Figure 7.1 details the balance for carbon in vegetative pea plants, and shows that 75 % of the net photosynthetic gain by the shoot is translocated to the root and nodules, approximately one-fifth of this eventually returning to the shoot in fixation products released from nodules, the remainder being utilized *in situ* for growth and respiration. Thus, at this stage of the life cycle the pea shoot derives as much as one-third of its non-respired carbon from materials passed through the nodules, a finding which stresses how important it is for adequate links to be established between root and shoot from the earliest stage in development.

The comparable flow pattern for nitrogen has been assembled from $^{15}NO_3^-$ feeding studies by Oghoghorie and Pate (1972, see Figure 7.3). One important feature is the inability of the distally-located root to obtain nitrogen directly from the nodules, and hence its dependence on nitrogen which has been processed in the shoot. Viewed together, the carbon and nitrogen cycling patterns highlight the interdependence of nodulated root and shoot for various organic materials, and raise the possibility of regulatory control of each part by the other through interchange of complementary sets of metabolites.

Balance studies for nitrogen during seed growth, using $^{15}NO_3^-$ as tracer,

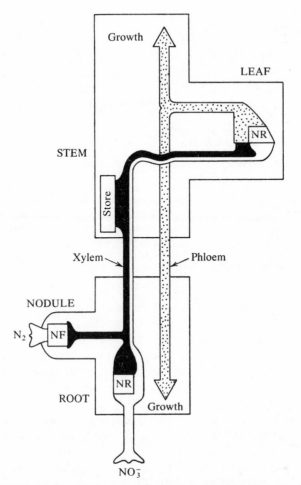

Figure 7.3. Scheme illustrating the main features of the assimilatory and transport systems for nitrogen in the field pea. The situation represented is a young vegetative plant, effectively nodulated and growing in the presence of a moderately high level of nitrate in the medium (after Oghoghorie and Pate, 1972).

□, nitrate; ■ aspartate, glutamine, homoserine, asparagine; ▦, various amino compounds; NR, nitrate reduction; NF, nitrogen fixation.

have shown that nitrogen assimilated at any time in the growth cycle is mobilized to the seeds with high efficiency (55–70 %). Thus, judging from the increase in nitrogen in the course of crop development, seeds are likely to be almost as dependent on nitrogen already present in the plant body at the time of flowering as they are on nitrogen entering later during reproduction (see Schilling and Schalldach, 1966; Pate and Flinn (1974)). This does not apply to carbon, however, for ^{14}C tracer studies have shown that carbon

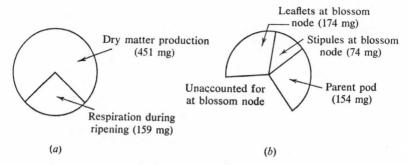

Figure 7.4. A balance sheet of carbon for seeds borne at the first fruiting node of field pea (*Pisum arvense* L.). (*a*) The requirement of the seeds for carbon estimated from losses of carbon from seeds in respiration and gains of carbon by the seeds in dry matter. (*b*) Sources of carbon translocated to the developing seeds from organs at the blossom node. The estimates include photosynthetically fixed carbon, carbon available through mobilization of dry matter, and, in the case of pods, carbon reassimilated from the respiring seeds.

assimilated during vegetative growth becomes 'immobilized' in structural matter (i.e. lignin and cellulose) close to the site of assimilation (Brennan, 1966). Consequently, fruit filling is very dependent on current photosynthesis. In *Pisum arvense*, for example, the seeds derive a large proportion of their carbon from photosynthesis by leaflets and stipules at the parent node (Flinn and Pate, 1970). Photosynthesis by the pod conserves and recycles the carbon respired by the ripening seeds (Figure 7.4).

There is evidence from several crop species that a consuming organ (sink) can exercise a controlling influence over the production and export of assimilates by 'source' organs such as photosynthesizing leaves. Lovell *et al.* (1972) have found that rate of ¹⁴C export from pea leaves can be greatly increased if, 20 hours before feeding, all other leaves are removed from the shoot. Since this increase in export is not evident if root or shoot apices are removed at the time of defoliation, it appears to be the demand for assimilates by these sinks which sets the tempo of export. The same applies to the mobilization of reserves from leaves since losses of protein and amino acids during leaf ageing can be greatly reduced if the shoot apex is cooled (Carr and Pate, 1967). On returning the apex to normal temperatures mobilization soon increases back to its previous rate.

In the sink-regulated situation described above, competition for assimilates is likely to result in organs of low competing power functioning at less than full capacity. Then, if a dominant sink be removed, assimilates are likely to become more readily available to less favoured organs. Evidence of such a diversion of assimilates has been obtained in the tracer studies on decapitated pea plants made by Husain (1967) and Morris and Thomas (1968), and evidence of benefit to root nodules has been demonstrated as a significant, albeit transient, increase in acetylene-reducing (nitrogen fixation) capacity immediately after the shoot tip has been removed (Halliday and Pate, unpublished). In the normal course of functioning of a plant there will be times when surplus, uncommitted photosynthate becomes available and it is of obvious advantage if this can be diverted to centres of storage, or harnessed usefully, say to root assimilation. Indeed, Roponen and Virtanen (1968) have shown that by removing the apical region of a shoot just before flowering and then allowing the vegetative axillary shoots to enlarge the photosynthetic canopy, a yield of fixed nitrogen can be obtained as much as eight times that of intact plants flowering and fruiting in a normal fashion. In the process, root, nodule and stem tissues of the vegetative plants accumulate massive levels of soluble forms of nitrogen.

There are certain situations, however, where the normal source–sink relationships for nitrogen seem to break down and, instead of storage organs of the shoot trapping nitrogen assimilated by a root, it is excreted to the rooting medium. This phenomenon was first observed with nodulated peas by Virtanen and co-workers in Finland (Virtanen and Laine, 1938; Virtanen, 1947), and sometimes occurred to such an extent that the shoots of the plants showed symptoms of nitrogen deficiency. It is now known that non-nodulated roots of peas also excrete amino compounds (Boulter *et al.* 1966; Ayers and Thornton, 1968), so the phenomenon is not restricted to nodules.

FLOWERING

First described as a genetically fixed characteristic by Gregor Mendel in 1865, time of flowering in pea is now regarded as resulting from a highly complex reaction of genotype to a range of interrelated physiological and environmental factors. The genetic system is under polygenic control (see Barber, 1959; Rowlands, 1964; Wellensiek, 1969; Murfet, 1971) and the modifying influence of environment is of a quantitative rather than qualitative nature (Haupt, 1969).

Early varieties characteristically show a node number of first flowering within the range 5–10. They are normally insensitive to day-length (Kopetz, 1938), unresponsive to vernalization (Barber, 1959), and may even initiate flowers in total darkness using cotyledon reserves (Haupt, 1969). Lack of response to environment is not surprising since inflorescence primordia may be developed only a few days after seed imbibition, often before the epicotyl has broken the soil surface (see Haupt, 1969). Removal of cotyledons immediately after germination delays flowering of early varieties by several nodes (Moore, 1964, 1965), and in so doing makes the seedling respond positively and quantitatively to photoperiod (earlier flowering in long days), or to added nutrients (nitrogen and carbohydrate additions advance flowering)

(Haupt, 1969). Auxin delays [flowering of the early variety 'Alaska' by several nodes (Leopold and Guernsey, 1954). Gibberellins tend to delay flowering of all pea varieties, whether tall or short, early or late flowering (Barber *et al.* 1958; Moore, 1965).

Flowering of late varieties of pea generally starts at a node number within the range 10–50. Late varieties are very responsive to environment, long days advancing flowering by 10–40 nodes, exposure to vernalization (1–7 °C for one to four weeks as a young seedling) usually lowering node number-to-first-flower by some two to three nodes (Barber, 1959; Highkin, 1960; Moore and Bonde, 1962). This vernalization effect is most noticeable in environments unfavourable for flowering, and it can be reversed by gibberellin treatment or by exposure to high temperatures given immediately after the cold experience (Moore and Bonde, 1962). Advancement of flowering in a late variety can also be achieved by depriving the seedling of minerals (Sprent, 1967) or by removing its cotyledons (Barber, 1959; Moore, 1964, 1965).

The vernalization and photoperiod responses of one late dwarf variety, 'Greenfeast', have been the subject of a most exhaustive series of studies. Using defoliation and leaf and stem masking techniques combined with long- or short-day treatments given at specific times in seedling life, it has been demonstrated that exposed cotyledons and the first three foliage leaves are sufficient to implement the photoperiodic response, and that once such induction is complete defoliation or short-day treatment fails to alter the plant's commitment to flower (Sprent, 1966; Paton, 1967, 1968, 1971). Vernalization in 'Greenfeast' acts additively and independently of the photoperiodic response, being presumed to condition the shoot apex to respond more quickly and positively to any floral stimulus exported from the leaves (Paton, 1971). Vernalization is supposed to act by leaching out or breaking down inhibitors present in the cotyledons (Amos and Crowden, 1969).

Since flowering in pea is sequential, the number of nodes initiated between first flowering and shoot apex senescence will be a major factor determining seed yield. It has already been mentioned that early nutrition of the seedling, the mass of vegetative material present when flowering begins, and the presence of developing fruits all affect the reproductive life span; but in some varieties genetic factors, responsive to day-length, may exercise an overriding influence. This is well demonstrated in the 'G_2' response type described by Marx (1968, 1969). Though insensitive to day-length as far as induction of flowering is concerned, plants of this type develop 50 or more reproductive nodes before apical senescence ensues under short days, but only four or so under long days. Genetic variability of this kind should enable the plant breeder to select for the number of reproductive nodes likely to be produced in a particular environmental situation.

FRUIT AND SEED DEVELOPMENT: ESTABLISHMENT OF
COTYLEDON RESERVES

The pea is self-fertile, pollination and fertilization being completed before the flower is fully open (Cooper, 1938). Although all ovules are usually fertilized some seeds, particularly those in terminal positions, may abort. Embryo failure in cv. 'Alaska' is reported to be as high as 30–50 % (Linck, 1961), though this undesirable character is not so frequently expressed in the modern, blunt podded varieties.

The pod increases first in length and width and then in wall thickness, attaining maximum fresh weight before the contained seeds become really active in laying down storage reserves (Flinn and Pate, 1968). After this pods lose dry matter and nitrogen steadily, final drying out being accompanied by a quite rapid loss of chlorophyll and photosynthetic capacity (Flinn and Pate, 1970). In their late life pods could supply up to 20–25 % of the seeds' total requirements for carbon and nitrogen (McKee *et al.* 1955; Raacke, 1957*b*; Flinn and Pate, 1968). But, as mentioned earlier, the pods also function in recycling carbon respired by the developing seed, and there is also the possibility that they might act in trapping, processing and exporting to the seeds nitrogen delivered from roots in the xylem. The continually changing pattern of proteins discernible in electrophoretograms of ageing pods (Flinn and Pate, 1968), and the changing composition of its pool of free amino acids (McKee *et al.* 1955; Flinn and Pate, 1968), are certainly suggestive of metabolic changes of some magnitude during pod development.

In contrast to some other legumes (e.g. see Rijven, 1972), the endosperm of pea is short lived and fails to reach a fully cellular condition or deposit insoluble reserves of any significance (Hayward, 1938; Kapoor, 1966). Nevertheless, its liquid contents are exceptionally rich in organic solutes, and balance sheets for the period 10–25 days after anthesis would indicate that the young embryo might obtain virtually all of its gross nutritional requirements by absorption of endospermic fluid and capture of solutes released from the seed coat (Raacke, 1957*a*; Flinn and Pate, 1968). Cytokinins of the endosperm may have special significance in early embryo growth in view of their presumed activity as regulators of cell division (see Burrows and Carr, 1970).

The timing of events in maturation of pea seeds has been the subject of many investigations (e.g. Bisson and Jones, 1932; Danielsson, 1956; Turner *et al.* 1957; Rowan and Turner, 1957; Bain and Mercer, 1966*a*; Flinn and Pate, 1968; Smith, 1974), and although effects due to environment and season make it difficult to relate results to any absolute time scale, there are sufficient similarities between results for different cultivars to suggest a general pattern for the species. This is summarized in Figure 7.5, using as time scale features such as size increases in the seed and changes in condition

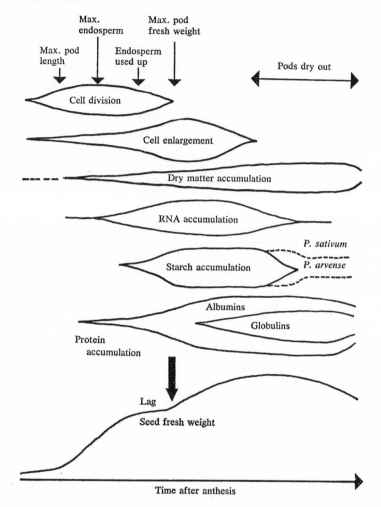

Figure 7.5. Schematic representation of the cardinal events in seed maturation of *Pisum* spp. The common time scale is the growth curve for seed fresh weight. All events depicted refer to rates within the developing cotyledons, divergent lines denoting periods of increasing rate, convergent lines times of decreasing rate. Differences between *P. sativum* and *P. arvense* are indicated. The solid arrow marks the approximate time when seeds of *P. sativum* are harvested for quick freezing (data compiled from sources listed in the text).

of the pod and endosperm. It can be seen that the phases of cell division and cell expansion in the cotyledons are virtually complete, the endospermic fluid has disappeared, and the pools of free sugars and amino acids have been established, before there is evidence of reserves of starch and protein being laid down in the cotyledons. In *Pisum arvense* (Flinn and Pate, 1968;

Burrows and Carr, 1970; Smith, 1974) and certain varieties of *P. sativum* (Carr and Skene, 1961) seed growth is clearly diauxic, the two phases of growth being separated by a short lag of a few days duration, coinciding with the time when the embryo fills the embryo sac. Carr and Skene (1961) regarded this lag as being caused by physical restriction of the embryo inside the seed coat, further growth of the seed then being possible only by stretching the seed coat. In *P. arvense* the lag also marks the cessation of cell division in the embryo (Smith, 1974), the appearance of storage globulins, and a sharp rise in the rate of starch accumulation in the cotyledons (Flinn and Pate, 1968).

Starch in peas is deposited as distinct grains, one per plastid (Bain and Mercer, 1966*a*). Round (smooth)-seeded varieties of *P. sativum* have on average 45 % of their mature seed dry weight as starch, the latter deposited in single grains of 38 % amylose: wrinkled-seeded varieties have less starch (on average 34 %) and grains of compound form with a higher percentage of amylose (see Yarnell, 1962). In *P. arvense*, Smith (1974) states that the full complement of grains is present in the cotyledon cells 20 days after anthesis, so that the main period of starch accumulation, over the period 20–35 days, is essentially a filling and maturation of already-formed grains. Thereafter, little further starch is laid down, although cell walls become impregnated with hemicelluloses (Flinn and Pate, 1968). In *P. sativum* starch accumulation usually continues until the seeds are almost fully ripe, and during the early stages of its synthesis a substantial drop occurs in the level of sucrose and other sugars in the seed (Bisson and Jones, 1932; Turner *et al.* 1957; Rowan and Turner, 1957).

The developing pea seed shows a changing spectrum of substances exhibiting cytokinin-like activity (Burrows and Carr, 1970) and at least three different gibberellins are present, including GA 5 and possibly GA 6 (Reinhard and Konopka, 1967).

It has been known for a long time that pea seed protein consists of two major fractions, a water-soluble 'albumin' fraction and a salt-soluble 'globulin' fraction (Osborne and Campbell, 1898; Osborne, 1926). Albumins accumulate progressively throughout seed development, and comprise many different proteins (Fox *et al.* 1964) whose relative amounts change during the maturation of the seed (Flinn and Pate, 1968). At maturity albumins are present in about two-thirds the quantity of the globulins (Varner and Schidlovsky, 1963; Beevers and Poulson, 1972).

Studies of the pea globulins by Danielsson (1949, 1952) substantiate Osborne's (1926) earlier finding of two major fractions – vicilin and legumin. Danielsson suggests that the 7S vicilin fraction has a molecular weight of 186000, the 12S legumin, 331000. The relative proportions of the two fractions vary during seed development, vicilin tending to accumulate in advance of legumin. Acrylamide gel electrophoresis of pea globulin resolves one major band (Boulter *et al.* 1967; Flinn and Pate, 1968) – the typical situation for members of the Vicieae (Boulter and Derbyshire, 1971). Compositional differences between vicilin and legumin, originally observed for sulphur content (see Osborne, 1926) and amino acid balance (Danielsson and Lis, 1952), are now fully substantiated (Boulter and Derbyshire, 1971). Vicilin and legumin have higher proportions of arginine and glutamyl residues, but less methionine and cystine than pea albumins, whilst pea seed proteins as a whole appear to be somewhat deficient in histidine and tryptophane in comparison with a 'standard' animal protein.

Certain of the protein reserves of the mature seed are located in membrane-bound 'protein bodies', roughly 2 μm in diameter and interspersed amongst the much larger, though less numerous starch grains (Varner and Schidlovsky, 1963; Bain and Mercer, 1966a). Protein bodies form just as the globulin proteins can first be detected, and since the time of their filling coincides with a period when labelling studies show globulins to be synthesized more rapidly than albumins (Beevers and Poulson, 1972), there is good reason for suggesting that these bodies are where the globulins are located in the seed.

Protein content of mature pea seeds usually lies within the range 15–35 %. There is some evidence that seed protein content is genetically controlled (see Pesola, 1955; Yarnell, 1962; Hynes, 1968), but effects due to environment may be found within a cultivar (Highkin, 1960; Robertson *et al.* 1962). *S*-triazines sprayed on plants before or during flowering can increase protein levels in legume seeds, effects being very noticeable in *Phaseolus* (Singh *et al.* 1972), but occurring also in peas (Wu, 1971). This group of chemicals apparently acts by altering the balance between enzymes of nitrogen and carbohydrate metabolism, so that any increase in protein accumulation by seeds is likely to be accompanied by a decrease in starch content.

CONCLUSION–THE BASES FOR IMPROVED CROP PERFORMANCE

So far in this chapter little mention has been made of the many problems facing the agronomist when trying to achieve consistently high yields from a pea variety under widely different soil and seasonal conditions. Many of these problems stem from the extreme sensitivity of most of the known varieties to factors within the environment, so that it should be a primary concern of the plant breeder to develop genetic stocks yielding well under a broad range of conditions (see Snoad, Arthur, Payne and Hobart, 1971), and of the crop physiologist to identify limiting processes and to devise means for ameliorating them.

Soil conditions during germination have already been discussed as a factor of importance in crop establishment, and possible improvements might now be made by selecting varieties resistant to pre-emergence attack by pathogens, and capable of germinating successfully in cold, damp soils (see Matthews and Dow, 1971). There is still much to be learned as to why certain seed provenances are poorer than others in field establishment (Carver, 1972): possibly conditions during ripening of the seed may provide the clue and should be studied accordingly.

Compared with other crop species little effort has been made so far to model specifically for canopy characteristics in the pea. The untidy habit of the crop and the highly variable leaf area index achieved under different situations (see Kornilov and Kostina, 1965; Eastin and Gritton, 1969; Meadley and Milbourn, 1970) may have militated against such study. Certainly, in the vining pea industry, where only a small fraction of the potential reproductive yield is utilized, factors other than the efficiency of light interception over the growth cycle appear to be paramount. Most important amongst these are sowing time, planting density, and plant habit in relation to harvesting. By judicious selection of varieties, sowing times and rates, highly accurate predictions of harvest times and yields can now be made, despite vagaries of season (see Younkin *et al.* 1950; Hope, 1962; van Dobben, 1963; King, 1966; Milbourn and Hardwick, 1968; Gritton and Eastin, 1968; Bland, 1971). The influences of soil water regime (Maurer *et al.* 1968), high temperature stress during flowering and fruiting (Lambert and Linck, 1958; Karr, 1959; Ormrod *et al.* 1970), soil compaction (Eavis and Payne, 1969) and weed growth (Marx and Hagedorn, 1961) have also received attention. Similarly, with the advent of quick freezing, quality control through attention to specific pathogens (e.g. Harland, 1948; Snoad, Payne and Hobart, 1971; Bland, 1971) and post-harvest physiology (Wager, 1954, 1957, 1964) has assumed much greater importance.

One cannot overestimate recent contributions by geneticists to the improvement of the species. A computer-based record system is available for retrieval of useful information concerning the many hundreds of pea varieties now known (see Johnson *et al.* 1971), and a '*Pisum* Genetics Association' has been formed to facilitate the exchange of information concerning the species and to foster preservation of valuable genetic stocks. Its publication *The* Pisum *Newsletter* provides up-to-date information on breeding studies on peas, much of this being of great interest to the crop physiologist. Varieties yielding high numbers of uniformly mature seeds suitable for freeze drying or processing have been developed. As a rule it is possible to harvest from only three or four reproductive nodes, so selection for more synchronous flowering (Marx, 1966), for more pods per node (Marx, 1966), and for more seeds per pod (Marx and Mishanec, 1962, 1967) offers a sensible approach to increasing useful yield. Similarly, for a vining pea it is of obvious benefit to devise genetic (Marx, 1968) or chemical (Herner, 1969) means of discouraging growth of the apex and axillary shoots once the optimum set of fruit has been obtained. These objectives are, in fact, the very opposite of those desirable in the market garden crop. Particular interest has been revived in the fasciation phenomenon in pea since it has been shown that vining varieties possessing this trait exhibit terminal and almost synchronous flowering (Scheibe, 1954, 1968; Marx and Hagedorn, 1962). A hidden benefit of fasciation may be that the bulky shoot axis is likely to hold much larger nutrient reserves for fruit ripening than in the case of conventional thin-stemmed varieties. Indeed, the ultimate barrier to higher yield of the vining pea may well be of a physiological nature, and relate to limitations imposed by the rapid and synchronous increase in requirements for nutrients by so many seeds, all in approximately the same stage of development. In such a situation the best variety may prove to be the one most efficient in mobilizing reserves and current assimilates to its fruits, and selection along these lines might well be possible.

In a world short of protein, it seems tragic that a large and increasing acreage of peas is now devoted to producing seeds harvested in unripe condition with a mere fraction of their potential seed reserves. Perhaps time will see a reversal of this trend through international recognition of a duty to use every crop to full advantage in terms of food production, even if this may mean a lower profit margin for the grower and, possibly, the loss of palatable commodities from our tables. The recent use of soybean (*Glycine max*) and field bean (*Vicia faba*) as protein sources for vegetable-based 'meat substitutes' for human consumption raises the question of whether pea could be used successfully in such a connection. At present, low and uncertain yields and low percentage of seed protein would argue against this, but there is no reason to suggest that these failings might not be corrected by suitable improvements in breeding and crop husbandry.

REFERENCES

Abbott, I. R. and Matheson, N. K. (1972). Starch depletion in germinating wheat, wrinkled-seeded peas and senescing tobacco leaves. *Phytochemistry* **11**, 1261–1272.

Amos, J. J. and Crowden, R. K. (1969). Effects of vernalization, photoperiod and the cotyledon on flower initiation in Greenfeast peas. *Aust. J. biol. Sci.* **22**, 1091–1103.

Austin, R. B. (1966). The influence of the phosphorus and nitrogen nutrition of pea plants on the growth of their progeny. *Plant and Soil* **24**, 359–368.

Ayers, W. A. and Thornton, R. H. (1968). Exudation of amino acids by intact and damaged roots of wheat and peas. *Plant and Soil* **28**, 193–207.

Bailey, L. H. (1949). *Manual of cultivated plants*. Macmillan, New York.

Bain, J. M. and Mercer, F. V. (1966a). Subcellular organization of the developing cotyledons of *Pisum sativum* L. *Aust. J. biol. Sci.* **19**, 49–67.

Bain, J. M. and Mercer, F. V. (1966b). Subcellular organization of the cotyledons in germinating seeds and seedlings of *Pisum sativum* L. *Aust. J. biol. Sci.* **19**, 69–84.

Barber, H. N. (1959). Physiological genetics of *Pisum*. II. The genetics of photoperiodism and vernalization. *Heredity* **13**, 33–60.

Barber, H. N., Jackson, W. D., Murfet, I. C. and Sprent, J. I. (1958). Gibberellic acid and the physiological genetics of flowering in peas. *Nature, Lond.* **182**, 1321–1322.

Beevers, L. and Poulson, R. (1972). Protein synthesis in cotyledons of *Pisum sativum* L. I. Changes in cell-free amino acid incorporation capacity during seed development and maturation. *Plant Physiol.* **49**, 476–481.

Beevers, L. and Splittstoesser, W. E. (1968). Protein and nucleic acid metabolism in germinating peas. *J. exp. Bot.* **19**, 698–711.

Bernstein, L. and Hayward, H. E. (1958). Physiology of salt tolerance. *Ann. Rev. Plant Physiol.* **9**, 25–46.

Bisson, C. S. and Jones, H. A. (1932). Changes accompanying fruit development in the garden pea. *Plant Physiol.* **7**, 91–106.

Bland, B. F. (1971). *Crop production: cereals and legumes.* Academic Press, London.

Blixt, S. (1970). *Pisum.* In *Genetic Resources in Plants,* eds O. H. Frankel and E. Bennett. *IBP Handbook* No. 11. Blackwell, Oxford, pp. 321–326.

Bond, L. (1948). Origin and developmental morphology of nodules of *Pisum sativum. Bot. Gaz.* **109**, 411–434.

Boulter, D. and Derbyshire, E. (1971). Taxonomic aspects of the structure of legume proteins. In *Chemotaxonomy of the Leguminosae.* Academic Press, London.

Boulter, D., Jeremy, J. J. and Wilding, M. (1966). Amino acids liberated into the culture medium by seedling roots. *Plant and Soil* **24**, 121–127.

Boulter, D., Thurman, D. A. and Derbyshire, E. (1967). A disc electrophoretic study of the globulin proteins of legume seeds with reference to their systematics. *New Phytol.* **66**, 27–36.

Bowen, M. R. and Wareing, P. F. (1971). Further investigations into hormone-directed transport in stems. *Planta* **99**, 120–132.

Brennan, H. (1966). Patterns of translocation in ageing plants of field pea (*Pisum arvense* L.). M.Sc. Thesis, Queen's University, Belfast.

Brouwer, R. (1962). Distribution of dry matter in the plant. *Neth. J. Agr. Sci.* **10**, 361–376.

Brouwer, R. and van Vliet, G. (1960). The influence of root temperature on growth and uptake in peas. *Meded. Inst. biol. scheik. Onderz. Landl-Gewass.* **108**, 23–26.

Bundy, J. W. (1971). Environmental limitations in crop production for the quick-freezing industry. In *Potential Crop Production,* eds P. F. Wareing and J. C. Cooper. Heinemann, London, pp. 351–361.

Burrows, W. J. and Carr, D. J. (1970). Cytokinin content of pea seeds during their growth and development. *Physiol. Plant.* **23**, 1064–1070.

Cameron, D. S. and Cossins, E. A. (1967). Studies of intermediary metabolism in germinating pea cotyledons. The pathway of ethanol metabolism and the role of the tricarboxylic acid cycle. *Biochem. J.* **105**, 323–331.

Canning, R. E. and Kramer, P. J. (1958). Salt absorption and accumulation in various regions of roots. *Amer. J. Bot.* **45**, 378–382.

Carr, D. J. and Burrows, W. J. (1966). Evidence of the presence in xylem sap of substances with kinetin-like activity. *Life Sciences* **5**, 2061–2077.

Carr, D. J. and Pate, J. S. (1967). Ageing in the whole plant. *Symp. Soc. exp. Biol.* **21**, 559–600.

Carr, D. J., Reid, D. M. and Skene, K. G. M. (1964). The supply of gibberellins from the root to the shoot. *Planta* **63**, 382–392.

Carr, D. J. and Skene, K. G. M. (1961). Diauxic growth curves of seeds, with special reference to French beans (*Phaseolus vulgaris* L.). *Aust. J. biol. Sci.* **14**, 1–12.

Cartwright, P. M. and Snow, D. (1962). The influence of foliar applications of urea on the nodulation pattern of certain leguminous species. *Ann. Bot., Lond.* **26**, 257–259.

Carver, M. F. F. (1972). Aspects of seed physiology which are influential in establishment problems of vegetable seeds. Ph.D. Thesis, Univ. Stirling.

Chin, L. T. Y., Poulson, R. and Beevers, L. (1972). The influence of axis removal on protein metabolism in cotyledons of *Pisum sativum. Plant Physiol.* **49**, 482–489.

Cole, S. (1961). *Neolithic Revolution*, 2nd edn. London, Brit. Mus. (Nat. Hist.).

Collander, R. (1941). Selective absorption of cations by higher plants. *Plant Physiol.* **16**, 691–720.

Cooper, D. C. (1938). Embryology of *Pisum sativum. Bot. Gaz.* **100**, 123–132.

Cossins, E. A. (1964). Formation and metabolism of lactic acid during germination of pea seedlings. *Nature, Lond.* **203**, 989–990.

Danielsson, C. E. (1949). Seed globulins of the Gramineae and Leguminosae. *Biochem. J.* **44**, 387–410.

Danielsson, C. E. (1952). A contribution to the study of the synthesis of the reserve proteins in ripening pea seeds. *Acta. chem. Scand.* **6**, 149–159.

Danielsson, C. E. (1956). Starch formation in ripening pea seeds. *Physiol. Plant.* **9**, 212–219.

Danielsson, C. E. and Lis, H. (1952). Differences in the chemical composition of some pea proteins. *Acta. chem. Scand.* **6**, 139–148.

Devadas, C. and Beck, C. B. (1972). Comparative morphology of the primary vascular systems in some species of Rosaceae and Leguminosae. *Amer. J. Bot.* **59**, 557–567.

Diener, T. (1950). Uber die Bedingungen der Wurzelknöllchenbildung bei *Pisum sativum* L. *Phytopath. Z.* **16**, 129–170.

Dilworth, M. J. (1966). Acetylene reduction by nitrogen-fixing preparations from *Clostridium pasteurianum. Biochim. Biophys. Acta* **127**, 285–294.

Dixon, R. O. D. (1964). The structure of infection threads, bacteria and bacteroids in pea and clover root nodules. *Arch. Mikrobiol.* **48**, 166–178.

Dobben, W. H. van (1963). The physiological background of the reaction of peas to sowing time. *Jaarb., Inst. biol. scheik. Onderz. Landb-Gewass.* pp. 41–49.

Dormer, K. J. (1946). Vegetative morphology as a guide to the classification of the Papilionatae. *New Phytol.* **45**, 145–161.

Dostal, R. (1967). On the lateral growth correlations exemplified by petioles and axillaries of *Pisum* cotyledons. *Biol. Plant.* **9**, 330–339.

Dostal, R. (1968). Right-left problem exemplified by *Pisum* seedlings. *Flora, Jena* **159**, 274–276.

Drennan, D. S. H. and Norton, C. (1972). The effect of ethrel on nodulation in *Pisum sativum* L. *Plant and Soil* **36**, 53–57.

Eastin, J. A. and Gritton, E. T. (1969). Leaf area development, light interception, and the growth of canning peas (*Pisum sativum* L.) in relation to plant population and spacing. *Agron. J.* **61**, 612–615.

Eavis, B. W. and Payne, D. (1969). Soil physical conditions and root growth. In *Root Growth*, ed. W. J. Whittington. Butterworths, London, pp. 315–338.

Federova, L. N. (1966). Influence of characteristics of pea varieties on intensity of nodule formation. *Agrochimya Poshvovedenie* **4**, 31–39.

Flentje, N. T. and Saksena, H. K. (1964). Pre-emergence rotting of peas in South Australia. III. Host–pathogen interactions. *Aust. J. biol. Sci.* **17**, 665–675.

Flinn, A. M. (1969). A nutritional study of fruit maturation in *Pisum arvense*. L. Ph.D. Thesis, Queen's University, Belfast.

Flinn, A. M. and Pate, J. S. (1968). Biochemical and physiological changes during maturation of fruit of the field pea (*Pisum arvense* L.). *Ann. Bot.* **32**, 479–495.

Flinn, A. M. and Pate, J. S. (1970). A quantitative study of carbon transfer from pod and subtending leaf to the ripening seeds of the field pea (*Pisum arvense* L.). *J. exp. Bot.* **21**, 71–82.

Fox, D. J., Thurman, D. A. and Boulter, D. (1964). Studies on the proteins of seeds of the Leguminosae. I. Albumins. *Phytochemistry* **3**, 417–419.

Gelin, O. and Blixt, S. (1964). Root nodulation in peas. *Agric. Hort. Genet.* **22**, 149–159.

Grasmanis, V. O. and Barley, K. P. (1969). The uptake of nitrate and ammonium by successive zones of the pea radicle. *Aust. J. biol. Sci.* **22**, 1313–1320.

Gritton, E. T. and Eastin, J. A. (1968). Response of peas (*Pisum sativum* L.) to plant population and spacing. *Agron. J.* **60**, 482–485.

Grobbelaar, N. and Steward, F. C. (1969). The isolation of amino acids from *Pisum sativum*. Identification of $L(-)$-homoserine and $L(+)$-O-acetylhomoserine and certain effects of environment upon their formation. *Phytochemistry* **8**, 557–559.

Guardiola, J. L. and Sutcliffe, J. F. (1971*a*). Control of protein hydrolysis in the cotyledons of germinating pea (*Pisum sativum*) seeds. *Ann. Bot.* **35**, 791–807.

Guardiola, J. L. and Sutcliffe, J. F. (1971*b*). Mobilization of phosphorus in the cotyledons of young seedlings of the garden pea (*Pisum sativum* L.). *Ann. Bot.* **35**, 809–823.

Guardiola, J. L. and Sutcliffe, J. F. (1972). Transport of materials from cotyledons during germination of the garden pea (*Pisum sativum* L.). *J. exp. Bot.* **23**, 322–337.

Hardy, R. W. F., Burns, R. C., Hebert, R. R., Holsten, R. D. and Jackson, E. K. (1971). Biological nitrogen fixation: a key to world protein. In *Biological Nitrogen Fixation in Natural and Agricultural Habitats*, eds T. A. Lie and E. G. Mulder. Martinus Nijhoff, The Hague, pp. 561–590.

Hardy, R. W. F., Holsten, R. D., Jackson, E. K. and Burns, R. C. (1968). The acetylene-ethylene assay for N_2 fixation: Laboratory and field evaluation. *Plant Physiol.* **43**, 1185–1207.

Harland, S. C. (1948). Inheritance of immunity to mildew in Peruvian forms of *Pisum sativum*. *Heredity* **2**, 263–269.

Harvey, D. M. (1971). Translocation of ^{14}C in *Pisum sativum* tissue. *Annual Report*, John Innes Institute, Norfolk, England, **62**, 35.

Haupt, W. (1969). *Pisum sativum* L. In *The Induction of Flowering*, ed. L. T. Evans. Macmillan, Melbourne, pp. 393–408.

Hawkes, J. and Wooley, L. (1963). *History of mankind*, vol. I. *Prehistory and the beginnings of civilization*. London, George Allen and Unwin.

Hayward, H. E. (1938). *The structure of economic plants*. Macmillan, New York.

Hector, J. M. (1936). *Introduction to the botany of field crops*, vol. II. *Non-cereals*. Central News Agency, Ltd, Johannesburg, S. Africa.

Hellmuth, E. O. (1971). The effect of varying air-CO_2 level, leaf temperature and illuminance on the CO_2 exchange of the dwarf pea, *Pisum sativum* var. Meteor. *Photosynthetica* **5**, 190–194.

Herner, R. C. (1969). Cytokinins and plant senescence of the garden pea (*Pisum sativum* L.). *Diss. Abstr.* **29**, 2316B.

Heywood, V. H. (1971). The Leguminosae – A sytematic purview. In *Chemotaxo-*

nomy of the Leguminosae, eds J. R. Harborne, D. Boulter and B. L. Turner. Academic Press, London, pp. 1–30.

Highkin, H. R. (1960). The effect of constant temperature environments and of continuous light on the growth and development of pea plants. *Cold Spring Harbor Symp. quant. biol.* **25**, 231–238.

Hope, G. W. (1962). Modification of the heat unit formula for peas. *Canad. J. Pl. Sci.* **42**, 15–21.

Husain, S. M. (1967). Studies on the phenomenon of apical dominance and its relationship to nutrient accumulation in *Pisum sativum* L. var. Alaska. *Diss. Abstr.* **27**, 4258B.

Husain, S. M. and Linck, A. (1966). Relationship of apical dominance to the nutrient accumulation pattern in *Pisum sativum* var. Alaska. *Physiol. Plant.* **19**, 992–1010.

Husain, S. M. and Linck, A. J. (1967). The effect of chilling of the physiological and simulated apex of 2- and 3-leaf plants of *Pisum sativum* L. cv. Alaska on lateral shoot growth, C^{-14} IAA movement and P^{-32}accumulation. *Physiol. Plant.* **20**, 48–56.

Hynes, M. J. (1968). Genetically controlled electrophoretic variants of a storage protein in *Pisum sativum*. *Aust. J. biol. Sci.* **21**, 827–829.

Janick, J., Schery, R. W., Woods, F. W. and Ruttan, V. W. (1969). *Plant science.* Freeman, San Francisco.

Jean, F. C. (1928). Root inheritance in peas. *Bot. Gaz.* **86**, 318–329.

Johnson, M. V., Snoad, B. and Davies, D. R. (1971). A computer based record system for *Pisum*. *Euphytica* **20**, 126–130.

Jones, V. M. and Boulter, D. (1968). Arginine metabolism in germinating seeds of some members of the Leguminosae. *New Phytol.* **67**, 925–934.

Juliano, B. O. and Varner, J. E. (1969). Enzymatic degradation of starch granules in the cotyledons of germinating peas. *Plant Physiol.* **44**, 886–892.

Kapoor, B. M. (1966). Contributions to the cytology of endosperm in Angiosperms – XII. *Pisum sativum* L. *Genetica* **37**, 557–568.

Karr, E. J. (1959). The effect of short periods of high temperature during day and night periods on pea yields. *Amer. J. Bot.* **46**, 91–93.

King, J. M. (1966). *Row width and plant populations in vining peas.* Pea Growing Research Organization, Ltd, England. Misc. Pub. No. 18, p. 4.

Kollöffel, C. (1970). Alcohol dehydrogenase activity in the cotyledons of peas during maturation and germination. *Acta. Bot. Neerl.* **19**, 539–545.

Kollöffel, C. and Sluys, J. V. (1970). Mitochondrial activity in pea cotyledons during germination. *Acta. Bot. Neerl.* **19**, 503–508.

Kopetz, L. M. (1938). Photoperiodische Untersuchungen an Pflückerbsen. *Gartenbauwissensch.* **12**.

Kornilov, A. A. and Kostina, V. S. (1965). The optimum leaf area for obtaining high pea yields. *Fiziol. Rast.* **12**, 551–553.

Lambert, R. G. and Linck, A. J. (1958). Effects of high temperature on yield of peas. *Plant physiol.* **33**, 347–350.

Larson, L. A. (1968). The effect soaking pea seeds with or without coats has on seedling coats. *Plant Physiol.* **43**, 255–259.

Larson, L. A. and Kyagaba, L. (1969). The effect of prolonged seed soaking on seedling growth of *Pisum sativum*. *Can. J. Bot.* **47**, 707–709.

Leblova, S., Zimakova, I., Sofrova, D. and Barthova, J. (1969). Occurrence of ethanol in pea plants in the course of growth under normal and anaerobic conditions. *Biol. Plant.* **11**, 417–423.

Lee, C. Y. and Shallenberger, R. S. (1969). Changes in free sugar during the germination of pea seeds. *Experientia* **25**, 692–693.

Leopold, A. C. and Guernsey, F. S. (1954). Flower initiation in the Alaska pea. II. Chemical vernalization. *Amer. J. Bot.* **41**, 181–185.

Lewis, O. A. M. and Pate, J. S. (1973). The significance of transpirationally-derived nitrogen in protein synthesis in fruiting plants of pea (*Pisum sativum* L.). *J. exp. Bot.* **24**, 596–606.

Lie, T. A. (1969). The effect of low pH on different phases of nodule formation in pea plants. *Plant and Soil* **31**, 391–406.

Lie, T. A. (1971). Temperature-dependent root-nodule formation in pea cv. Iran. *Plant and Soil* **34**, 751–752.

Linck, A. J. (1961). The morphological development of the fruit of *Pisum sativum* var. Alaska. *Phytomorphology* **11**, 79–84.

Linck, A. J. and Sudia, T. W. (1962). Translocation of labelled photosynthate from the bloom node leaf to the fruit of *Pisum sativum*. *Experientia* **18**, 69–70.

Linck, A. J. and Swanson, C. A. (1960). A study of several factors affecting the distribution of phosphorus-^{32} from the leaves of *Pisum sativum*. *Plant and Soil* **12**, 57–68.

Linke, R. D. and Marinos, N. G. (1970). Effects of a pregermination pulse treatment with morphactin on *Pisum sativum*. *Aust. J. biol. Sci.* **23**, 1125–1131.

Lockhart, J. A. and Gottschall, V. (1961). Fruit-induced and apical senescence in *Pisum sativum* L. *Plant Physiol.* **36**, 389–393.

Lovell, P. H. (1969). Interrelationships of cotyledonary shoots in *Pisum sativum*. *Physiol. Plant.* **22**, 506–515.

Lovell, P. H. and Lovell, P. J. (1970). Fixation of CO_2 and export of photosynthate by the carpel in *Pisum sativum*. *Physiol. Plant.* **23**, 316–322.

Lovell, P. H., Oo, H. T. and Sagar, C. R. (1972). An investigation into the rate and control of assimilate movement from leaves in *Pisum sativum*. *J. exp. Bot.* **23**, 255–266.

McIntyre, G. I. (1971). Water stress and apical dominance in *Pisum sativum*. *Nature, Lond.* **230**, 87–88.

McKee, H. S., Nestel, L. and Robertson, R. N. (1955). Physiology of pea fruits. II. Soluble nitrogenous constituents in the developing fruit. *Aust. J. biol. Sci.* **8**, 467–475.

Makasheva, R. Kh. (1971). Species of the genus *Pisum*. *Trudy prikl. Bot. Genet. Selek.* **44**, 86–104.

Marx, G. A. (1966). Uniformity of maturity in peas. *Farm Research* **32**, 12–13.

Marx, G. A. (1968). Influence of genotype and environment on senescence in peas, *Pisum sativum* L. *Bioscience* **18**, 505–506.

Marx, G. A. (1969). Some photo-dependent responses in *Pisum*. I. Physiological behaviour. *Crop Sci.* **9**, 273–276.

Marx, G. A. and Hagedorn, D. J. (1961). Plant population and weed growth relations in canning peas. *Weeds* **9**, 494–496.

Marx, G. A. and Hagedorn, D. J. (1962). Fasciation in *Pisum*. *J. Hered.* **53**, 31–43.

Marx, G. A. and Mishanec, W. (1962). Inheritance of ovule number in *Pisum sativum* L. *Proc. Amer. Soc. Hort. Sci.* **80**, 462–467.

Marx, G. A. and Mishanec, W. (1967). Further studies on the inheritance of ovule number in *Pisum*. *Crop Sci.* **7**, 236–239.

Matthews, P. and Dow, P. (1971). Pea pathogens. *Annual Report*, John Innes Institute, Norfolk, England, **62**, 31–32.

Maurer, A. R., Ormrod, D. P. and Fletcher, H. F. (1968). Response of peas to environment. IV. Effects of soil water regimes on growth and development of peas. *Can. J. Pl. Sci.* **48**, 129–137.

Meadley, J. T. and Milbourn, G. M. (1970). The growth of vining peas. 2. The effect of density of planting. *J. agric. Sci.* **74**, 273–278.

Mendel, G. (1865). Versuche über Pflanzen-Hybriden. *Verhandl. Naturforsch. Verein Brünn* **4**, 3.

Milbourn, G. M. and Hardwick, R. C. (1968). The growth of vining peas. 1. The effect of time of sowing. *J. agric. Sci.* **70**, 393–402.

Minchin, F. R. and Pate, J. S. (1973). The carbon balance of a legume and the functional economy of its root nodules. *J. exp. Bot.* **24**, 259–271.

Mishustin, E. N. and Shilnikova, V. K. (1971). *Biological fixation of atmospheric nitrogen.* Macmillan, London.

Mitchell, D. J. and Bidwell, R. G. S. (1970). Compartments of organic acids in the synthesis of asparagine and homoserine in pea roots. *Can. J. Bot.* **48**, 2001–2007.

Moore, T. C. (1964). Effects of cotyledon excision on the flowering of five varieties of *Pisum sativum*. *Plant Physiol.* **39**, 924–927.

Moore, T. C. (1965). Effects of gibberellin on the growth and flowering of intact and decotylized dwarf peas. *Nature, Lond.* **206**, 1065–1066.

Moore, T. C. and Bonde, E. K. (1962). Physiology of flowering in peas. *Plant Physiol.* **37**, 149–153.

Morris, D. A. and Thomas, E. E. (1968). Distribution of ^{14}C-labelled sucrose in seedlings of *Pisum sativum* L. treated with indoleacetic acid and kinetin. *Planta* **83**, 276–281.

Mulder, E. G. (1948a). Importance of molybdenum in the nitrogen metabolism of micro-organisms and higher plants. *Plant and Soil* **1**, 94–119.

Mulder, E. G. (1948b). Investigations on the nitrogen nutrition of pea plants. *Plant and Soil* **1**, 179–212.

Mulder, E. G., Bakema, K. and Veen, W. L. van (1959). Molybdenum in symbiotic nitrogen fixation and in nitrate assimilation. *Plant and Soil* **10**, 319–334.

Murfet, I. C. (1971). Flowering in *Pisum*. Three distinct phenotypic classes determined by the interaction of a dominant early and a dominant late gene. *Heredity* **26**, 243–257.

Nakamura, E. (1964). Effect of decapitation and IAA on the distribution of radioactive phosphorus in the stem of *Pisum sativum*. *Plant Cell Physiol.* **5**, 521–524.

Nakano, M. and Asahi, T. (1972). Subcellular distribution of hydrolase in germinating pea cotyledons. *Plant Cell Physiol.* **13**, 101–110.

Nawa, Y. and Asahi, T. (1971). Rapid development of mitochondria in pea cotyledons during the early stage of germination. *Plant Physiol.* **48**, 671–674.

Nutman, P. S. (1956). The influence of the legume in root nodule symbiosis. A comparative study of host determinants and functions. *Biol. Rev.* **31**, 109–151.

Oghoghorie, C. G. O. (1971). The physiology of the field pea-*Rhizobium* symbiosis in the presence and absence of nitrate. Ph.D. Thesis, Univ. Belfast.

Oghoghorie, C. G. O. and Pate, J. S. (1971). The nitrate stress syndrome of the nodulated field pea (*Pisum arvense* L.). Technique for measurement and evaluation in physiological terms. In *Biological nitrogen fixation in natural and agricultural habitats. Plant and Soil (special volume)*, pp. 185–202.

Oghoghorie, C. G. O. and Pate, J. S. (1972). Exploration of the nitrogen transport system of a nodulated legume using ^{15}N. *Planta* **104**, 35–49.

Oinuma, J. (1948). Cytological and morphological studies on root nodules of garden peas. *Pisum sativum* L. *Seibutu* **3**, 155–161.

Ormrod, D. P., Maurer, A. R., Mitchell, G. and Eaton, G. W. (1970). Shoot apex development in *Pisum sativum* L. as affected by temperature. *Can. J. Pl. Sci.* **50**, 201–202.

Osborne, T. B. (1926). *The Vegetable Proteins*, 2nd edn. Longmans Green and Co., London.

Osborne, T. B. and Campbell, G. F. (1898). The proteins of the pea, lentil, horse bean and vetch. *J. Am. Chem. Soc.* **20**, 410–419.

Pahlich, E. and Joy, K. W. (1971). Glutamate dehydrogenase from pea roots. Purification and properties of the enzyme. *Can. J. Biochem.* **49**, 127–138.

Pate, J. S. (1956). The physiology of nodule development of some Leguminosae. Ph.D. Thesis, Queen's University, Belfast.

Pate, J. S. (1958). Nodulation studies in legumes. I. The synchronization of host and symbiotic development in the field pea *Pisum arvense* L. *Aust. J. biol. Sci.* **11**, 366–381.

Pate, J. S. (1962). Root-exudation studies on the exchange of C^{14}-labelled organic substances between the roots and shoot of the nodulated legume. *Plant and Soil* **17**, 333–356.

Pate, J. S. (1966). Photosynthesizing leaves and nodulated roots as donors of carbon to protein of the shoot of the field pea (*Pisum arvense* L.). *Ann. Bot.* **30**, 93–109.

Pate, J. S. (1973). Uptake, assimilation and transport of nitrogen compounds by plants. *Soil Biol. Biochem.* **5**, 109–120.

Pate, J. S. and Flinn, A. M. (1973). Carbon and nitrogen transfer from vegetative organs to ripening seeds of field pea (*Pisum arvense* L.). *J. exp. Bot.* **24**, 1090–1099.

Pate, J. S. and Greig, J. M. (1964). Rhythmic fluctuations in the synthetic activities of the nodulated root of the legume. *Plant and Soil* **21**, 163–184.

Pate, J. S. Gunning, B. E. S. and Briarty, L. G. (1969). Ultrastructure and functioning of the transport system of the leguminous root nodule. *Planta* **85**, 11–34.

Pate, J. S. and O'Brien, T. P. (1968). Microautoradiographic study of the incorporation of labelled amino acids into insoluble compounds of the shoot of a higher plant. *Planta* **78**, 60–71.

Pate, J. S., Walker, J. and Wallace, W. (1965). Nitrogen-containing compounds in the shoot of *Pisum arvense* L. II. The significance of amides and amino acids released from roots. *Ann. Bot.* **29**, 475–493.

Pate, J. S. and Wallace, W. (1964). Movement of assimilated nitrogen from the root system of the field pea *Pisum arvense* L. *Ann. Bot.* **28**, 83–99.

Paton, D. M. (1967). Leaf status and photoperiodic control of flower initiation in a late variety of pea. *Nature, Lond.* **215**, 319–320.

Paton, D. M. (1968). Photoperiodic and temperature control of flower initiation in the late pea cultivar Greenfeast. *Aust. J. biol. Sci.* **21**, 609–617.

Paton, D. M. (1971). Photoperiodic induction of flowering in the late pea cultivar Greenfeast: The role of exposed cotyledons and leaves. *Aust. J. biol. Sci.* **24**, 609–618.

Perry, D. A. and Harrison, J. G. (1970). The deleterious effect of water and low temperature on germination of pea seed. *J. exp. Bot.* **21**, 504–512.

Pesola, V. A. (1955). Protein content of field pea seeds as a varietal character. *Acta. agr. Fenn.* **83**, 125–132.

Phillips, D. A. (1971). Abscisic acid inhibition of root nodule initiation in *Pisum sativum*. *Planta* **100**, 181–190.

Raacke, I. D. (1957*a*). Protein synthesis in ripening pea seeds. II. Development of embryos and seed coats. *Biochem. J.* **66**, 110–113.

Raacke, I. D. (1957b). Protein synthesis in ripening pea seeds. III. Study of the pods. *Biochem. J.* **66**, 113–116.

Reed, H. S. (1942). The relation of zinc to seed production. *J. agric. Res.* **64**, 635–644.

Reinhard, E. and Konopka, W. (1967). Ein Beitrag zur Analyse der Gibberelline in Samen von *Pisum sativum* L. *Planta* **77**, 58–76.

Reynolds, J. D. (1955). Marsh spot of peas: a review of present knowledge. *J. Sci. Fd. Agric.* **6**, 725–734.

Reynolds, J. D. (1960). Manuring the pea crop. *Agriculture* **66**, 509–513.

Rijven, A. H. G. C. (1972). Control of the activity of the aleurone layer of fenugreek *Trigonella foenum-graecum* L. *Acta. Bot. Neerl.* **21**, 381–386.

Roberts, E. H. and Roberts, D. L. (1972). Viability nomographs. In *Viability of seeds*, ed. E. H. Roberts. Chapman and Hall, London.

Robertson, R. N., Highkin, H. R., Smydzuk, J. and Went, F. W. (1962). The effect of environmental conditions on the development of pea seeds. *Aust. J. biol. Sci.* **15**, 1–15.

Roponen, I. E., Valle, E. and Ettala, T. (1970). Effect of temperature of the culture medium on growth and nitrogen fixation of inoculated legumes and rhizobia. *Physiol. Plant.* **23**, 1198–1205.

Roponen, I. E. and Virtanen, A. I. (1968). The effect of prevention of flowering on the vegetative growth of inoculated pea plants. *Physiol. Plant.* **21**, 655–667.

Rowan, K. S. and Turner, D. H. (1957). Physiology of pea fruits. V. Phosphate components in the developing seed. *Aust. J. biol. Sci.* **10**, 414–425.

Rowlands, D. G. (1964). Genetic control of flowering in *Pisum sativum* L. *Genetica* **35**, 75–94.

Sachs, T. (1966). Senescence of inhibited shoots of peas and apical dominance. *Ann. Bot.* **30**, 447–456.

Sachs, T. (1968a). The role of the root in the induction of xylem differentiation in peas. *Ann. Bot.* **32**, 391–399.

Sachs, T. (1968b). On the determination of the pattern of vascular tissue in peas. *Ann. Bot.* **32**, 781–790.

Sachs, T. and Thimann, K. V. (1967). The role of auxins and cytokinins in the release of buds from dominance. *Amer. J. Bot.* **54**, 136–144.

Salter, P. J. and Drew, D. H. (1965). Root growth as a factor in the response of *Pisum sativum* L. to irrigation. *Nature, Lond.* **206**, 1063–1064.

Schalldach, I. and Schilling, G. (1966). Effect of N applications at various times on N fixation by *Pisum sativum*. *Albrecht-Thaer-Arch.* **10**, 829–839.

Scheibe, A. (1954). Der *fasciata*-Typus bei *Pisum*, seine pflanzenbauliche und züchterische Bedeutung. *Pfl. Zücht.* **33**, 23–30.

Scheibe, A. (1968). The fasciata type of pea and seed certification. *Saatgut Wirt* **20**, 126–128 (*Field crop abstracts* **21**, 2599).

Schilling, G. and Schalldach, I. (1966). Translocation, incorporation and utilization of late applied N in *Pisum sativum*. *Albrecht-Thaer-Arch.* **10**, 895–907 (*Field crop abstracts* **20**, 1089).

Schöllhorn, R. and Burris, R. H. (1966). Study of intermediates in nitrogen fixation. *Fedn. Proc. Fedn. Am. Socs. exp. Biol.* **58**, 213.

Schwinghamer, E. A., Evans, H. J. and Dawson, M. D. (1970). Evaluation of effectiveness in mutant strains of *Rhizobium* by acetylene reduction relative to other criteria of N_2 fixation. *Plant and Soil* **33**, 192–212.

Sebanek, J. (1965). The interaction of endogenous gibberellin with cotyledons and axillary buds in the pea (*Pisum sativum* L.). *Biol. Plant.* **7**, 194–198.

Seth, A. K. and Wareing, P. F. (1964). Interaction between auxins, gibberellins and kinins in hormone-directed transport. *Life Science* 3, 1483–1487.

Seth, A. K. and Wareing, P. F. (1967). Hormone-directed transport of metabolites and its possible role in plant senescence. *J. exp. Bot.* **18**, 65–77.

Shain, Y. and Mayer, A. M. (1968). Activation of enzymes during germination: amylopectin-1,6-glucosidase in peas. *Physiol. Plant.* **21**, 765–776.

Simon, E. W. and Harun, R. M. (1973). Leakage during seed imbibition. *J. exp. Bot.*, **23**, 1076–1085.

Singh, B., Campbell, W. F. and Salunkhe, D. K. (1972). Effects of *S*-triazines on protein and fine structure of cotyledons of bush beans. *Amer. J. Bot.* **59**, 568–572.

Smillie, R. M. (1962). Photosynthetic and respiratory activities of growing pea leaves. *Plant Physiol.* **37**, 716–721.

Smith, D. L. (1974). Nucleic acid, protein and starch synthesis in developing cotyledons of *Pisum arvense* L. *Ann. Bot.* **37**, 795–804.

Smith, D. L. and Flinn, A. M. (1967). Histology and histochemistry of the cotyledons of *Pisum arvense* L. during germination. *Planta, Berl.* **74**, 72–85.

Snoad, B., Arthur, A. E., Payne, A. and Hobart, J. (1971). Biometrical analyses of yield components in peas. *Annual Report*, John Innes Institute, Norfolk, England, **62**, 28–31.

Snoad, B., Payne, A. and Hobart, J. (1971). Biological control of flax tortrix moth attack in peas. *Annual Report*, John Innes Institute, Norfolk, England, **62**, 31.

Snow, R. (1929). The young leaf as the inhibiting organ. *New Phytol.* **28**, 345–358.

Snow, R. (1931). Experiments on growth and inhibition. II. New phenomena of inhibition. *Proc. Roy. Soc. B.* **108**, 305–316.

Snow, R. (1937). On the nature of correlative inhibition. *New Phytol.* **36**, 283–300.

Sorokin, H. and Sommer, A. L. (1940). Effects of calcium deficiency upon the roots of *Pisum sativum*. *Amer. J. Bot.* **27**, 308–318.

Sorokin, H. P. and Thimann, K. V. (1964). The histological basis for inhibition of axillary buds in *Pisum sativum*, and the effects of auxins and kinetin on xylem development. *Protoplasma* **59**, 326–350.

Sprent, J. I. (1966). Role of the leaf in flowering of late pea varieties. *Nature, Lond.* **209**, 1043.

Sprent, J. I. (1967). Effects of nutrient factors, water supply and growth-regulating substances on the vegetative growth pattern of peas and their relationship to node of first flower. *Ann. Bot.* **31**, 607–618.

Sprent, J. I. (1968*a*). The inability of gibberellic acid to stimulate amylase activity in pea cotyledons. *Planta* **82**, 299–301.

Sprent, J. I. (1968*b*). Effects of benzyladenine on cotyledon metabolism and growth of peas. *Planta* **81**, 80–87.

Stanfield, B., Ormrod, D. P. and Fletcher, H. F. (1966). Response of peas to environment. 2. Effects of temperature in controlled environment cabinets. *Can. J. Pl. Sci.* **46**, 195–203.

Thimann, K. V. and Skoog, F. (1934). On the inhibition of bud development and other functions of growth substance in *Vicia faba*. *Proc. Roy. Soc. Lond. B.* **114**, 317–339.

Torfason, W. E. and Nonnecke, I. L. (1959). A study of the effects of temperature and other factors upon the germination of vegetable crops. 2. Peas. *Can. J. Pl. Sci.* **39**, 119–124.

Turner, J. F., Turner, D. H. and Lee, J. B. (1957). Physiology of pea fruits. IV. Changes in sugars in the developing seed. *Aust. J. biol. Sci.* **10**, 407–413.

Varner, J. E., Balce, L. V. and Huang, R. C. (1963). Senescence of cotyledons of germinating peas. Influence of axis tissue. *Plant Physiol.* **38**, 89–92.

Varner, J. E. and Schidlovsky. (1963). Intracellular distribution of proteins in pea cotyledons. *Plant Physiol.* **38**, 139–144.

Vartiovaara, U. (1937). Investigations on the root-nodule bacteria of leguminous plants. XXI. The growth of the root-nodule organisms and inoculated peas at low temperatures. *J. agric. Sci.* **27**, 626–637.

Virtanen, A. I. (1947). The biology and chemistry of nitrogen fixation by legume bacteria. *Biol. Rev.* **22**, 239–269.

Virtanen, A. I. and Berg, A. M. (1954). γ-glutamyl-alanine in pea seedlings. *Acta. chem. Scand.* **8**, 1089–1090.

Virtanen, A. I., Berg, A. M. and Kari, S. (1953). Formation of homoserine in germinating pea seeds. *Acta. chem. Scand.* **7**, 1423–1424.

Virtanen, A. I., Jorma, J., Erkama, J. and Linnasalmi, A. (1947). On the relation between nitrogen fixation and leghaemoglobin content of leguminous root nodules. *Acta. chem. Scand.* **1**, 90–111.

Virtanen, A. I. and Laine, T. (1938). Biological synthesis of amino acids from atmospheric nitrogen. *Nature, Lond.* **141**, 748–749.

Virtanen, A. I. and Miettinen, J. K. (1963). Biological nitrogen fixation. In *Plant Physiology*, vol. 3, ed. F. C. Steward, pp. 565–668.

Wager, H. G. (1954). The effect of artificial wilting on the sugar content and respiration rate of maturing green peas. *New Phytol.* **53**, 354–363.

Wager, H. G. (1957). The effect of artificial wilting on the carbon dioxide production of developing pea seeds. *New Phytol.* **56**, 230–246.

Wager, H. G. (1964). Physiological studies on the storage of green peas. *J. Sci. Fd. Agric.* **15**, 245–252.

Wallace, T. (1961). *The diagnosis of mineral deficiencies in plants by visual symptons.* 3rd edn. H.M.S.O., London.

Wardlaw, I. F. and Mortimer, D. C. (1970). Carbohydrate movement in pea plants in relation to axillary bud growth and vascular development. *Can. J. Bot.* **48**, 229–237.

Wellensiek, S. J. (1969). The physiological effects of flower forming genes in peas. *Z. Pflanzenphysiol.* **60**, 388–402.

Wieringa, K. T. (1958). Transport of amino acids in leguminous plants. In *Nutrition of the Legumes*, ed. E. C. Hallsworth. Butterworths, London, pp. 256–265.

Wieringa, K. T. and Bakhuis, J. A. (1957). Chromatography as a means of selecting effective strains of *Rhizobia. Plant and Soil* **8**, 254–260.

Wong, P. P. and Evans, H. J. (1971). Poly-β-hydroxybutyrate utilization by soybean (*Glycine max* Merv.) nodules and assessment of its role in maintenance of nitrogenase activity. *Plant Physiol* **47**, 750–755.

Woodbridge, C. G. (1969). Boron deficiency in pea *Pisum sativum* cv. Alaska. *Proc. Ann. Soc. hort. Sci.* **94**, 542–544.

Wu, M. T. (1971). The regulatory effects of certain S-triazine compounds on the nutritive composition of beans and seeds of peas (*Pisum sativum* L.) and sweet corn (*Zea mays* L.) and responses of metabolic systems. Ph.D. Thesis, Utah State University, Logan.

Yarnell, S. H. (1962). Cytogenetics of the vegetable crops. III. Legumes. A. Garden peas, *Pisum sativum* L. *Bot. Rev.* **28**, 465–537.

Yorke, J. S. and Sagar, G. R. (1970). Distribution of secondary root growth potential in the root system of *Pisum sativum. Can. J. Bot.* **48**, 699–704.

Younkin, S. D., Hester, J. B. and Hoadley, B. D. (1950). Interaction of seeding rates and nitrogen levels on yield and sieve size of peas. *Proc. Am. Soc. hort. Sci.* **55**, 379–383.

8

Potato

J. MOORBY AND F. L. MILTHORPE

Many thousands of clones of the 150 or so known species of tuber-bearing *Solanum* are no doubt cultivated in restricted areas of the Andean region of Peru and Bolivia, the centre of origin of the genus. With the development of a few superior cultivars, these are being gradually supplanted; almost all the cultivars now grown throughout the world belong to the species *tuberosum* and most to the sub-species *tuberosum*, one of the five or six recognized groups of this species (Burton, 1966).

This chapter will be concerned almost exclusively with *Solanum tuberosum* s.sp. *tuberosum* and particularly with those cultivars grown in Europe, North America and Australia. These have all been developed from a very narrow gene pool (Simmonds, 1962). Nevertheless, what little information has been published on other species suggests that, apart from responses such as specific short-day requirements for tuberization, they behave similarly to *S. tuberosum* (Alvey, 1965; Ezeta and McCollum, 1972). Species other than *tuberosum* often possess valuable properties such as disease and nematode resistance but often have undesirable tuber qualities; nevertheless, high-yielding strains of *S. stoloniferum* and *S. andigena* 'neo-tuberosum' have been developed.

The cultivars which are our concern are grown from the tropics to the sub-polar regions, and comprise a major food crop in most countries (see Table 1.1). We will consider here, in an ontogenetic order, those aspects of the physiology of individual processes and of community interrelationships which are particularly relevant to the potato. As potatoes are propagated vegetatively, there is a continued passage of a vast number of cells from one generation to the next without genetic change or any sharply defined boundary. Nevertheless, it is convenient to regard the life of an individual as beginning with the initiation of a tuber on a stolon of the mother plant. The resulting mature tuber is both the propagule disseminated in time and space for subsequent crops, and the harvested commodity. We will commence by considering those aspects of the differentiation and development of tubers relevant to their behaviour as a propagule and then follow in greater detail the crop they produce throughout its ontogeny.

DIFFERENTIATION OF THE MOTHER TUBER

Each tuber arises as a swelling between the terminal bud and the penultimate expanding internode of a diageotropic underground stem or 'stolon'. The visible swelling reflects a change in the polarity of cell expansion from the longitudinal to the radial dimension (Booth, 1963). The terminal bud of the stolon at this stage bears about 12 primordial leaves, the older six to eight of these subtending axillary buds (Milthorpe, 1963). It continues to produce new leaves, buds and stem tissue at a rate which keeps very much in step with the expansion of the internodes some 12 leaves removed from the apical meristem. Extensive secondary division occurs throughout the primary tissue and continues during the whole growth period of the tuber, being mainly responsible for the increase in volume (Plaisted, 1957). The tuber and its apical bud (i.e. the terminal bud of the stolon) are therefore part of the primary axis of the stolon and the main buds on the tuber are secondary axes. Each main bud usually has two associated third-order buds to form the 'eye'. The total number of eyes per tuber, N, is a function of tuber size and can usually be described by

$$N - N_0 = (cA)^k, \qquad (8.1)$$

where N_0 is a lower limit of N, A is the surface area of the tuber, and c and k are constants. In the cultivar Majestic, c, k, and N_0 were 0.0412, 1.64 and 3.6, respectively with A in cm^2 (Bleasdale, 1965). A less exact approximation is one eye per cm length of tuber.

The second- and higher-order buds cease growth in acropetal succession along the primary axis. Some evidence suggests that growth of the tuber tissue beneath an eye may cease soon after the buds of that eye stop growing; certainly, the apical bud and the tuber as a whole cease growth at the same time. The growth of a population of tubers ceases only with the death of the above-ground parts (the 'haulm' or 'vine') but individual tubers may cease before this (p. 239).

It seems likely that the cessation of growth of lateral buds is due to correlative inhibition and that of the apical bud to lack of photosynthate. At harvest, however, all buds are usually dormant, 'dormancy' being defined as that state where there is no cell division when tubers or isolated buds are put in environments favourable for growth. Alternative definitions of dormancy are often used (e.g. Burton, 1963; Emilsson and Lindblom, 1963; Rappaport and Wolf, 1969); these are usually based on the presence or absence of visible elongation growth (as distinct from radial growth with continued activity of the apical meristem), irrespective of whether this is due to innate physiological mechanisms or an unfavourable environment. Such definitions appear to be more concerned with agricultural convenience than physiological clarity!

STORAGE PHASE OF GROWTH

Loss of dormancy

With the passage of time following harvest, mitotic activity in the apical meristems of the buds is gradually resumed and the buds elongate as 'sprouts'. The length of the dormant period, i.e. from harvest to this resumption of growth, varies from 1 to 15 weeks between varieties but is little influenced by environmental conditions during either the growth or the storage of the tubers (Burton, 1966). As it has a strong genetic component, it has been possible to select varieties with especially short dormant periods for use in areas where two crops per season are possible. The usual index is based on visible growth and therefore is confounded with effects on the rate of sprout elongation; it is adequate for general purposes but more exact work requires detection of the resumption of mitotic activity or, better, elucidation of the earliest biochemical changes involved.

Although the processes leading to the resumption of bud growth have no rigid environmental control, they are accelerated by treatment with ethylene chlorhydrin, glutathione, thiourea, carbon disulphide, cytokinins, gibberellins, wounding and, as long known to potato farmers, by providing an ample supply of water (Slomnicki and Rylski, 1964; Goodwin, 1966; Lascarides, 1967; Madec and Perennec, 1969; Bruinsma and Swart, 1970). Some of these substances are used commercially to break dormancy and to ensure even emergence, especially in virus-testing programmes and in areas with two crops per season. Some care must be exercised as the normal apical dominance relationships (see below) may be disrupted leading to unwanted effects (Holmes *et al.* 1970). The onset of growth may be delayed by substances such as inhibitors of DNA, RNA, and protein syntheses, abscisic acid, and amyl alcohol. Although ethylene and other volatile substances are produced naturally by tubers, they do not appear to be of major importance in controlling dormancy (Burton and Meigh, 1971).

There is a reasonable amount of evidence suggesting that in dormant buds protein synthesis is blocked at some level; indeed, there is little synthesis of DNA, RNA or protein, cell division is suppressed and there is no evidence of cell expansion (Rappaport and Wolf, 1969). The end of dormancy appears to be associated with a decrease in the internal concentration of abscisic acid and an increase in gibberellins although it has been suggested that these do not act at the same point in the system.

Numbers of sprouts established

Following the loss of dormancy, buds commence to grow at a rate determined largely by the temperature. The apical bud is the first to start growing and the others follow in basipetal succession. As growth progresses, apical

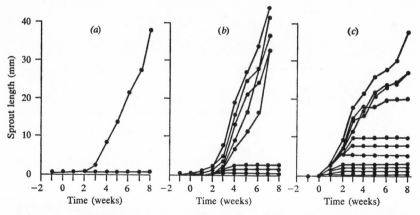

Figure 8.1. Length of each individual sprout on a tuber in relation to time from first visible growth. Tubers were of the early cultivar 'Arran Pilot' and were stored in darkness from lifting at temperatures of (a) 15 °C, (b) 10 °C, (c) at 1 °C until one month after sprouts in (b) had commenced to grow and then transferred to 10 °C (Goodwin, 1963).

dominance is established within the sprout population with the larger sprouts inhibiting the smaller (Goodwin, 1963, 1967). This is well illustrated in Figure 8.1: where tubers are stored from lifting at 15 °C or above, the apical sprout grows at a rate sufficient to inhibit all other buds at a small size. At a less favourable temperature, many buds continue to grow for a long time. Where the tubers are kept at a 'non-growing' temperature until dormancy of all sprouts is lost, and then transferred to a 'growing' temperature, they all start to grow but the smaller sprouts are gradually suppressed by the larger. Similar patterns are found in the main-season British cultivars and the American cultivars 'Kennebec' and 'Sebago'. It is possible, therefore, to arrange pre-determined populations of growing sprouts by manipulation of the storage temperature. If the apically dominant sprout is damaged or removed, all inhibited sprouts resume growth until another becomes dominant. Care must therefore be taken when planting sprouted tubers, because damage may lead to the emergence of a large number of sprouts without apical dominance being re-established and, hence, undesirably intense competition.

Simple relationships exist between the number of sprouts produced, S, and the tuber weight, W. These may be expressed in the form (Bleasdale, 1965)

$$S - bN_0 = (dW)^{k'}, \qquad (8.2)$$

where b and d are constants, $k' = 0.67k$, and k and N_0 have the meaning and values shown in (8.1). The values of the parameters, however, vary greatly with environmental history and conditions, as will be evident from what has been said above. Moreover, the numbers of sprouts which emerge after planting are more closely related to the numbers of sprouts which are growing at planting than to the total number of sprouts present. Satisfactory relationships which allow these to be predicted as a function of tuber size and storage history have yet to be established.

Chemical changes during sprout growth

The growing sprouts produce a stimulus which leads to the breakdown of starch and proteins in the general mass of the tuber. This stimulus is almost certainly hormonal, is rapidly transmitted throughout the whole tuber via the network of internal phloem, depends on the presence of the growing sprout, and can be replaced by gibberellic acid (Edelman and Singh, 1966; Edelman *et al.* 1969; Morris, 1966, 1967*a*). Many other chemical changes occurring during this time are documented by Emilsson and Lindblom (1963). The breakdown of starch appears to take place through sucrose phosphate synthetase, leading to the production of sucrose (Pressey, 1970) and its hydrolysis to glucose and fructose by invertase (Pressey and Shaw, 1966; Moll, 1968). The concentrations of invertase and reducing sugars increase during sprouting, and at low temperatures (< 5 °C), and the activity of an inhibitor of invertase decreases. It also seems that the synthesis of invertase is regulated in part by the concentration of sucrose (Moll, 1970). As do Hardy and Norton (1968), Moll suggests that the hexoses are stored in separate compartments, possibly the vacuoles, and as the concentration in these rises there is a feed-back effect inhibiting invertase synthesis. All these changes are reversed on transfer to high temperature (say, 18 °C) and can be manipulated in the direction required simply by transfer to the required temperature. They are almost certainly involved in the 'reconditioning' of stored tubers for processing. The amyloplast membranes also disintegrate when tubers are stored at 4 °C, even if storage is in nitrogen or 5 % carbon dioxide in air, but not when stored at 25 °C; changes in lipid content and protein composition of the membranes have also been recorded (Ohad *et al.* 1971).

Rate of sprout growth

Although the number of growing sprouts per tuber is almost certainly controlled by the apical-dominance mechanism described above, the rate at which the sprouts grow seems to be determined by the rate of supply of some substrate – at least, at temperatures above 10 °C. At any given temperature, the rate of increase of dry weight of sprouts on single-sprout tubers is constant over a long period and is proportional to tuber size; with multi-sprout tubers, it is an inverse function of the number of growing sprouts per tuber

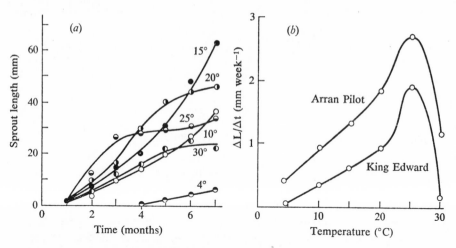

Figure 8.2. Growth of the apical bud in relation to temperature. (*a*) Changes with time in 'Arran Pilot' at different temperatures. (*b*) Variation in the initial rate of elongation with time in the early cultivar 'Arran Pilot' and the main-season cultivar 'King Edward' (Sadler, 1961).

as well as being positively related to the tuber weight and is independent of the spacing between sprouts (Headford, 1962; Morris, 1966). The limiting substrate does not appear to be carbohydrate as large amounts of starch accumulate in the sprouts themselves and the concentration of sugars in the tubers increases with time (mainly sucrose at temperatures above and hexoses at temperatures below 5 °C, as discussed above). The changes in nitrogen compounds are less well documented and sprout growth may well be restricted by the rate of breakdown of proteins. Some evidence suggests that the only amino acid which moves freely from the tuber to the sprout is proline; this accumulates in dark-grown sprouts but disappears when they are exposed to light (Breyhan *et al.* 1962). The predominance of proline among the free amino acids suggests, by analogy with other systems, that this is a 'stress' situation; there is no doubt that the whole system is retarded, directly or indirectly, by shortage of water as evidenced by the rapid increase in growth rate following its supply (cf. Figure 8.5).

Elongation growth is curvilinear with time, initially being fastest at about 25 °C (Figure 8.2). Gradually, a point is reached where the main axis elongates no further due to death of the apical meristem. The time to this 'tip-death' is an inverse function of the rate of growth of the sprout and appears to be due to calcium deficiency associated with the low concentrations in the tubers and difficulties of calcium transport to the sprouts (Baker and Moorby, 1969). Complete absence of light induces etiolation but differences in intensity and duration of visible radiation above daily values of about 0.5 W m^{-2}

have little effect (Headford, 1962). The application of gibberellic acid also leads to sprouts of an etiolated nature but the effect of darkness does not seem to be associated with endogenous gibberellins (Morris, 1967 *b*).

During growth in storage, the sprouts continue to differentiate new primordial leaves, axillary buds, diageotropic stolons, root primordia and, eventually, the basal buds grow out as negatively geotropic branches. Flowers may also be initiated in well-developed sprouts. The length of the sprout is a good index of its degree of development; for example, in cv. 'Arran Pilot', about one stolon is differentiated for every 5 mm length, the mean length of stolons is about half that of the sprout, and the dry weight of the sprout, W mg, is related to its length, L mm, irrespective of cultivar or temperature by $W \approx 0.22 \, L^{2.2}$ (Sadler, 1961).

It is possible, therefore, by manipulating the tuber size, storage temperature and duration of storage to control the number, size and degree of differentiation of the sprouts at the time of planting. As will be discussed later, these have a large influence on the subsequent performance of the crop. Despite much speculation on the effect of the environment in which the seed tubers were grown on subsequent performance, this seems to be negligible and any initial differences can readily be eliminated by suitable storage treatments (Goodwin *et al.* 1969 *a, b*).

Storage for subsequent consumption

We are primarily concerned here with crop production and the above has been written in that context. However, most storage practice is aimed at the prevention of sprouting and the maintenance of the tubers in a condition as close as possible to that at harvest. The issues and procedures involved are discussed fully by Burton (1966). Briefly, the tubers are usually stored at temperatures of 8–10 °C, where respiration is near minimum (Figure 8.3). Temperatures below 5 °C lead to high concentrations of hexose sugars which are undesirable and also result in higher respiration losses. Respiration rate also changes with time, being about 9 mg CO_2 (kg fresh weight)$^{-1}$ h^{-1} at harvest and falling to about one-third of this value during the first month of storage; it increases with sprout growth, the total respiratory loss being approximately equal to the dry weight of sprouts produced. Because the temperatures used for storage allow reasonable rates of sprout growth, use is made of various suppressants such as isopropyl-N-phenylcarbamate (van Vliet and Sparenberg, 1970). Ventilation is usually at a rate adequate to prevent condensation but not sufficient to cause excessive losses of water.

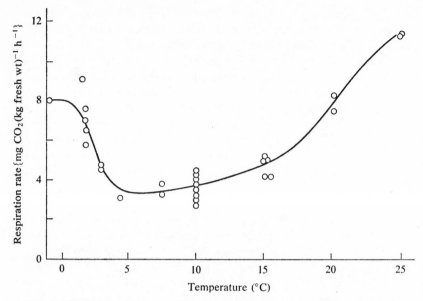

Figure 8.3. Respiration rates of potato tubers stored at various temperatures. The points relate to a number of varieties and investigations (Burton, 1966).

GROWTH IN THE FIELD

The season of the year chosen for the final, prominent and productive phase of growth depends on the climates. Two regions may be distinguished – the cool-temperate regions in which one crop only per annum is possible, and warmer areas in which two crops may be arranged (Figure 8.4). This figure was constructed from data of two specific experiments and climatic parameters of the two regions; nevertheless, it summarizes to a surprising degree some of the main features of potato growth. (It is suggested that the reader pause, compare the various curves, and make his interpretations; the general themes illustrated will emerge during the progress of this chapter.) Within each crop, three phases may be recognized: (i) that from planting to the establishment of a leaf surface of some 200–300 cm² per plant, when the plant depends on substrates from the mother tuber; (ii) the first period of autotrophic growth, in which production of the haulm is predominant; and (iii) the period of tuber growth, during which the haulm gradually senesces and dies.

   In the cool-temperate regions, the length of the growing season is usually limited by temperature, planting being determined by temperatures adequate for growth and the need to avoid late frosts whereas the end of the season is determined by the need to escape early frosts and the ravages of *Phytophthora* blight. In the warmer regions, the same considerations apply, the period

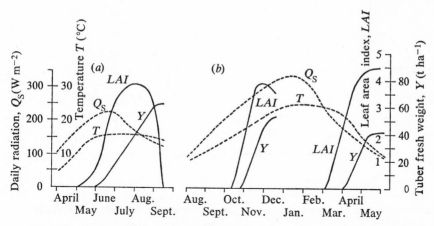

Figure 8.4. Changes of the main components of a potato crop during its growing season in (*a*) a region allowing one summer crop only (Midlands, England), and (*b*) a region with spring and autumn crops (M.I.A., Australia). Data in (*a*) due mainly to Ivins and Bremner (1964) and in (*b*) to Sale (1973).

between the two crops usually being one of escape from high temperatures which retard tuberization and from high rates of evaporation. Within these environmental restraints, the grower also adapts to marketing considerations – especially to meeting periods of shortages with high prices and avoiding periods of excess production.

The pre-emergence phase of growth

As explained above, it is possible to prepare for planting tubers bearing sprouts over a wide array of developmental states. These may range from tubers with all buds still dormant to those with one or more large well-developed growing sprouts and a number of others correlatively inhibited at different stages. On planting, those which have stopped growing do not recommence growth but the rate of growth of the growing sprouts increases rapidly (Figure 8.5) and the rate of respiration per tuber increases eight- to ten-fold. Apical dominance does not appear to be significant at this stage; all sprouts growing at the time of planting finally emerge, irrespective of environmental conditions, and indeed some lateral branches may also emerge. The number of above-ground stems is, therefore, usually greater than the number of growing sprouts but may be less than the total number of sprouts on the planted tuber. (In some situations, such as the side of a ridge, a diageotropic stolon may also emerge; exposure to light results in accelerated growth of the primordial leaves of its terminal bud and it changes to a normal negatively geotropic stem. Stolons on advanced sprouts may also develop in this way during storage, especially under high light intensities.)

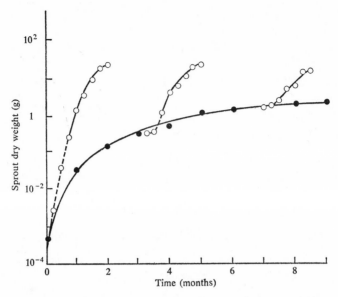

Figure 8.5. Growth of sprouts of cv. 'King Edward' in storage (closed circles) and following planting in the field (open circles). The broken lines indicate pre-emergence growth (Headford, 1962).

The number of stems appearing above the ground, as well as the rate of emergence, is increased by increasing the concentration of mineral ions in the soil solution. The rate of absorption of mineral nutrients during even the earliest stages of emergence growth appears to be significant (Moorby, 1967, 1968; Morris, 1967a), although there is often a loss from the whole system (cf. Figure 8.6) amounting to 20% of the potassium and 50% of the phosphorus from the plant as a whole during the first 10–15 days following planting (Carpenter, 1957; Moorby, 1968). This represents some kind of poorly understood leaching mechanism; nevertheless, at the same time nutrients are being absorbed by the sprouts from the soil. Eventually, all but about 20% of the content of nitrogen, phosphorus and potassium in the mother tuber are lost or transported from the mother tuber, the net transfer occurring at an exponentially declining rate. Similarly, about 80% of the dry weight of the tuber is transferred, leaving little except cell walls and membranes enclosing dilute solutions of sugars – but also some active enzyme systems.

Marked competition between sprouts for water and nutrients may occur at this time, irrespective of whether the sprouts are on the same or separate tubers. The number of shoots emerging above the ground, S, and the mean weight, W, of these shoots are related to the population per unit area, D, by the usual competition functions

$$S = aD^b; \quad W = a'D^{-b}, \tag{8.3}$$

where *a* and *b* vary with variety and $0 < b < 1$ over the range from full to no competition (Moorby, 1967). Such relationships hold where all growing sprouts are of the same size. More usually, a range of sizes may be expected; in which case the larger sprouts emerge first and tend to dominate the smaller sprouts, differences between the largest and the smallest being accentuated with time. Because of the apical dominance phenomenon in this system, however, the range in sizes between growing sprouts is usually not great and (8.3) often holds.

The rate of emergence, i.e. the reciprocal of the time from planting to appearance above the soil, is increased by increase in soil temperature, soil water (over the range -1.5 to -0.3 kJ kg^{-1}), mineral nutrient concentration, length of longest sprout at planting and shallowness of planting. In one experiment at constant temperatures and with the top of the longest sprout placed 5 cm below the surface (Sadler, 1961), the rate of emergence, G day^{-1}, was linearly related to temperature over the range 0°–25 °C, increasing by 0.0067 day^{-1} °C^{-1}, and to length of the longest sprout, L mm, at 25 °C, by

$$G = 0.167\{1 - 0.82 \exp(-0.065L)\}. \qquad (8.4)$$

That is, the time to emergence varied from 33 days with immeasurably short sprouts to six days with very long sprouts.

Although sprout development prior to planting leads to earlier establishment and subsequent growth, it also encourages two abnormalities which seriously impair later development. One is premature tuberization ('little potato') which is more likely the older the mother tuber, the better-developed the sprouts, and the lower the soil temperature. The other is 'coiled sprout', in which the elongating sprouts lose all geotropic control, form loops and coils, often become fasciated, and frequently grow long distances before emerging. The more developed the sprout the more prone it is to coil, with medium-sized sprouts having a greater tendency at low than higher temperatures (Moorby and McGee, 1966). There are associated effects of depth of planting and degree of soil compaction (Lapwood *et al.* 1967) and infection by *Verticillium nubilum* (Pitt *et al.* 1965). Dr A. H. Catchpole has shown that growing sprouts produce ethylene and that exposure to ethylene induces incipient coiling. Both of these conditions result from useful procedures being applied over-enthusiastically and can be minimized by careful husbandry (Cox, 1970).

Haulm, root and stolon growth

Four types of shoots, depending on their origin relative to the mother tuber, may be recognized: (i) Main stems arising from first- and higher-order buds differentiated during growth of the mother tuber; (ii) branches arising from below-ground nodes on these main stems; (iii) slender leafy branches developing infrequently from stolons; and (iv) above-ground axillary branches. Types (i) and (ii) develop a virtually independent existence soon after emergence and form the main framework of the plant (Milthorpe, 1963).

As 12–15 leaves are present in an undeveloped bud (about 26 on a sprout 2 cm long) and possibly 40 is the maximum produced on any one axis, differentiation of new leaves from apical meristems plays a less important part than in species grown from seed. The development of the leaf surface depends mainly on the expansion of leaves already present and on the production of axillary branches. As these commence growth soon after emergence, expansion of the leaf surface is relatively more rapid than in many other species. The minimum and optimum temperatures for leaf expansion are about 7 °C and 20 °C, respectively, whereas the optimum for stem elongation and branch production is about 25 °C (Borah and Milthorpe, 1962; Bodlaender, 1963). Haulm growth is hastened by high amounts of radiation and is due primarily to effects on axillary branch formation. Adequate quantitative functions describing these relationships have yet to be established, the overall effects being further complicated by a marked inverse relationship between the rate of stem elongation and intensity of radiation. This is even more distinct under a source of visible radiation other than normal daylight and appears to involve a highly developed phytochrome response. (Controlled environment facilities require supplementary infra-red radiation, for example, before potatoes can be grown satisfactorily.)

Branch production is generally inversely related to the rate of stem elongation but, superimposed on this effect, is a marked stimulation of branching with increase in radiation. There also appear to be differences in the capacity for growth between buds at different positions, fewest branches developing from the middle region of the stem (Lovell and Booth, 1969).

Increase in nitrogen supply particularly stimulates branch production and increases the areas of individual leaves but hastens senescence of the lower leaves (Watson, 1963). The increase in cell number per leaf ascribed to mineral nutrients by Humphries and French (1963) is surprising since increased cell expansion only is the usual response. The overall effect of supplying increasing amounts of fertilizer is to increase leaf area, leaf area duration, rate of haulm growth and the time to tuber initiation, with phosphorus and potassium having smaller and less persistent effects than nitrogen (cf. Figure 8.8).

Although there has been little work directly on the effect of water deficits on haulm growth, the plants appear to respond in much the same way as other crop plants, with a decrease in growth arising from reduced photosynthesis due to increased stomatal resistance. The stomatal resistance appears to follow much the same pattern as that of most C_3 plants, showing no distinctive feature such as the marked sensitivity of sorghum or the insensitivity of lucerne to light. Observations in England showed variations from 0.13 to 10 s cm^{-1} with only small differences through the canopy (Burrows, 1969).

Stolons are differentiated early in the growth of a sprout and several are usually present at planting. These extend rapidly after planting and new main stolons grow out from the underground nodes. The number of stolon-bearing nodes per sprout is usually an inverse function of the number of growing sprouts per tuber and is positively related to the tuber weight. The numbers of primary, and also of secondary, stolon axes formed and their lengths are not greatly influenced by radiation but are greater at low than at high temperatures and are increased by spacing and increase in nitrogen supply. Growth of the individual stolons appears to be rather complex; they grow at different rates and these rates relative to each other change with time (Lovell and Booth, 1969). In this respect, they resemble the growth of individual tubers (p. 239).

Very little is known about root growth. Observations based on the absorption of ^{32}P injected into the soil indicated that the roots of well-established plants are well distributed both between and below the ridges in which they are planted. Post-planting cultivations are often excessive in some traditional practices and lead to lower yields than are obtained from cultivation-free techniques; however, it has not been possible to ascribe this to root disturbance (Newbould *et al.* 1968).

Tuber initiation

Tubers may be initiated by plants over a very wide range of developmental states, varying from very early in plants from tubers with well-developed sprouts to late in plants with excessive haulm growth. (Indeed, in more leisurely by-gone times, the fortunate in Europe often produced 'new' potatoes for Christmas from the former; and Claver (1971) has produced four generations of tubers without the intercession of foliage. Now, the formation of 'little potato' in storage or before haulm emergence is accepted as an unfortunate abnormality, seriously impairing the yield potential.) Possibly, by analogy with flowering, the older the plant the more prone it is to form tubers. Tubers usually form at the stolon tips indicating that the reactions in these tissues are more biassed towards tuber formation – but this is not invariable as sometimes small aerial tubers form from axillary buds or directly on sprouts.

Superimposed on the effect of age, the changes which occur in the stolon tip giving rise to tubers are hastened by the environmental conditions of short days, low temperature, high daily radiation and low mineral nutrient supply (Slater, 1963, 1968). There would seem to be two sets of reactions favouring tuberization – one hormonal and associated with photoperiod, the other of a substrate nature interpretable in the sense that all conditions which increase the concentration of photosynthate in the stolon tips (high production and low use in growth by other parts) promote tuber formation.

In *tuberosum* cultivars, there is no strict requirement for short days but there is a substantial variation in response between them (Bodlaender, 1963; Krug, 1963). Generally, early varieties show less response to photoperiod than late-season varieties; with both, the longer the photoperiod the longer the time from emergence to tuber initiation but with increasing day-lengths late varieties require progressively longer times than do early varieties. This facultative response is valuable as higher yields are obtained the greater the haulm growth at the time of tuber initiation.

The response to day-length is usually greatly modified by other factors, especially fertilizer supply. High levels of fertilizer, especially nitrogen, which encourage extensive haulm growth greatly delay tuber formation. If other seasonal factors allow, such crops will eventually give the highest yields, but often they do not. The greater the degree of sprout development prior to planting the shorter the time between planting and tuber initiation, due mainly to the shorter time to emergence. There may also be a small effect associated with the age of the sprout *per se*. Less marked effects arise from low temperatures: the lower the temperature the more rapid is tuber initiation and the more tubers formed; raising plants in a greenhouse to a reasonable size and then transplanting to the lower temperatures of the field also leads to rapid tuber initiation (Burt, 1964*a*, 1965).

The metabolic changes, which occur in the pith and cortical cells of the expanding internodes of the stolon and cause them to grow radially and induce secondary cell division, still remain virtually unknown (Claver, 1964; Gregory, 1964; Madec, 1963, 1964; Slater, 1963). Much discussion has centred around the formation of tuber-promoting substances in the mother tuber; it seems more likely that the responses represent changes in the sprout and its associated stolons as the tendency to form tubers persists when the sprouts are excised and grown aseptically (Bodlaender and Marinus, 1969). There appears to be an accumulation of sugars in apices immediately prior to tuber initiation (Slater, 1968) and Lovell and Booth (1967) maintain that starch synthesis in the stolons is the first detectable sign of tuberization, occurring before there is any histological evidence.

The application of gibberellic acid to the haulm retards tuberization, whereas growth inhibitors such as (2-chloroethyl)-trimethylammonium chloride (CCC) and *N*-dimethylaminosuccinamic acid (B9) stimulate tuberization (Gifford and Moorby, 1967; Humphries and Dyson, 1967), but these could be indirect effects arising from effects on haulm growth. We have followed the concentrations of gibberellins (GA) in stolon tips, finding high concentrations prior to and at the earliest sign of tuberization, and very low values soon after. Smith and Rappaport (1969) find much the same. This suggests that gibberellins are not involved in the initial stimulus but may be associated during radial growth. Smith and Rappaport invoked an inhibitor which was not abscisic acid (ABA) and, unlike El-Antably, *et al.* (1967), were unable to find any effect of ABA on the tuberization of whole plants or cultured stolons. McCorquodale (1971), using aseptically cultured stolon apices, found that tubers were initiated in media with very low concentrations of GA, there being no effect of CCC, B9, ABA, auxins or kinins (but see Palmer and Smith, 1969). Ethylene has also been shown to promote tuberization (Catchpole and Hillman, 1969; Garcia-Torres and Gomez-Campo, 1972) and changes in nucleic acid fractions in stolon apices and leaves have been reported (Oslund and Li, 1972).

It is clear, therefore, that there is as yet no satisfactory physiological explanation of tuberization. McCorquodale (1971) suggested that the continued synthesis of GA in the shoots and/or roots and its transport to the stolons maintains the elongation growth of stolons and, if this decreases, tubers form. However, our findings mentioned above indicate that GA concentrations in tubers are determined locally and are not involved in the initial stimulus.

Tuber growth

Detailed descriptions of the anatomy, growth and compositional changes of the tuber have been given by Artschwager (1924), Plaisted (1957), Booth (1963) and Reeve *et al.* (1969) and will not be discussed further here.

Most of the tubers which grow to harvestable size are formed within a period of two weeks. However, stolons often continue to grow and branch after this time and new tubers are being continually formed and resorbed. There appears never to be any shortage of sites for tuber formation and the main issues concern the growth of these early-formed tubers.

Soon after tuber initiation there is a two- to three-fold increase in the assimilation of $^{14}CO_2$ and the proportion of assimilates exported from the leaves is doubled, most of this going to the tubers (Moorby, 1968). When tubers are harvested some 24 h after exposing plants to $^{14}CO_2$, there is very little correlation between the ^{14}C-content and fresh weights of the individual tubers (Lovell and Booth, 1967; Moorby, 1968); those with the higher contents also have a much higher proportion of the labelled carbon in the ethanol-insoluble fraction, presumably starch. The patterns fluctuate in such a way as to suggest that not all the larger tubers grow at the same time and the largest at any one time does not necessarily have the highest growth rate at that time (Moorby, 1967, 1968). There is a far better correlation between ^{14}C-content and tuber weight if the tubers are sampled a few weeks after exposure to $^{14}CO_2$. About 50 % of the labelled photosynthate is exported from the leaves within 24 hours. The remainder is converted into ethanol-insoluble compounds, mainly starch, and these are broken down and transported during the following weeks (Moorby, 1970). The half-time of this slower turnover is about 3.5 weeks.

There is no satisfactory explanation for these variations in growth rate between tubers. Although particular groups of leaves may supply particular tubers, there is no good evidence of this nor any reason why their outputs should fluctuate violently. It is conceivable that transport may impose some limitation. Taking the data on which Figure 8.4*b* was based, there was a maximum tuber growth rate of $25 \text{ g m}^{-2} \text{ day}^{-1}$ and 60 tubers m^{-2}. Our observations suggest a mean cross-sectional area of phloem per stolon of 1.76 mm^2 and, allowing 30 % for respiration, this gives a mean rate of transport into each tuber over the 24 hours of $1.3 \text{ g (cm}^2 \text{ phloem)}^{-1} \text{ h}^{-1}$. This is about one-third of the usually accepted figure (Canny, 1971). Transport should not prevent all tubers from growing simultaneously, therefore, but if only one-third grow at any one time for other reasons it would be approaching the maximum for those tubers. Diurnal fluctuations in transport are small (Baker and Moorby, 1969).

Rhythmical diurnal changes in fresh weight and volume of a growing tuber show there is considerable movement of water out of the tuber in response to transpiration during the midday hours, which is made good at night. This movement interrupts the flow of ions through the xylem to the tubers but flow through the phloem appears to be uninterrupted (Baker and Moorby, 1969). A considerable proportion of the nitrogen, phosphorus and potassium reaches the tubers via the leaves through phloem transport, either within a few hours of absorption or more slowly as a consequence of senescence. Calcium, however, can move only in the xylem and enters the tubers mainly during the dark periods. This may be one explanation for the much lower concentrations of calcium in tubers than in the leaves.

The rate of growth of the tubers as a whole – the rate of 'bulking' – is exponential for the first two to three weeks but then becomes almost linear. Soon after assumption of this constant rate of bulking, the rate of axillary-branch production decreases and this leads gradually to a decrease in the rate of production of new leaves (Milthorpe, 1963). There is also a gradual increase in the rate of senescence of the older leaves, a substantial net migration of nitrogen, phosphorus, and potassium from the haulm to the tubers (Figure 8.6), and a decrease in the dry weight of the haulm. At this time, the rate of increase of dry weight of the tubers often exceeds that of the total dry weight made by the plants; this transfer of previously assimilated material rarely amounts to more than about 25 % of that moving at any one time and is about 10 % of the total dry weight of the tubers. The senescence of the haulm accelerates as the tuber bulking increases and eventually becomes complete (Figure 8.4*a*). Frequently, however, the haulm is artificially destroyed before this time.

Once the bulking rate enters its constant phase it appears to be very

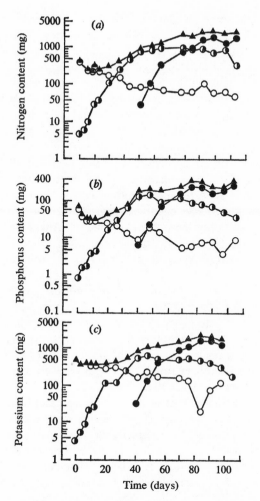

Figure 8.6. Changes in content of the main mineral nutrients in different parts of potato plants grown in perlite supplied with Long Ashton nutrient solution from planting. Triangles denote the total system, open circles the mother tuber, closed circles new tubers, and pied circles the remainder of the plant (Moorby, 1968).

insensitive to short-term fluctuations of temperature and radiation. It may continue in this constant phase for many weeks and only start to decrease when the leaf area index has decreased to about one (cf. Figure 8.8 and most bulking curves; also Dyson and Watson (1971) for contrary evidence). Very high temperatures (> 30 °C) or water deficits do inhibit tuber growth, often without influencing stolon growth, so that tuber buds sometimes grow out as stolons (Lugt *et al.* 1964). When the stress is relieved, the new stolons initiate tubers ('chain tuberization'); if no stolons have grown out, regions of the old tuber underlying some of the buds may resume growth producing 'knobby' tubers. In both situations, there is no resumption of starch synthesis by the non-growing part (or whole) of the primary tuber. A water deficit (in the leaves) of about -500 J kg^{-1} will induce this condition (Sabalvoro, 1965; McCorquodale, 1966). There appear to be no changes in the activities of the enzymes associated with starch synthesis (cf. Figure 8.7), but there is increased stomatal resistance and a reduced rate of photosynthesis. The primary effect appears to be a reduction in the supply of sugars to the tubers or some change associated with this – further evidence, perhaps, that the 'message' for tuberization coming from the leaves is closely tied to sucrose transport. A similar effect is found when plants with growing tubers are sprayed with gibberellic acid: tubers cease to grow and the buds grow out as stolons. The proportion of ^{14}C exported from the leaves is reduced from 60–70% to 55–60% and only 15% of this mobile ^{14}C enters the stolon-tuber system compared with 85–90% in the normal plant (Lovell and Booth, 1967). Which of the several possible explanations for this result holds is unknown.

There has been no work on the reasons why normal tuber growth and starch synthesis are not resumed when the stress is removed. The rate of conversion of assimilate entering the primary tuber is much less than in the growing secondary tuber even though it has to pass through the latter. Sometimes, the starch already present in the primary tuber is broken down and transferred to the secondary tubers, leaving a mass of cells containing a dilute solution of sugars – the so-called 'glassy' tuber.

There is reasonable evidence that sucrose is the sugar translocated and that it enters the tuber through the phloem. There is some evidence that, in the tomato, downward translocation occurs in the external phloem, but it is not known if this is so in the potato. Mrs H. Armstrong has shown that the phloem, which forms two concentric cylinders of tissue in the proximal parts of the stolons, has many anastomoses at the stolon nodes and gradually decreases in cross-sectional area, finally breaking into separate strands as it enters the tuber. There is, therefore, ample opportunity for the sugar to be transferred throughout the whole tuber.

Sucrose appears to enter the tuber cells without being initially hydrolysed (Hardy and Norton, 1968) and is then converted to starch. The biochemical pathways of starch synthesis are not fully substantiated but are probably similar to those shown in Figure 8.7. The activity of sucrose synthetase is low in small tubers but increases rapidly to a peak, declining rapidly at the end of the growing season (Pressey, 1969). The activity of invertase declines throughout most of the growing season but there is an increase in activity of an invertase inhibitor which masks this enzyme. The sucrose-6-phosphate synthetase activity is initially much lower than that of sucrose synthetase but it gradually increases to 100-fold as the tuber grows (Pressey, 1970). It is not

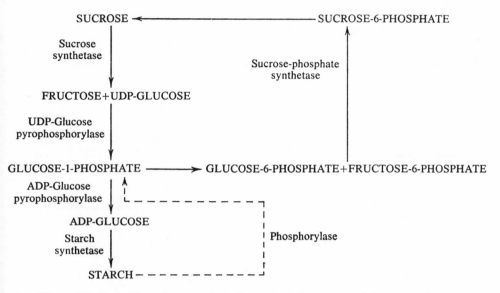

Figure 8.7. Probable pathways of starch synthesis in potato tubers. Note that all reactions are reversible and only the most likely pathways are shown.

yet certain whether any of these enzymes provide 'bottlenecks' in loading starch into the tuber cells. Present evidence would suggest that they do not; that the limitations to growth of any tuber comes rather from the supply of sucrose to it (or from more subtle reactions, associated with sucrose supply, controlling the radial expansion and secondary cell division of the tuber-forming cells).

SPECIAL FEATURES OF CROP GROWTH

Factors determining yield and tuber sizes

The most striking feature of the potato crop is its tremendous plasticity, much of which can be manipulated by the farmer. Crops can be arranged giving yields varying from 15–20 t ha^{-1} in eight weeks to 70–100 t ha^{-1} (*ca* 20 % dry matter) in five to six months. The basic underlying physiological feature is that, with any one cultivar, the smaller the leaf area at the time of tuber initiation then the slower is the rate of bulking and the lower the final yield – but acceptable low yields are achieved earlier in the season (Figure 8.8). Significant marketable yields early in the season, when the market price may be four to five times that holding at the time main-season crops are harvested, are obtained by planting seed tubers with well-advanced sprouts and using little fertilizer, especially nitrogen. If seed tubers are kept in an unsprouted condition until planting and large amounts of nitrogen are

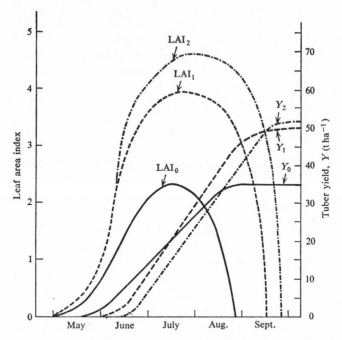

Figure 8.8. Change with time in leaf area and tuber yield of cv. Majestic grown at three rates of nitrogen supply (0, 1, and 2) in England (Ivins and Bremner, 1964).

applied, tuber initiation is delayed until a fairly large leaf area has been established. Such plants usually have a large number of main stems per seed tuber and the number of daughter tubers set per main stem remains fairly constant (for any one variety). The rate of bulking is then relatively high and continues for a much longer time – mainly because a significant leaf area is maintained for a longer time. Care must then be exercised to ensure that other contingencies of the growing season (infection by *Phytophthora* blight, frosts, etc.) allow the potential to be realized.

Compromises are sought between opposing tendencies of various growth processes, especially between the tubers and the haulm. Encouragement of haulm growth retards tuber initiation but once tubers have eventually formed, a large haulm surface maintains tuber growth at a high rate for a long time. The tubers, however, gradually assume ascendancy and retard the production of new leaf surface, eventually leading to the complete senescence of the haulm and thereby to cessation of tuber growth. A further interrelationship, of little significance in tuber production, concerns flower and fruit development. Flowers are initiated by most cultivars but these usually abscind at an early stage – sometimes before opening, sometimes after anthesis, and infrequently fruits are even set and develop – depending mainly on cultivar and partly on environment. This abscission appears to be associated with tuber growth; plant breeders frequently remove the tubers to obtain an adequate production of seed. Such plants produce an excessive growth of haulm.

These opposing interrelationships confound the application of simplified growth analysis (Watson, 1952). In the pre-tuber stage, there is a close linear relationship between crop growth rate and leaf area up to a leaf area of about 5 m² m⁻²; if higher leaf areas are produced by increasing the density of planting or application of nitrogen, interference with light and nutrient supplies occur (Bremner and Taha, 1966; Bremner and Radley, 1966). However, at least in European climates, the other features discussed preclude the use of large quantities of fertilizer, so competition is rarely a serious issue (see below). In the post-tuber phase, the crop growth rate changes from one dominated by the haulm to one consisting solely of tuber growth. The constant rate of bulking usually continues during a period of declining leaf area – until it reaches about 1 m² m⁻²; so over this phase the crop growth rate is negatively related to leaf area and there is an increasing net assimilation rate. The net assimilation rate when plotted over the whole season follows a broad-U shape (Milthorpe, 1963). These features mitigate against the use of these standard parameters in comparing different treatments.

The long period with a constant rate of bulking in any one crop, and the variation of this rate between different crops, deserve further comment. This rate appears to be determined around the time of tuber initiation and is probably some function of the numbers of growing tubers set and the amount of leaf area then supplying them with photosynthate. It has not yet been possible to predict this rate with any certainty. Once established, its apparent insensitivity to fluctuations in light supply and temperature (but not to water) suggests that it controls the rate of photosynthesis. Much evidence supports this contention. Removal of tubers or reducing their growth by lowering the soil temperature leads to lower rates of net assimilation and of photosynthesis (Burt, 1964*b*, 1966; Nösberger and Humphries, 1965) and partial defoliation has no effect on rate of tuber growth (Humphries, 1969). The dominant effect of the tuber 'sink' on the rate of photosynthesis – seen also where tomato scions are grafted on potato stocks (Khan and Sagar, 1969) – is rather surprising, especially as the photosynthetic capacity would be expected to decline with export of mineral ions to the developing tubers.

The numbers of tubers set per stem varies with variety, decreases to some extent with the number of stems per seed tuber, but is little influenced by the number of stems per unit ground area. With increase in stem density, there is a decrease in mean tuber size at harvest (Bremner and Taha, 1966; Bleasdale, 1965; Sharpe and Dent, 1968; and others). Whether this arises from reduced foliage area (conflicting with the evidence given above) or from interference with nutrient supplies is not certain. The mean tuber size often reflects the larger proportion of tubers in the smaller size grades, the number of large tubers per unit area remaining reasonably constant. The relationship between yield of tubers and density appears to follow the usual parabolic function relevant to particular organs (Donald, 1963). With main-season crops, maximum total and 'ware' (commercially desirable size grade) yields are obtained with about 85–90 thousand plants ha^{-1}, and about four times this number of main stems. The desirable commercial grade consists of tubers within the range of about 70–350 g, although they will be marketed usually in a number of size grades within this range.

Quality factors

We are unable in this chapter to discuss the many aspects of quality, i.e. the suitability of a tuber for a particular purpose. Different purposes have different requirements: for cooking, size, texture, colour, flavour and sound-ness are important with variations depending on the particular table use; for processing, which takes an increasing proportion of the crop, other factors such as dry-matter content and concentration of reducing sugars become pre-eminent, again with variations depending on the end-product. Many of these aspects are discussed by Burton (1966). It might be mentioned here, however, that requirements to meet certain standards of quality conflict with the direct aim of simply higher yields; with increasing demand for specific processing purposes, there is also need for the grower to produce tubers of increasingly stringent quality standards. This in turn requires a deeper understanding of the functioning of the crop than is yet available as well as detailed information on tuber biochemistry (Swain *et al.* 1963).

The micro-environment and exchange processes

Three interrelated components are pre-eminent: the temperature influencing the rates of various physiological processes, the water status of the plant influencing rates of leaf expansion and photosynthesis, and the rate of photosynthesis. These are determined by the local climate, arising from the large-scale circulation and the local flux of radiation (Rose, 1966; Monteith, 1973), and by crop parameters, particularly the extent and disposition of the leaves, the leaf resistance to diffusion and the mesophyll resistance in photo-synthesis (Milthorpe and Moorby, 1974). Here, we will follow standard treatments and examine only those aspects peculiar to the potato – the crop parameters. Among these, we should recognize that potatoes are normally grown on ridges, some 40–90 cm apart, and that cover is rarely complete until about the time of tuber initiation (say, height 60–70 cm and leaf area about 3 m^2 m^{-2}).

(i) *Water status*

The water status of a crop at any instant depends on the rate of evaporation and the rate of supply of water to the plant, the latter being mainly a function of the depth and density of roots and the soil water content. Following a widely used approximate procedure (Penman, 1948; Monteith, 1965; Cowan and Milthorpe, 1968) we may write

$$E = \frac{\epsilon(Q_N - G)/\lambda + \rho(q' - q)/R_a}{\epsilon + 1 + R_l/R_a},$$ (8.5)

where E is the evaporation from a homogeneous extended surface, ϵ the increase of latent heat content with increase of sensible heat content of saturated air at the ambient temperature, Q_N the net radiation absorbed by the surface, G the net heat flux to the soil, λ the latent heat of evaporation of water, ρ the density of air, q' and q the saturation and actual specific humidities respectively of the ambient air, R_a an atmospheric diffusion resistance and R_l a surface diffusion resistance.

Working at a level which gives acceptable estimates when averaged over a week, and taking Q_s as measured incoming short-wave radiation,

$$Q_N \doteq 0.66 \, Q_s - 2.1 \text{ MJ m}^{-2} \text{ day}^{-1}$$ (8.6)

and $0.28 \geqslant G/Q_N \geqslant 0.10$ from bare soil to full cover (Szeicz *et al.* 1969). Equation (8.5) applies to bare soil as well as to all degrees of cover by potatoes because of compensating effects between differences in reflection and heating of the surface. The atmospheric resistance, R_a, can be calculated as

$$1/R_a = k^2 u/[\ln\{(z-d)/z_0\}]^2 \text{ cm s}^{-1}$$ (8.7)

where k is von Karman's constant ($= 0.4$), u the wind speed at height z, d the zero plane displacement and z_0 the roughness length. With potatoes, d is constant at about 20 cm, z_0 decreases curvilinearly with wind speed from about 9.5 cm at 0.5 m s^{-1} to 4 cm at 4 m s^{-1} and is linearly related to crop height, h, by $z_0 = 0.105 \, h$ (Szeicz *et al.* 1969). Values for the surface resistance, R_l, are less certain; for moist bare soil, R_l appears to be about 1.3 s cm^{-1} rising to infinity with the evaporation of the available water in the top 10–15 cm soil (say, with the loss of 20–25 mm water). With vegetation present, irrespective of degree of cover, it seems a reasonable approximation to take $R_l = r_l/L$, where r_l is the leaf resistance and L the leaf area per unit area of soil (LAI) (Monteith, 1965). Some evidence (Szeicz and Long, 1969) indicates that R_l is independent of soil water potential in the top 15 cm, ψ_s, over the range from field capacity to about -400 J kg^{-1} and then increases linearly with decrease in potential. It would seem to be not unreasonable from the evidence of Burrows (1969) to take

$$r_1 = 1 \text{ s cm}^{-1}, \quad \text{where } \psi_s > -400 \text{ J kg}^{-1},$$

$$r_1 = -0.02(\psi_s + 400), \quad -400 > \psi_s > -2000 \text{ J kg}^{-1} \qquad (8.8)$$

and adjust for L to obtain R_1.

At a more approximate level – but probably adequate for forecasting irrigation need in a climate where the mean daily evaporation does not exceed 3–4 mm d^{-1} – constant values for R_a and R_1 may be used in (8.5), together with the arbitrary rule that the calculated rate applies until the soil water deficit reaches about 25 mm; it then falls sharply to zero (cf. Penman, 1963). This gives an 'adjusted transpiration rate'.

Treatments at a more detailed level – in the context of diurnal changes and a full appreciation of the exchange processes – are still inadequate in respect of the potato crop. The main difficulties reside in the partitioning of radiation fluxes between soil and vegetation, and between latent and sensible heat, during the period when cover is incomplete. An initial investigation (Graetz, 1972) of these exchanges in an analogous system – a vineyard – has indicated that surface and physiological parameters are far more important than is implicit in the above treatment. There is need to extend to these patterned communities the type of treatment developed by Cowan (1968) and Thom (1972) for horizontally uniform canopies in which the disparities in surface transfers of momentum, heat and mass are taken into account.

(ii) *Temperature*

The radiation flux and its partition between latent and sensible heat leads to temperature profiles within and above the crop of the type shown in Figure 8.9. In this well-watered situation with complete cover, canopy air temperatures were as much as 3 °C higher during the day and 1–1.5 °C lower during the night than screen temperatures. During the day the highest temperatures occurred near the soil surface – temperature is usually highest in the leaf layers of greatest density – and at night the lowest temperatures were at the top of the canopy, where the 20–24 h vapour pressure profile indicated dew deposition. (Most dew is formed in the upper half of the canopy and is, of course, a significant feature in development of *Phytophthora* blight.) Observations made in Israel under more intense daily radiation (*ca* 25 MJ m^{-2} d^{-1}) indicated differences between leaf and screen temperatures of 1° with wet soil to 5 °C or more in dry soil (Lomas *et al.* 1972).

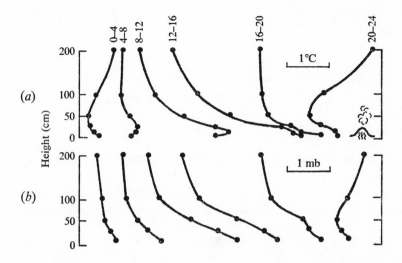

Figure 8.9. (*a*) Mean temperature and (*b*) vapour pressure profiles over 4-hourly periods in and above a potato crop. The mean crop height is indicated (Long and Penman, 1963).

(iii) *Canopy photosynthesis*

Appreciable progress has been made in recent years in delineating the light interception pattern and predicting the photosynthesis of crops with relatively homogeneous horizontal distribution (Eastin *et al.* 1969; Loomis *et al.* 1971; Šetlík, 1970). A number of difficulties arise in applying these approaches to the potato. First, as with total radiation, there is the lack of a suitable treatment to describe the light relations during the period of incomplete cover. During the period of complete cover, although there are grounds for using standard photosynthesis-light response curves for typical temperate C_3 plants, light-extinction coefficients are not well defined. Moreover, 'sink' effects, as described before, appear to be particularly important; procedures for taking these into account are still rudimentary.

Nevertheless, one procedure along these lines has been used successfully with three varieties in a number of seasons and water regimes in the Netherlands by Rijtema and Endrödi (1970). They calculated a potential rate applying to any C_3 crop with complete cover as determined by the visible radiation flux for the particular centre and adjusted this for variation of R_a, of the R_l appropriate to carbon dioxide, and degree of cover. We believe this relationship may be modified by the degree of sprout development at planting; it also requires estimates of the degree of cover which cannot yet be readily predicted. At a more approximate level, Penman (1963) used the adjusted transpiration rate, defined above, to forecast yields of tubers. Satisfactory relationships during the ontogeny of any one crop were obtained but there was an appreciable scatter (about $\pm 30\%$) with different crops in different seasons.

Limitations to yield

Maximum yields of 19 tonnes tuber dry matter ha^{-1} (about 95 t marketable produce ha^{-1}) have been recorded (Milthorpe and Moorby, 1974). This is appreciably greater than the productivity of any cereal crop but does not approach the maximal yield of sugar beet roots grown in an extended season. However, the world yields are only 13.4 t ha^{-1} (about 2.7 t dry matter ha^{-1}) and, even in a reasonably suitable climate like the United Kingdom, the average tuber yield is about 20 t ha^{-1}, although many main-season crops would yield two to three times as much. Probably many reasons contribute to this, including a high-price market early in the season, the need to obtain tubers in a restricted and reasonably uniform size range, and as yet poorly defined, although recognized, requirements for quality. Nevertheless, it seems that there has been less success in selecting new high-yielding varieties of potatoes than of cereals; for example, until very recently, over 70 % of the cultivars grown in the United Kingdom had been introduced at the beginning of this century. It is not easy on current understanding to pin-point the physiological restrictions to yield although, as we have emphasized through-out, the marked mutual opposition of haulm and tuber growth, the determina-tion of tuber set and the factors controlling the growth of individual tubers, which do not seem to include photosynthetic capacity of the haulm, appear to be important.

Towards predicting the responses of a crop

It seems to us that with the potato, as with all crops, we would obtain a much clearer understanding of the above issues, as well as a satisfactory procedure for predicting yields, from the construction of a quantitative simulation model. Current interest in this activity and the degree of success achieved in defining the light relationships, photosynthesis, and water relations (Šetlík, 1970) augur well for accelerating success although many issues have yet to be faced in handling the supply of mineral nutrients and of generating the plant (Milthorpe and Moorby, 1974). The large degree of plasticity shown by the potato makes this a more formidable challenge than for many other crop species. Much of that written above has had this task in mind: it seemed to us that we should first appreciate the major issues and, even though we have not formulated intermediate relationships adequately, a number of these and the degree of variation shown by the parameters can be extracted from the references cited. An excellent beginning has been provided by Raeuber and Engel (1966).

We attach a very cursory and preliminary flow diagram as the basis of one approach towards this goal (Figure 8.10). It does little more than arrange the

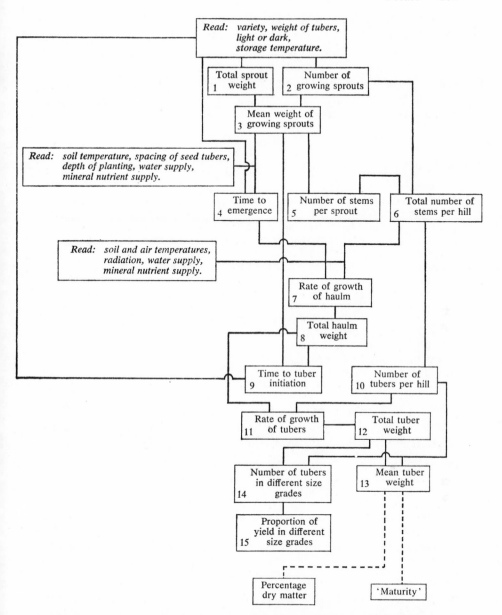

Figure 8.10. Outline of flow diagram of a possible simulation model to describe the growth of a potato crop. The plant characteristics required are written in roman script; required information and intermediate parameters are in italics. Numbers indicate the sequence of operations.

main plant parameters in an order in which they are to be dealt with and leaves unstated the many connecting functions. Much lies hidden in the boxes with italic script, although a great deal of this can be translated from other crops. We believe it is possible for this to be developed to an acceptable degree in the foreseeable future; when this is achieved, we will understand the potato crop much more clearly and a future chapter would need to be only a fraction of the length of this!

REFERENCES

Alvey, N. G. (1965). The effect of temperature on two tuberous *Solanum* species. *Eur. Potato J.* **8**, 1–13.

Artschwager, E. (1924). Studies on the potato tuber, *J. agric. Res.* **27**, 809–835.

Baker, D. A. and Moorby, J. (1969). The transport of sugar, water and ions into developing potato tubers. *Ann. Bot.* **33**, 729–741.

Bleasdale, J. K. A. (1965). Relationships between sett characters and yield in main-crop potatoes. *J. agric. Sci., Camb.* **64**, 361–366.

Bodlaender, K. B. A. (1963). Influence of temperature, radiation and photoperiod on development and yield. *Proc. 10th Easter Sch. agric. Sci. Univ. Nottingham*, 199–210.

Bodlaender, K. B. A. and Marinus, J. (1969). The influence of the mother tuber on growth and tuberization of potatoes. *Neth. J. agric. Sci.* **17**, 300–308.

Booth, A. (1963). The role of growth substances in the development of stolons. *Proc. 10th Easter Sch. agric. Sci. Univ. Nottingham*, 99–113.

Borah, M. N. and Milthorpe, F. L. (1962). Growth of the potato as influenced by temperature. *Indian J. Pl. Physiol.* **5**, 53–72.

Bremner, P. M. and Radley, R. W. (1966). Studies in potato agronomy. II. The effects of variety and time of planting on growth, development and yield. *J. agric. Sci., Camb.* **66**, 253–262.

Bremner, P. M. and Taha, M. A. (1966). Studies in potato agronomy. I. The effects of variety, seed size and time of planting on growth, development and yield. *J. agric. Sci., Camb.* **66**, 241–252.

Breyhan, T., Fischnich, O. and Heilinger, F. (1962). Stoffwechsel- und enwicklungs-physiologische Untersuchungen an Kartoffelknollen und -keimen. *Landbau-forsch. Völkenrode* **12**, 78–80.

Bruinsma, J. and Swart, J. (1970). Estimation of the course of dormancy of potato tubers during growth and storage, with the aid of gibberellic acid. *Potato Res.* **13**, 29–40.

Burrows, F. J. (1969). The diffusive conductivity of sugar beet and potato leaves. *Agric. Meteorol.* **6**, 211–216.

Burt, R. L. (1964a). Influence of short periods of low temperature on tuber initiation in the potato. *Eur. Potato J.* **7**, 197–208.

Burt, R. L. (1964b). Carbohydrate utilization as a factor in plant growth. *Aust. J. biol. Sci.* **17**, 867–877.

Burt, R. L. (1965). The influence of reduced temperatures after emergence on the subsequent growth and development of the potato. *Eur. Potato J.* **8**, 104–114.

Burt, R. L. (1966). Some effects of temperature on carbohydrate utilization and plant growth. *Aust. J. biol. Sci.* **19**, 711–714.

Burton, W. G. (1963). Concepts and mechanism of dormancy. *Proc. 10th Easter Sch. agric. Sci. Univ. Nottingham*, 17–40.

Burton, W. G. (1966). *The potato*. Veenman and Zönen, Wageningen.

Burton, W. G. and Meigh, D. F. (1971). The production of growth-suppressing volatile substances by stored potatoes. *Potato Res.* **14**, 96–101.

Canny, M. J. (1971). Translocation: mechanism and kinetics. *Ann. Rev. Pl. Physiol.* **22**, 237–260.

Carpenter, P. N. (1957). Mineral accumulation in potato plants. *Bull. Me. agric. Exp. Sta.* **562**, 1–22.

Catchpole, A. H. and Hillman, J. (1969). Effect of ethylene on tuber initiation in *Solanum tuberosum* L. *Nature, Lond.* **223**, 1387.

Claver, F. K. (1964). Estudio sobre la tuberizacion de plantas y brotes de papa. *Rev. Fac. Agron. La Plata* **40**, 171–183.

Claver, F. K. (1971). Formacion de cuatro generaciones de tuberculos sin foliaje en '*Solanum tuberosum*' L. *Rev. Fac. Agron. La Plata* **47**, 65–73.

Cowan, I. R. (1968). Mass, heat and momentum exchange between stands of plants and their atmospheric environment. *Quart. J. roy. Meteorol. Soc.* **94**, 523–544.

Cowan, I. R. and Milthorpe, F. L. (1968). Plant factors influencing the water status of plant tissues. In *Water deficits and plant growth*, ed. T. T. Kozlowski. Academic Press, New York, pp. 137–193.

Cox, A. E. (1970). Coiled sprout survey in south-west Cornwall, England, 1968 and 1969. *Potato Res.* **13**, 332–341.

Donald, C. M. (1963). Competition among crop and pasture plants. *Adv. Agron.* **15**, 1–118.

Dyson, P. W. and Watson, D. J. (1971). An analysis of the effects of nutrient supply on the growth of potato crops. *Ann. appl. Biol.* **69**, 47–63.

Eastin, J. D., Haskins, F. A., Sullivan, C. Y. and van Bavel, C. H. M. (1969). *Physiological aspects of crop yield*. Amer. Soc. Agronomy, Madison.

Edelman, J., Jefford, T. G. and Singh, S. P. (1969). Studies on the biochemical processes in the potato tuber: the pathway and control of translocation from the tuber. *Planta* **84**, 48–56.

Edelman, J. and Singh, S. P. (1966). Studies on the biochemical basis of physiological processes in the potato tuber: changes in carbohydrates in the sprouting tuber. *J. exp. Bot.* **17**, 696–702.

El-Antably, H. M. M., Wareing, P. F. and Hillman, J. (1967). Some physiological responses to *d,l*,abscisin (dormin). *Planta, Berl.* **73**, 75–90.

Emilsson, B. and Lindblom, H. (1963). Physiological mechanisms concerned in sprout growth. *Proc. 10th Easter Sch. Agric. Sci. Univ. Nottingham*, 45–62.

Ezeta, F. N. and McCollum, R. E. (1972). Dry-matter production and nutrient uptake and removal by *Solanum andigena* in the Peruvian Andes. *Amer. Potato J.* **49**, 151–163.

Garcia-Torres, L. and Gomez-Campo, C. (1972). Increased tuberization in potatoes by ethrel (2-chloro-ethyl-phosphonic acid). *Potato Res.* **15**, 76–78.

Gifford, R. M. and Moorby, J. (1967). The effect of CCC on the initiation of potato tubers. *Eur. Potato J.* **10**, 235–238.

Goodwin, P. B. (1963). Mechanism and significance of apical dominance in the potato tuber. *Proc. 10th Easter Sch. agric. Sci. Univ. Nottingham*, 63–71.

Goodwin, P. B. (1966). The effect of water on dormancy in the potato. *Eur. Potato J.* **9**, 53–63.

Goodwin, P. B. (1967). The control of branch growth on potato tubers. II. The pattern of sprout growth. *J. exp. Bot.* **18**, 87–99.

Goodwin, P. B., Brown, A., Lennard, J. H. and Milthorpe, F. L. (1969*a*). Effect of centre of production, maturity and storage treatment of seed tubers on the growth of early potatoes. I. Sprout development in storage. *J. agric. Sci., Camb.* **73**, 161–166.

Goodwin, P. B., Brown, A., Lennard, J. H. and Milthorpe, F. L. (1969*b*). Effect of centre of production, maturity and storage treatment of seed tubers on the growth of early potatoes. II. Field growth. *J. agric. Sci., Camb.* **73**, 167–176.

Graetz, R. D. (1972). A micrometeorological study of the contributions of the individual elements of a community of plants to the momentum, heat and vapour fluxes from that community as a whole. Ph.D. thesis, Australian National University.

Gregory, L. E. (1964). Physiology of tuberization in plants. (Tubers and tuberous roots) *Encyl. Pl. Physiol.* **15** (1), 1328–54. Springer Verlag.

Hardy, P. J. and Norton, G. (1968). Sugar absorption and compartmentation in potato tuber slices. *New Phytol.* **67**, 139–143.

Headford, D. W. R. (1962). Sprout development and subsequent plant growth. *Eur. Potato J.* **5**, 14–22.

Holmes, J. C., Lang, R. W. and Singh, A. K. (1970). The effect of five growth regulators on apical dominance in potato seed tubers and on subsequent tuber production. *Potato Res.* **13**, 342–352.

Humphries, E. C. (1969). The dependence of photosynthesis on carbohydrate sinks: current concepts. *Proc. Int. Symp. Tropical Root Crops, Univ. of West Indies* **1**, ii, 34–45.

Humphries, E. C. and Dyson, P. W. (1967). Effect of a growth inhibitor, *N*-dimethyl-aminosuccinamic acid (B9), on potato plants in the field. *Eur. Potato J.* **10**, 116–126.

Humphries, E. C. and French, S. A. W. (1963). The effects of nitrogen, phosphorus, potassium and gibberellic acid on leaf area and cell division in Majestic potato. *Ann. appl. Biol.* **52**, 149–162.

Ivins, J. D. and Bremner, P. M. (1964). Growth, development and yield in the potato. *Outl. Agric.* **4**, 211–217.

Khan, A. and Sagar, G. R. (1969). Alteration of the pattern of distribution of photosynthetic products in the tomato by manipulation of the plant. *Ann. Bot.* **33**, 753–762.

Krug, H. (1963). Zum Einfluss von Temperatur und Tageslichtdauer auf die Entwicklung der Kartoffelpflanze (*Solanum tuberosum* L.) als Grundlage der Ertragsbildung. *Gartenbauwiss.* **28**, 515–564.

Lapwood, D. H., Hide, G. A. and Hirst, J. M. (1967). An effect of soil compaction on the incidence of potato coiled sprout. *Pl. Path.* **16**, 61–63.

Lascarides, D. L. (1967). Shortening the dormant period of spring-grown seed potatoes for midsummer planting. *Eur. Potato J.* **10**, 100–107.

Lomas, J., Schlesinger, E., Zilka, M. and Israeli, A. (1972). The relationship of potato leaf temperatures to air temperature as affected by overhead irrigation, soil moisture and weather. *J. appl. Ecol.* **9**, 107–119.

Long, I. F. and Penman, H. L. (1963). The micro-meteorology of the potato crop. *Proc. 10th Easter Sch. agric. Sci. Univ. Nottingham*, 183–190.

Loomis, R. S., Williams, W. A. and Hall, A. E. (1971). Agricultural productivity. *Ann. Rev. Pl. Physiol.* **22**, 431–468.

Lovell, P. H. and Booth, A. (1967). Effects of gibberellic acid on growth, tuber formation and carbohydrate distribution in *Solanum tuberosum*. *New Phytol.* **66**, 525–537.

Lovell, P. H. and Booth, A. (1969). Stolon initiation and development in *Solanum tuberosum* L. *New Phytol.* **68**, 1175–1185.

Lugt, C., Bodlaender, K. B. A. and Goodijk, G. (1964). Observations on the induction of second-growth in potato tubers. *Eur. Potato J.* 7, 219–227.

McCorquodale, A. J. (1966). Second-growth of potato tubers in relation to water deficit and the movement of photosynthate. M.Sc. thesis, Univ. of Nottingham.

McCorquodale, I. A. (1971). Tuberization of potato stolons in aseptic culture. Ph.D. thesis, Univ. of Nottingham.

Madec, P. (1963). Tuber-forming substances in the potato. *Proc. 10th Easter Sch. agric. Sci. Univ. Nottingham*, 121–130.

Madec, P. (1964). Les développements les plus recents dans le domaine de la physiologie de la pomme de terre. *Proc. 2nd Trienn. Conf. EAPR*, 1963, 36–59.

Madec, P. and Perennec, P. (1969). Levée de dormance de tubercles de pomme de terre d'âge différent: action de la rindite, de la gibberelline et l'oeilletonnage. *Eur. Potato J.* 12, 96–115.

Milthorpe, F. L. (1963). Some aspects of plant growth. *Proc. 10th Easter Sch. agric. Sci. Univ. Nottingham*, 3–16.

Milthorpe, F. L. and Moorby, J. (1974). *An introduction to crop physiology*. Cambridge University Press, Cambridge.

Moll, A. (1968). Die Saccharase im Stoffwechsel der Kartoffelknolle. I. Die Rolle der Saccharase bei der Ausbildung des typischen Zückergehaltes der Knollen. *Flora, Jena* 159, 277–292.

Moll, A. (1970). Die Saccharase in Stoffwechsel der Kartoffelknollen. III, IV. *Biochem. Physiol. Pflanzen* 161, 74–80, 81–90.

Monteith, J. L. (1965). Evaporation and environment. *Symp. Soc. exp. Biol.* 19, 205–234.

Monteith, J. L. (1973). *Principles of environmental physics*. Arnold, London.

Moorby, J. (1967). Inter-stem and inter-tuber competition in potatoes. *Eur. Potato J.* 10, 189–205.

Moorby, J. (1968). The influence of carbohydrates and mineral nutrient supply on the growth of potato tubers. *Ann. Bot.* 32, 57–68.

Moorby, J. (1970). The production, storage and translocation of carbohydrates in developing potato plants. *Ann. Bot.* 34, 297–308.

Moorby, J. and McGee, S. (1966). Coiled sprout in the potato: the effect of various storage and planting conditions. *Ann. appl. Biol.* 54, 159–170.

Morris, D. A. (1966). Intersprout competition in the potato. I. *Eur. Potato J.* 9, 69–85.

Morris, D. A. (1967a). Intersprout competition in the potato. II. *Eur. Potato J.* 10, 296–311.

Morris, D. A. (1967b). The influence of light, gibberellic acid and CCC on sprout growth and mobilization of tuber reserves in the potato (*Solanum tuberosum* L.). *Planta* 77, 224–232.

Newbould, P., Howse, K. R. and Elliott, J. G. (1968). Effect of soil type and post-planting cultivation on the yield of tubers and the zones of soil from which nutrients are absorbed by potatoes. *Agric. Res. Council Letcombe Lab. Ann. Rep.* 19, 51–52.

Nösberger, J. and Humphries, E. C. (1965). The influence of removing tubers on dry-matter production and net assimilation rate of potato plants. *Ann. Bot.* 29, 579–588.

Ohad, I., Friedburg, I., Ne'eman, Z. and Schramm, M. (1971). Biogenesis and degradation of starch. *Pl. Physiol., Lancaster* 47, 465–477.

Oslund, C. R. and Li, P. H. (1972). Metabolism of nucleic acids in potato stolon tips at the onset of tuberization induced by long nights and low temperature. *Potato Res.* **15**, 32–40.

Palmer, C. E. and Smith, O. E. (1969). Cytokinins and tuber initiation in the potato *Solanum tuberosum* L. *Nature, Lond.* **221**, 279–280.

Penman, H. L. (1948). Natural evaporation from open water, bare soil and grass. *Proc. Roy. Soc. A.* **193**, 120–145.

Penman, H. L. (1963). Weather and water in the growth of potatoes. *Proc. 10th Easter Sch. agric. Sci. Univ. Nottingham*, 191–198.

Pitt, D., Hardie, J. L., Hall, T. D. and Graham, D. C. (1965). *Verticillium nubilum* Pethybr. as a cause of the coiled sprout disorder of potatoes. *Pl. Path.* **14**, 14–22.

Plaisted, P. H. (1957). Growth of the potato tuber. *Pl. Physiol., Lancaster* **32**, 445–453.

Pressey, R. (1969). Potato sucrose synthetase: purification, properties, and changes in activity associated with maturation. *Pl. Physiol., Lancaster* **44**, 759–764.

Pressey, R. (1970). Changes in sucrose synthetase and sucrose phosphate synthetase activities during storage of potatoes. *Amer. Potato J.* **47**, 245–251.

Pressey, R. and Shaw, R. (1966). Effect of temperature on invertase, invertase inhibitor, and sugars in potato tubers. *Pl. Physiol. Lancaster* **41**, 1657–1661.

Raeuber, A. and Engel, K.-H. (1966). Untersuchungen über den Verlauf der Massenzunahme bei Kartoffeln (*Sol. tuberosum* L.) in Abhängigkeit von Umwelt- und Erbguteinflüssen. *Abhand. d. Meteorol. Dienstes d. Deut. Demökrat. Repub.* No. **76**.

Rappaport, L. and Wolf, N. (1969). The problem of dormancy in potato tubers and related structures. *Symp. Soc. exp. Biol.* **23**, 219–240.

Reeve, R. M., Hautala, E. and Weaver, M. L. (1969). Anatomy and compositional variations within potatoes. I, II. *Amer. Potato J.* **46**, 361–373, 374–386; cf. also *ibid.* **47**, 148–162.

Rijtema, P. E. and Endrödi, G. (1970). Calculation of production of potatoes. *Neth. J. agric. Sci.* **18**, 26–36.

Rose, C. W. (1966). *Agricultural physics*. Pergamon, Oxford.

Sabalvoro, E. G. (1965). The effect of moisture stress on second-growth of potato tubers. M.Sc. thesis, Univ. of Nottingham.

Sadler, E. (1961). Factors influencing the development of sprouts of the potato. Ph.D. thesis, Univ. of Nottingham.

Sale, P. J. M. (1973). Productivity of vegetable crops in a region of high solar input. I, II. *Aust. J. agric. Res.* **24**, 733–749, 751–762.

Šetlík, I. (1970). *Prediction and measurement of photosynthetic productivity*. Pudoc, Wageningen.

Sharpe, P. R. and Dent, J. B. (1968). The determination and economic analysis of relationships between plant population and yield of main crop potatoes. *J. agric. Sci., Camb.* **70**, 123–129.

Simmonds, N. W. (1962). Variability in crop plants, its use and conservation. *Biol. Rev.* **37**, 442–465.

Slater, J. W. (1963). Mechanisms of tuber initiation. *Proc. 10th Easter Sch. agric. Sci. Univ. Nottingham*, 114–120.

Slater, J. W. (1968). The effect of night temperature on tuber initiation of the potato. *Eur. Potato J.* **11**, 14–22.

Slomnicki, I. and Rylski, I. (1964). Effect of cutting and gibberellin treatment on autumn-grown seed potatoes for spring planting. *Eur. Potato J.* **7**, 184–192.

Smith, O. E. and Rappaport, L. (1969). Gibberellins, inhibitors, and tuber formation in the potato, *Solanum tuberosum*. *Amer. Potato J.* **46**, 185–191.

Swain, T., Hughes, J. C., Linehan, D., Mapson, L. W., Self, R. and Tomalin, A. W. (1963). Biochemical aspects of the quality of potatoes. *Proc. 10th Easter Sch. agric. Sci. Univ. Nottingham*, 160–178.

Szeicz, G., Endrödi, G. and Tajchman, S. (1969). Aerodynamic and surface factors in evaporation. *Water Resources Res.* **5**, 380–394.

Szeicz, G. and Long, I. F. (1969). Surface resistance of crop canopies. *Water Resources Res.* **5**, 622–663.

Thom, A. S. (1972). Momentum, mass and heat exchange of vegetation. *Quart. J. roy. Meteorol. Soc.* **98**, 124–134.

van Vliet, W. F. and Sparenberg, H. (1970). The treatment of potatoes with sprout inhibitors. *Potato Res.* **13**, 223–227.

Watson, D. J. (1952). The physiological basis of variation in yield. *Adv. Agron.* **4**, 101–145.

Watson, D. J. (1963). Some features of crop nutrition. *Proc. 10th Easter Sch. agric. Sci. Univ. Nottingham*, 233–247.

9

Sugar beet

G. W. FICK, R. S. LOOMIS AND W. A. WILLIAMS

Very few crop plants yet possess, after thousands of generations of domestication, combinations of morphological and physiological traits which we would judge to be ideally suited for agriculture. Even with intensive plant breeding, plant improvement will continue to be slow unless rather specific goals can be identified. Donald (1968), Swaminathan (1972), and Bingham (1972) have discussed their concepts of such ideal forms or 'ideotypes' of wheat. But new ideotypes may differ so radically from present forms in behaviour and in their production systems that it is difficult to encourage plant breeders and others to accept them as goals unless practical experience or an adequate means for quantitative prediction is at hand.

Plant physiology serves to explain the basis of plant behaviour and the issue is whether it can also serve to suggest new ideotypes and new systems of production. To do this, we must deal with the integrative problems which ideotypes present; e.g. given three alternative levels for a key enzyme, or morphological trait, what is the consequence to community behaviour? One approach is through whole-plant research but there are only a few examples where information on organ and tissue level physiology has been translated into explanations and predictions of field behaviour (e.g. Ludwig *et al.*'s (1965) study of light adaptation in cotton communities). Such research with whole plants is extremely difficult and we lack the skill and genetic materials for rapid progress. Perhaps also, an important element of the system is missing – a structural hypothesis of the physiology of the community system.

It is our contention that explanatory models of community behaviour can be structured from physiological information and should go hand-in-hand, as an integrative tool, with cellular, tissue and whole-plant physiology. To fulfill this research role, the models should predict field or phytotron performance of whole plants and communities. That is, they should be multi-level, with clear structural and quantitative correspondence to the real biological system; they should deal with the dynamics of development over time; and they should explain how changes occur in the system. The single-level, associative systems models of technology have a role in management, but they fail as scientific research tools for integrative studies of the complex interactions of morphology, physiology, and environment.

[259]

Examples of relatively simple, explanatory models of crop growth include Curry's (Curry, 1971; Curry and Chen, 1971) corn model and Patefield and Austin's (1971) sugar beet model. De Wit's group (de Wit et al. 1970) have proposed a more detailed approach in their Elementary Crop Growth Simulator (ELCROS). Our own SUBGRO model, to be presented here, has its roots in Duncan's (Duncan et al. 1967) photosynthesis model and in ELCROS, and represents a preliminary, integrative hypothesis of sugar beet growth (Fick, 1971).

Model development begins with a study of the real system and our first consideration is a description of the sugar beet crop. The structure of the model and its principal physiological themes are then examined. Finally, we will illustrate how the model synthesizes physiological opinion in a way that permits the testing of complex hypotheses about crop growth.

THE SUGAR BEET CROP

The genus Beta (family Chenopodiaceae) has been a domesticated companion of man for many centuries. Mediterranean cultures extending to at least 3000 B.C. used descendants of the wild Beta maritima for food and medicinal purposes. Through the dark ages it served as a basic vegetable for humans. Despite its long lineage, types with appreciable sweetness were not reported until the 1500s and it was not until nearly 1800 that commercial sugar was produced in Europe from selections of fodder beets (mangel-wurzel). At present the sugar beet, after less than 200 years of development, supplies about 40% of the world's commercial sucrose (Table 1.1).

The Beta vulgaris species is extremely versatile. Leafy chard, table beet, sugar beet and fodder beet are included and are sometimes distinguished as botanical varieties. The sugar beet, as a product of political contingencies, has been the subject of intensive technological development and there is a fair amount known about its physiology. Rather surprisingly, there has been very little study, except by Watson's group (Watson and Baptiste, 1938; Watson and Selman, 1938; Watson and Witts, 1959), of the comparative physiology of the diverse types within the species – a potentially fruitful approach. Although genetic discontinuities obviously exist, much of what is said about one of the types presumably will apply in a general way to the others.

This versatility may, like the wide adaptation of the vulgaris cultivars, relate in part to our use of their vegetative leaves and roots, rather than reproductive parts. Reproduction depends upon a more complex sequential pattern in development (vegetative growth, flowering and fruit set, growth and maturation) and thus is vulnerable at more decision stages to failures or limitations imposed by environment on growth and development.

Both annual and biennial types are found within the species and some closely related wild *Beta*'s are short-lived perennials. The sugar beet has been highly selected for biennial behaviour but is managed as an annual for sugar production. The induction of flowering requires extended exposure to low temperatures (0–10 °C; Stout, 1946) and is hastened if this is followed by long days. Varieties with very strong induction requirements (and hence 'bolting resistant') are used for early plantings and in mild climates where sugar beet may be kept in the field over-winter. Our focus will be on aspects of vegetative growth relevant to sugar production, which precede flowering.

Sugar beet embryos tend to be rather small with seed weights (including testa, embryo and perisperm) seldom exceeding 5 mg (Savitsky, V. and Savitsky, H. 1965). The small size of seed dictates the need for shallow planting and a fine-textured seed bed. It also means that sugar beet is especially subject to problems in emergence such as crusting. Emergence is epigeous with the hypocotyledonary arch serving as the penetrating structure. The photosynthetic area of the expanded cotyledons is small (*ca* 1 to 2 cm²) and foliage leaves are initiated slowly. As a result, seedling development is generally slow until after four to six foliage leaves have been formed, which may be one month after planting even under favourable conditions. The large size of mature sugar beets requires a relatively large space per plant (*ca* 0.1 m²) compared with its initial leaf area.

Aggregate, 'multigerm', fruits with one to several seeds each are common in *Beta*. More recently, 'monogerm' varieties have been developed following the discovery of appropriate germ plasm (Savitsky, V. 1950) and genetic selection for a larger seed size is now possible (this was not so with multigerm lines). But little progress has been made, perhaps because large seeds are associated with thick ovary walls and thus with concomitant problems in hard-seededness and germination inhibitors (Hogaboam and Snyder, 1964). Slower germination offsets any initial size advantage from the large seed.

Watson (1952) pointed out long ago that one could work around the problem of slow seedling growth by early planting of varieties adapted to grow well in cool spring temperatures so that full cover of the ground is achieved before the seasonal peak in insolation. Variability exists among sugar beet lines in their ability to grow in cool climates but it has not been exploited. In fact, Ulrich (1961) has shown that certain disease-resistant lines widely used in the United States, after being developed through mass selections made in a warm climate, were less adapted to cool climates than their disease-susceptible parents.

As the seedling plants develop, the leaves form a rosette with a 5/13 phyllotaxy on a compact, unelongated stem. The first 20 leaves constitute a juvenile group (Bouillenne *et al.* 1940; Loomis and Nevins, 1963) with size of blade and length of petiole increasing with successive leaves. Leaves 15 to 20 are usually the largest in area, and subsequent leaves are somewhat smaller and relatively constant in size and shape. With favourable conditions, the blade may exceed 3 dm² in area and, along with its petiole, 1 m in length (1 to 2 dm² and 0.5 m in length are more common sizes). The length and area of leaves are strongly influenced by variety, number of leaves on the plant, climate, and nutrient supply. Blades are usually dark green in colour, broad, and rounded in shape. Petioles are short and thick in cool climates, whereas in hot climates petioles are long and thin and blades paler green in colour and more elongate in shape (Ulrich, 1956). Larger leaves of intermediate colour and shape are produced in a 'moderate' climate (23 °C day, 17 °C night). Blade expansion after appearance is due mostly to cell enlargement (Terry, 1968) and follows a saturation curve with maximal area being attained in 10 to 20 days. Mature blade size and weight per unit area seem to be determined in part by intraplant competition, since the mature blades retain a capability for expansion up to the onset of senescence and can be released by the removal of other leaves (Morton and Watson, 1948).

After the juvenile phase, new leaves unfold from the single apical meristem at the centre of the rosette at a more or less constant rate characteristic of the variety. Rates of leaf initiation in favourable environments range from two (Humphries and French, 1969) to four or five leaves per week (Loomis and Nevins, 1963). The rate is less with low light, low temperature or nitrogen deficiency. The longevity of leaves after unfolding ranges, inversely with temperature, from five to eleven weeks (Ulrich, 1956). Initially, new leaves appear more rapidly than old leaves die, but in a constant environments an equilibrium number is eventually reached (Ulrich, 1954; Terry, 1968). Ulrich (1954) showed that the sugar beet will continue to form new leaves and associated stem tissue without elongation for several years if floral induction does not occur.

The sugar beet is tap-rooted and secondary vascular meristematic activity in the upper portion of this root and in the hypocotyl and stem lead to the formation of the 'storage beet'. The ontogeny of the cambia is unclear. Artschwager (1926) concluded that they arise in the secondary phloem as it is laid down by a primary cambium, while Hayward (1938) suggested that they arise in proliferated pericycle tissue. Artschwager, and earlier authors, concluded that the cambia, which number four to twelve or more according to the environment, are initiated in quick succession and a root of 0.5 cm may already possess its full complement. The extent of cambial activity in the root declines acropetally producing a tapered storage root. Only limited secondary thickening occurs in lateral roots.

Cambial activity in the root continues throughout the first growing season and is known to extend into a second year when the crop is over-wintered in mild climates. The plants which Ulrich (1954) maintained in a vegetative state for several years made extensive storage root growth during the first year; thereafter the storage 'beets' continued to increase rapidly in size only through the growth of stem tissue. It is not known whether the root cambia have only a limited life or whether internal substrate supplies become limiting. Under field conditions, plant densities are such that each plant achieves much less than its potential root size (0.5 to 1 kg rather than 10 kg fresh wt or more) and hence the potential for cambial activity does not limit storage.

Little is known about the controls on root enlargement; Winter (1954) found that the formation of cambia in decapitated beet plants could be stimulated by applications of auxin. In radish, which enlarges through the activity of a single primary cambium, cambial activity can be initiated with auxin but sugar and cytokinin must also be supplied for the activity to be maintained (Loomis and Torrey, 1964). The transport of both auxin and cytokinin in this system is acropetally polar. Similar controls probably operate in beet.

The anatomy of the upper hypocotyl and the stem are more complex than in the storage root. The cambial rings are joined in a complex anastomosing system with the leaf traces so that all of the leaves are connected with other leaves and with all portions of the root. The expanding leaves represent a strong sink for translocates. A mature leaf normally delivers labelled substrates only to young leaves and portions of the storage root on the same side of the plant (Joy, 1964; Belikov and Kostetskii, 1964). Translocate from old leaves may also enter adjacent young leaves, but a greater proportion of their translocate is directed to adjacent parts of the storage root. However, if the source leaves are removed from one side of the plant, the translocate from the remaining source leaves is spread more uniformly throughout all the expanding leaves and all portions of the storage root. Removal of leaves in selected patterns does not affect the gross morphology of the storage beet.

Sucrose is the principal carbohydrate found in the root and its concentration may exceed 75% of the dry weight and 18% of the fresh weight of storage beet. Topographically, the greatest concentrations of sucrose are associated with the vascular rings while the concentration in the intercambial 'storage' parenchyma generally is 1–2% less. Each vascular cambial ring consists of a series of vascular bundles separated by parenchymatous rays (pericyclic rays according to Hayward, 1938); the distribution of sucrose among the tissues of particular vascular rings has not been studied.

The mean sucrose concentration of the entire root is a very stable parameter (low coefficients of variation) for a particular environment and variety. Sugar beet varieties have sometimes been grouped into 'yield' (large roots and moderate sucrose concentration) or 'sugar' (smaller roots and greater sucrose concentration) types. The sugar types have lower capabilities for growth, and current commercial varieties are largely of the yield type. Rapidly growing plants of this type with adequate supplies of water and nitrogen will have root sucrose concentrations ranging from 10 to 14% fresh wt. Night temperature and nitrogen supply are the dominant environmental factors controlling sucrose concentration with low night temperature and nitrogen deficiency singly or in combination causing greater concentrations (Ulrich, 1955). A simple interpretation has been that growth restrictions allow more photosynthate to accumulate as sucrose in the root. However, there must also be either increases in turgor or decreases in other soluble constituents such that the balance of ψ_{solute} and turgor remains constant (Loomis and Worker, 1963).

Considering an example from our own work, nitrogen deficiency increased sucrose concentration from 12.4 to 16.8 % fresh wt and from 75.0 to 77.8 % dry wt. Relative to the total water content, sucrose molality changed from 0.43 to 0.63 corresponding to about 5 bars difference in ψ_{solute}. If turgor remained constant, 0.20 moles of soluble, non-sucrose materials must have been exported from the cells. It is known that the content of K^+, Na^+, NO_3^-, Cl^- and amino and organic acids decreases sharply with nitrogen deficiency (Houba et al. 1971; Snyder, 1971). Taking the mean osmole weight of such substances in typical ratios as 50 g mole^{-1}, the export of 0.20 moles would account entirely for the slight shift from 75.0 to 77.8 % dry weight. The same assumption of constant osmolality explains the variation in sucrose concentration with nitrogen supply in other data sets (e.g. those of Snyder, 1971) and agrees with direct observations by Goodban and McCready (1968) on the osmolality of sap expressed from high and low nitrogen roots. However, Milford and Watson (1971), from observations that root cell size and water content increased with nitrogen supply, were led to conclude that the effect of nitrogen on sucrose percentage fresh weight was simply due to the change in water content. This implies that turgor increased to offset the increase in sucrose osmolality. A relative consistency of turgor seems more likely, but root cell water relations obviously need more study.

In the harvest of commercial fields, the living leaves are beaten from the thickened spherical stem (crown), or part of the stem is cut away from the rest of the beet. In either case the lower (older) part of the stem is free of leaves because of leaf senescence and is treated as part of the storage beet. Stem tissue generally has lower sucrose concentrations and higher concentrations of soluble non-sucrose materials than does the storage root. Therefore sugar beets have been selected genetically to have a small 'crown' relative to size of root.

Relatively little is known about the fibrous root system of sugar beet. Soil excavations made in the field (Jean and Weaver, 1924; Andrews, 1927) reveal that the fibrous roots reach depths of over 2 m under favourable conditions. But their development is slow and four to six weeks usually elapse after planting before soil moisture extraction is uniform throughout the first metre of soil (Doneen, 1942; Taylor and Haddock, 1956). The fibrous root system will continue to increase in depth through the growing season if water is available at depth. We have had to work with plants grown in pots to learn something about the morphology of the fibrous roots. In vermiculite-nutrient cultures, fibrous root weights increased continuously reaching 14 g plant^{-1} 113 days after emergence, with a length of 286 m g^{-1} and a surface area of 6.48 cm^2 m^{-1} (Fick *et al.* 1972). At that time, the bulk of the absorptive surface area was on quaternary branches with a mean diameter of 1.12 mm. There were about 150 branches m^{-1} of fibrous root length. The kind of rooting medium and the confining influence of the pots used in such experiments probably influences fibrous root development: e.g. Thorne *et al.* (1967) observed a peak of 9 g plant^{-1}. While we know something about the gross morphology of the root system, we have little information about the distribution of physiological activities (water and nutrient uptake for example) within the system.

During the juvenile phase the crop achieves a full leaf canopy and an extensive fibrous root system, but only a slight enlargement of the storage root occurs even though the root cambia are initiated very early. Extensive root enlargement appears to begin when root sucrose concentrations approach 5 to 8 % fresh wt (Ulrich, 1952). The juvenile phase of leaf development and the lag in storage root enlargement can be interpreted as resulting from intraplant competition for substrate, with the growth of leaves and fibrous roots having higher priorities than storage root growth (Terry, 1968; Ulrich, 1952). But this opens intriguing questions of how the fibrous roots maintain a priority over storage roots when the substrate is transported through the storage roots.

The period of vegetative growth with full foliage which follows the juvenile phase is characterized by a stable sucrose concentration in the roots and rapid growth rates. In California, vigorous crops commonly yield more than 60 mt ha^{-1} of fresh tops and 80 mt ha^{-1} of fresh storage beets in a five to six month growing season. Total dry matter (tops plus storage beets) would exceed 20 mt ha^{-1} with 40 to 50 % of that amount being sucrose. Beet and sucrose yields are greater with longer growing seasons: the commercial records in California are 132 mt ha^{-1} of fresh beets (at 13 % sucrose), and 19.1 mt ha^{-1} of sucrose (115 mt ha^{-1} beets at 16.5 % sucrose; dry matter estimated at 26.7 mt in beets and 18 mt in tops in 240 days). For the latter crop, the mean crop growth rate is estimated to have been 18.5 g m^{-2} day^{-1}, and the overall rate of sucrose storage 10.7 g m^{-2} day^{-1}.

The sugar beet crop is usually managed so that the harvest is made during cool, autumn weather after a period of nitrogen deficiency and, perhaps, moisture stress. These conditions favour a greater sucrose concentration and a smaller concentration of soluble non-sucrose materials. This contributes to more efficient processing but generally does not influence sucrose yields, if the stresses are not prolonged. It is common to speak of this as a 'maturation' or 'ripening' period but no equivalent of biological maturation or ripening is involved. In mild winter areas (the warm Mediterranean climates), where sugar beet may be grown as a winter annual for late spring or summer harvest, low night temperatures may not occur and the principal focus then is on nitrogen management.

Since the main vegetative growth phase is indeterminate until floral induction occurs, harvests can be made whenever economically adequate yields are attained. Processing plant capacities, and whether or not temperatures are low enough to allow the beet roots to be accumulated at the factories in storage piles, are also determining factors. In California, temperatures are generally too warm for extended storage (post-harvest respiration and decay losses become excessive) and the harvest schedule is geared rather closely to processing plant capacity. In order to accommodate the entire crop and to extend the plant investment over a long season, some fields must be harvested early at low tonnages, and average yields for the area are well below full-season biological potentials but sufficient for an adequate economic return to the grower. A recent monograph (Johnson *et al.* 1971) provides an extensive treatment of sugar beet management practices.

A SIMULATION MODEL: SUBGRO

Basic approach

The central issue in any model of plant growth is the manner in which photosynthates are allocated to growth, storage and respiration. There are two basic approaches to the problem: (1) to incorporate in the model information on the distribution patterns which have been observed for real plants, or which might hypothetically be observed; and (2) to predict the distribution pattern from information about the physiological processes which determine source–sink relations.

The first approach has been used by Monsi and Murata (1970) to demonstrate, first, that slight variations in distribution can lead over time to very large differences in plant growth, and second, that allocation ratios for selected rice varieties are rather conservative (similar) over a range of environments. In another example, Patefield and Austin (1971) used time-varying allometric equations in constructing their simple explanatory model of sugar beet growth. But the uniformity shown by the rice varieties is not the general case for all rice varieties, nor within other species. As examples, simple distribution ratios for tiller production could not apply over a range of cereal plant densities, nor to both determinate and indeterminate varieties of tomatoes or beans. Thus, single-level models are generally limited to specific problem solutions; Patefield and Austin conclude that explanatory models based on 'the component processes of translocation, cell division and expansion, leaf initiation, and their dependence on the environment' would be much more powerful.

Most of the events of development and differentiation which control partitioning clearly involve genome triggering, frequently through the agency of plant hormones. Thus, Heslop-Harrison (1969) builds his conceptual models of development and differentiation with the hormonal milieu as the principal control. But the capacity of present-day computers and gaps in our information limit the degree of detail in explanatory models. Generally speaking, only about three levels of organization may be incorporated in complex models according to the range of relaxation (transition) times involved (de Wit, 1970). A transition may be mimicked by advancing time in some fraction of the transition time. Thus a physiological event occurring in a time span of 15 minutes (like stomatal closure) might be mimicked by advancing time in the model at one-minute intervals. But this would require 144000 iterations for a 100-day growing season, and a high degree of program detail. The model would likely exceed present computer memory capacities and would also require inordinate amounts of computer time. A one-hour time interval (2400 iterations for 100 days) will mimic diurnal as well as seasonal events with between two and three orders of magnitude range in transition times. Thus, molecular level information about hormones and enzyme induction is less appropriate than tissue or organ level information, if crop behaviour in the field is to be the output; and leaf initiation rates would be used as input rather than hormonal limitations to cell division in the apical meristem.

What this means is that, at our lowest level of organization in a community model, we must frequently depart from an explanatory mode and introduce black-box type associative information (e.g. leaf initiation as function of temperature). Tissues are assigned certain usual capabilities for relative growth rates (RGR, the logarithmic growth rate constant) thus combining aspects of cell division and enlargement. This means, for example, that the qualitative and quantitative effects of hormones are integrated into opinions about the capability of the root tip to grow as a function of age, reserves, nitrogen supply, water or temperature. An alternative which we have used only with the photosynthesis section, is to build sophisticated submodels covering rapid transitition events, and to use general summary tables or relations from these as inputs to the main model.

The need for an associative, integrative nature of many of the inputs would be true regardless of the starting level of the model. Fortunately, it is also theoretically sound when used in multilevel models since associative relations at one level became explanatory at higher levels.

Translating these ideas into a model description, it means that the photosynthate source will be estimated from leaf photosynthesis and that information about the capacity of various sinks will serve as a basic means of determining distribution. Thornley (1972*a*) has approached this by introducing resistances to phloem transport between the leaf source and the stem and between stem and root. This effectively varies substrate concentration in each part so that the allocation priorities are leaves > stems > roots. In ELCROS, Brouwer and de Wit (1969) developed the same set of priorities by varying the input information regarding the growth of each tissue as a function of a single pool of reserves. In either approach, respiration may be calculated in proportion to substrate supply, the amount of growth, and the amount of living tissue. Brouwer and de Wit's method simplifies programming in the model but it confounds the influences of position, translocation and growth as a function of reserves. Our first versions of SUBGRO use an approach similar to that of ELCROS. Sucrose movement in the sugar beet is rapid (*ca* 50 cm h^{-1}, Geiger *et al.* 1969) and does not follow simple gradient patterns between sources and sinks (Mortimer, personal communication). Certain stages of substrate movement apparently involve active transport against gradients (Geiger and Christy, 1971) and portions of the sucrose may be compartmentalized and thus outside the gradients. Whatever the reasons, we do not know enough as yet about the transport and accumulation of sucrose in the beet to employ some other approach, and Thornley's resistance hypothesis appears unsound for sugar beet.

Simple partitioning of reserves by any of the above methods leaves out certain features of root–shoot relations. Root growth is dependent upon photosynthates from the leaves, but leaf growth in turn is dependent upon the amount of fibrous roots and the water and nutrients which they acquire. Thornley (1972b) constructed a root:shoot functional balance from nitrogen and photosynthate supply considerations, while ELCROS (Brouwer and de Wit, 1969) and SUBGRO (Fick, 1971) use a functional balance based on water and photosynthate supply.

Model structure

The SUBGRO model is written in CSMP (IBM, 1967, 1969) which, like other simulation languages (Gordan, 1969), is specifically designed with built-in timing and integration routines to deal with time-varying phenomena. The model has 16 semi-independent sections concerned with major aspects of the system. These are connected by information pathways to allow interaction. The first section translates daily weather station data into hourly microclimate variables as a basis for calculating the current rates of various biological and physical processes. Several sections are devoted to community processes including transpiration, water uptake, photosynthesis and respiration. Another group of growth sections serves to calculate growth rates, weights, sizes, numbers or other attributes of leaves, fibrous roots, storage roots, stored sugar and reserves. These 'state' variables serve to describe the crop at any point in time. Energy content has not been considered (like most herbaceous materials other than oil seeds, sugar beet tissues show little variation in energy content per g dry wt). With initial conditions set for two-leaf stage or older plants, a particular latitude, and day of the year, the program follows daily weather and cycles during each hour from calculating growth rates to updating the weights and surface areas based on these rates. Soil water and nutrient supply are considered to be non-limiting and a soil environment section needs to be developed before problems on the practical management of these factors can be considered.

SUBGRO is one of the longest CSMP programs in existence, testifying to the complexity of biological problems. The size of the model and the number of interactions required has precluded stochastic treatment of input variables; otherwise this could be done easily through use of the Gaussian generator in CSMP. Like real plants, SUBGRO includes in its explanatory mechanisms homeostatic controls which reduce the need for considering variability.

In the model, photosynthates pass through a pool called reserves and are then partitioned to respiration, growth or storage. The partitioning controls involve three descriptions: (1) the basic *growth rate* of the sink as a function of temperature and the amount of tissue capable of growth; (2) the effect of *reserve supply* in limiting growth; and, (3) the effect of *internal water status* of the plant in limiting growth. We shall examine each of these topics in some detail.

Growth rates

The main physiological parameters needed as input for SUBGRO are the potential or maximum possible growth rates for each plant part. The actual hourly growth rates are computed from these values by first making an adjustment for the temperature and second, for the supply of limiting factors.

The maximum possible growth rate can be estimated either on the basis of maximum observed relative growth rates (RGR) or maximum observed absolute growth rates (AGR). Both approaches are used in SUBGRO. Radford (1967) has described in detail the practical and mathematical considerations in measuring RGR, and de Wit *et al.* (1970) discussed the difficulties of estimating the unrestrained RGR for a leaf growth model of corn. RGR expressed as a material flux is the change in weight per unit of weight already present in the growing plant or organ (g g^{-1} day^{-1}). In growth chamber studies, the maximum RGR for sugar beet that we measured, in seedling shoots, was 0.44 g g^{-1} day^{-1} (Fick, 1971) while Terry (1970) reported values up to 0.5 g g^{-1} day^{-1}. RGR is generally positive and declines as the plants develop and the proportion of mature, non-growing tissue increases. However, the high values measured in early growth are probably maintained by the tissue that is actually capable of further growth. Using the approach of de Wit *et al.* (1970) for estimating the weight of tissue capable of growth, the maximum possible growth rate then is the product of the maximum RGR and the weight of the tissue that can grow.

In the present versions of SUBGRO, we have introduced certain AGR limits. Theoretically, the product of RGR as a function of temperature and as constrained by reserves, water, and the amount of tissue capable of maximum growth rate should prove self-limiting. But in the fibrous root section of SUBGRO this did not prove sufficiently accurate. The problem indicates a need for further research on the appropriate maximum RGR for each tissue, on the portion of tissue capable of maximum RGR, on the relative effects of water and reserves in limiting storage and fibrous root growth and perhaps in this case on transport. ELCROS deals with determinate growth and includes a developmental stage control which serves to provide boundary limits. But vegetative growth in the sugar beet is indeterminate. The AGR limits may be viewed as quantitative developmental constraints (such as might be imposed by hormones) which provide organismal and community level (associative) boundaries to the course of the simulations. Such limits are easily derived from the literature for most parts of the sugar beet; data for fibrous roots were not available and so were found from experiments (Fick *et al.* 1972) to be 0.2 g dry matter plant^{-1} day^{-1}.

There are thus two estimates of the maximum possible growth rate: one based on the maximum RGR; and one based on the maximum AGR per plant, per growing unit like a leaf, or per unit of land area. SUBGRO selects the smaller of these at any given time. The maximum growth rate typically increases exponentially according to the RGR model as constrained by resources and environment and then becomes constant at the appropriate maximum AGR.

Perhaps the most sophisticated formulation of the AGR component of SUBGRO is found in the top growth section where the development of individual leaves is modelled (Figure 9.1). The final weight of individual leaves shows considerable variability and is strongly influenced by the environment. Ulrich (1952) showed that plants grown in low light have individual leaves weighing only 0.9 g, while at natural intensities and day-lengths they may weigh 3.3 g. He also showed that plants grown at 2 °C had individual leaves that averaged 1.3 g in dry weight, compared with 2.6 g at 26 °C. SUBGRO incorporates an upper limit on the dry weight of individual leaves, 3 to 3.5 g, that is approached only under the most favourable conditions.

The maximum AGR for individual leaf growth was based on experiments (Loomis and Nevins, 1963; Kelley and Ulrich, 1966; and Fick *et al.* 1971) which showed that sugar beet leaves continue to increase in area under good conditions until just a few days before senescence. At any one time, no more than 10 leaves per plant have been observed to have near maximum growth rates. Rapid growth for a particular leaf lasts no longer than 20 days. SUBGRO sets the maximum number of growing leaves at 20. These are treated in four classes of five leaves each with maximum absolute growth rates, starting with the youngest, of 0.11, 0.11, 0.16, and 0.05 g leaf^{-1} day^{-1}.

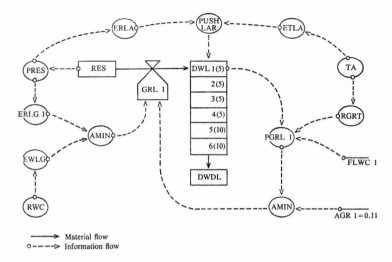

Figure 9.1. Relational diagram of SUBGRO leaf growth as a transfer of material from reserves (RES). The dry weight of the youngest age group (DWL 1) times the fraction of the leaf weight capable of growing (FLWC1) along with the relative growth rate (RGRT) as a function of air temperature (TA) determines the possible growth rate of the youngest leaves (PGRL1). The minimum of PGRL1 and the absolute growth rate (AGR1) is then modified by the more limiting effect of reserves (PRES) on leaf growth (ERLG1) and of relative water content (RWC) on leaf growth (EWLG). The result is the actual growth rate of the youngest leaves (GRL1). A similar process establishes the leaf growth rates for the other age classes (DWL2 to DWL6). Individual leaves advance through the 'push-down table' to a pool of dead leaves (DWDL) as push signals arise from a leaf appearance generator (LAR) controlled by the effect of temperature (ETLA) and the effect of reserves (ERLA) on leaf appearance rate.

With optimum conditions, a leaf could grow rapidly for 21 days, would cease growth in 28.5 days, and reach a maximum weight of 3.07 g.

The amount of simulated time that an individual leaf is in a particular growth rate class is determined by a model for the rate of leaf appearance. Analysis of leaf count data (Ulrich, 1955, 1956; Thorne *et al.* 1967; and Loomis and Nevins, 1963) indicates a ceiling leaf appearance rate of almost 0.7 leaves day^{-1} after 8 to 10 leaves have unfolded (Figure 9.2). The scatter in these data after 10 leaves have appeared is removed by assuming that the variability is due to temperature and variety effects (Figure 9.2). In SUBGRO the actual leaf appearance rate is the product of the maximum possible value of 0.7 leaves day^{-1} and the relative rate of appearance as a function of air temperature, shown in Figure 9.2*b*. The computer language used, CSMP, has routines to interpolate between *x,y*-co-ordinates, so a functional relationship like that of Figure 9.2*b* can be used directly without description by a mathematical equation.

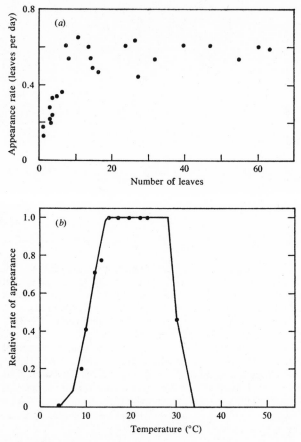

Figure 9.2. (a) The appearance rate of sugar beet leaves as a function of the number of leaves that have appeared. Data are for the varieties GW 304 (Ulrich, 1955), US 22/3 (Ulrich, 1956), and Klein E. (Thorne *et al.* 1967). (b) The relative rate of leaf appearance as a function of temperature. The data are from (a), after 10 leaves have appeared, and have been transformed using maximum leaf appearance rates of 0.61 for GW 304, 0.56 for US 22/3, and 0.50 for 'Klein E'.

A model of leaf senescence has not yet been included in SUBGRO; rather, there is an arbitrary limit of 40 living leaves per plant, consistent with observations of a relatively stable leaf number per plant. An equilibrium between leaf appearance and senescence after the stable leaf number is reached must be the cause of this stability, but an explanatory mechanism is needed to improve the model and to permit simulation of genotypes with widely divergent patterns of foliage development. The present formulation limits the dry weight of living tops per plant to 123 g. This is near the maximum value of 134 g reported by Loomis and Ulrich (1959), and it is much

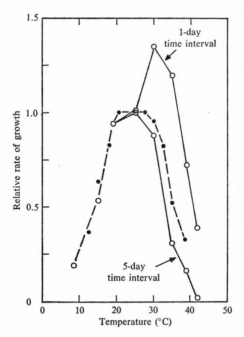

Figure 9.3. Rate of growth as a function of temperature as measured with the tops of sugar beet seedlings at 1- and 5-day intervals (open circles; Fick, 1971) and with beet roots of older plants at 6- and 13-week intervals (closed circles, maximum observations; Radke and Bauer, 1969). The broken line is the functional relationship used in SUBGRO and the ordinate values are relative to the maximum of this curve.

larger than most values reported in the literature. Though the top growth model has certain quantitative deficiencies, it gives a very good approximation of the growth of real sugar beet canopies.

We noted that the maximum possible growth rates at optimum temperature for each leaf-age class are adjusted first for the effect of temperature. In practice, we encounter problems with this since the optimum temperature for growth may depend on the time interval over which the measured response is followed. Since we lack the detailed experimental data for fully modelling the transitions involved, the alternative is to use time intervals long enough so that transitional phenomena (enzyme synthesis and degradation?) have stabilized at new equilibria as revealed by the temperature response curve.

As an example, the general effect of temperature on storage beet growth was derived from yields obtained after long periods of growth with controlled root temperatures (Radke and Bauer, 1969; and Ulrich, in Loomis *et al.* 1971) (Figure 9.3). Relative values from these yield curves were introduced into SUBGRO as multipliers for the effect of temperature on rate of growth, with the optimum temperature given the value of 1.0. Using soil temperature, the maximum possible growth rate of storage beets is adjusted for the effect of temperature. A similar approach is used for the growth rates of leaves and fibrous roots. With the curve used in SUBGRO, the multiplier always has a value between 0.0 and 1.0.

The final and key adjustment made in the computation of the actual growth rates is for limiting factors. Again, the actual growth rate is the product of the maximum possible growth rate adjusted for temperature and a multiplier designated as relative rate of growth selected for the most limiting factor. The two basic physiological parameters used in SUBGRO as regulators of computed growth rates, reserve level as a percent of total dry weight (PRES) and the relative water content of leaves (RWC), are handled as limiting factors. We shall study them further, but for now we need only note that a multiplier is selected for each growing part of the plant for the effects of both PRES and RWC (Figure 9.4). The smaller of these, as the more limiting factor, is used as the final rate-determining multiplier. Additional factors could be handled in the same way, or as simultaneous limitations. Justification of the relationships shown in Figure 9.4 is given later (p. 282), as are several changes in these first assumptions relating to beet growth and sugar storage.

After the actual growth rate is computed, it is used to update the weights of the part of the plant to which it applies (as a transfer from reserves). In the case of tops, the size of each growing leaf is computed separately and the weights of all the leaves on a plant are summed to give the total top weight per plant. This value is multiplied by the population density to give the top dry matter yield at any given time. These values are used further to estimate the height of the canopy (Figure 9.5*a*) and the leaf area index (Figure 9.5*b*), thus providing information needed to compute transpiration and photosynthesis.

Reserve supply

The internal supply of reserves is the integral of the rate of photosynthesis minus the rates of respiration, growth, and storage. The total weight of reserves is assumed to be distributed among all the plant parts and, at any point in time, the concentration of reserves, expressed as a percent of the accumulated total dry matter (PRES, Figure 9.4) is the same for all plant parts. In starved plants, PRES may assume negative values representing catabolism. PRES occupies a central position in the sugar beet model and, as we shall see, the tentative nature of our present views indicates the need for additional research.

The photosynthesis component of the reserve integral was calculated (Duncan *et al.* 1967) from a description of the daily solar radiation totals,

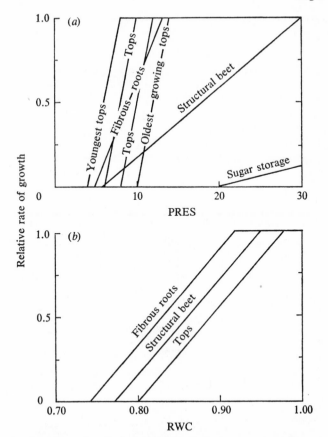

Figure 9.4. (*a*) The relative effect of reserves as a percentage of total dry weight (PRES) (as used in version I of SUBGRO), and (*b*) the relative effect of relative water content of the leaves (RWC) on the rate of growth of several growing parts of a sugar beet crop as used in SUBGRO. PRES and RWC represent limiting factors on growth rate and their combined effect regulates dry matter distribution in new growth (Fick *et al.* 1973).

the structure of the leaf canopy and the photosynthetic response of sugar beet leaves to light. Vertical distribution of leaf area index (LAI) and frequency distributions of leaf angles along the vertical profile were determined at Davis for various stages of development and for several densities (for an example, see Loomis *et al.* 1971). The light response curve was taken from Hall and Loomis (1972) for 290 ppm CO_2 and 25 °C. The output of the Duncan model was summarized in photosynthesis tables for clear and overcast skies as a function of LAI and input to SUBGRO (Table 9.1). This approach assumes that canopy architecture is fixed for each level of LAI. In practice this has not created a problem; the rather crude associative relationship

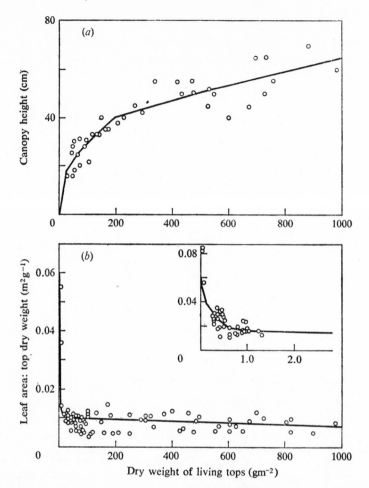

Figure 9.5. (*a*) Data points showing the relationship between the dry weight of sugar beet tops and the height of the canopy as derived from studies by Bouillenne *et al.* (1940), Loomis and Ulrich (1962), and R. S. Loomis (previously unpublished). The eye-fitted set of successive straight lines is the relationship used as a linear-function generator in SUBGRO to estimate canopy height from top dry weight. (*b*) Data points show the relationship between the dry weight of sugar beet tops and the leaf area to top dry weight ratio as derived from studies by Watson and Baptiste (1938), Kuiper (1962), Campbell and Viets (1967), and R. S. Loomis (previously unpublished). The eye-fitted set of successive straight lines is the relationship used as a linear-function generator in SUBGRO to compute leaf area index from dry weight of living tops. The inset shows the relationship for small plants.

Table 9.1. *Gross photosynthesis rates* (in $g \, CH_2O \, m^{-2} \, h^{-1}$) *of sugar beet canopies in the field as a function of leaf area index* (LAI, $m^2 \, m^{-2}$) *and the sine of the height of the sun* (SINE) *for clear* (C) *and overcast* (O) *skies. Calculated with the Duncan photosynthesis simulation program* (*Duncan et al. 1967*)

SINE LAI	0.0	0.1	0.2	0.3	0.4	0.5	0.6	0.7	0.8	0.9	1.0	Sky
0.	0.00	0.00	0.00	0.00	0.00	0.00	0.00	0.00	0.00	0.00	0.00	C
	0.00	0.00	0.00	0.00	0.00	0.00	0.00	0.00	0.00	0.00	0.00	O
1.	0.00	0.26	0.81	1.30	1.75	2.12	2.44	2.70	2.89	3.01	3.12	C
	0.00	0.07	0.18	0.28	0.39	0.51	0.63	0.74	0.87	0.98	1.06	O
2.	0.00	0.32	0.96	1.66	2.34	3.07	3.75	4.17	4.48	4.67	4.80	C
	0.00	0.09	0.24	0.39	0.55	0.72	0.89	1.04	1.22	1.37	1.48	O
3.	0.00	0.34	1.02	1.76	2.55	3.42	4.28	4.89	5.27	5.63	5.92	C
	0.00	0.11	0.28	0.44	0.62	0.81	1.01	1.17	1.39	1.57	1.70	O
4.	0.00	0.34	1.03	1.79	2.63	3.56	4.54	5.25	5.80	6.24	6.59	C
	0.00	0.11	0.29	0.46	0.65	0.85	1.06	1.24	1.46	1.64	1.78	O
5.	0.00	0.34	1.04	1.81	2.67	3.61	4.62	5.40	6.01	6.52	6.98	C
	0.00	0.11	0.30	0.48	0.67	0.88	1.09	1.27	1.50	1.69	1.84	O
6.	0.00	0.34	1.04	1.82	2.69	3.65	4.69	5.51	6.16	6.77	7.30	C
	0.00	0.12	0.31	0.49	0.68	0.89	1.11	1.28	1.51	1.70	1.85	O
7.	0.00	0.35	1.05	1.82	2.70	3.68	4.73	5.61	6.30	6.94	7.51	C
	0.00	0.12	0.31	0.49	0.68	0.89	1.11	1.29	1.52	1.71	1.86	O
8.	0.00	0.34	1.05	1.82	2.70	3.68	4.75	5.65	6.34	6.97	7.60	C
	0.00	0.12	0.31	0.49	0.68	0.89	1.11	1.30	1.53	1.72	1.87	O
9.	0.00	0.34	1.05	1.82	2.70	3.68	4.76	5.67	6.37	6.99	7.66	C
	0.00	0.12	0.31	0.49	0.69	0.90	1.12	1.31	1.54	1.73	1.89	O
10.	0.00	0.33	1.05	1.82	2.70	3.68	4.77	5.68	6.39	7.00	7.71	C
	0.00	0.12	0.31	0.49	0.69	0.91	1.13	1.32	1.56	1.75	1.91	O

between LAI and dry weight of tops (Figure 9.5*b*) has proved to be more limiting to the range of possible simulations.

SUBGRO computes for each hour of the day, the sky condition (amount of cloudiness and hence direct and diffuse light distributions, as described by de Wit, 1965), the solar elevation, and the LAI, and then finds the appropriate photosynthesis rate by interpolation in the photosynthesis tables. The potential rate of net photosynthesis is then adjusted for the effect of temperature and the effect of leaf water status (RWC). Again, this adjustment is done with multipliers that represent the relative effect on the absolute rate.

We used the data of Nevins and Loomis (1970) to define the effect of temperature on sugar beet photosynthesis. Ito (1965), Hofstra and Hesketh (1969), and Hall (1970) all reported somewhat different relationships indicating the need for further refinement of experimental procedures. Data on the

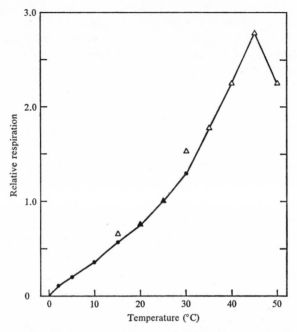

Figure 9.6. The relative rate of respiration in leaves as a function of temperature, set equal to 1.0 at 25 °C. Measurements were made on sugar beet leaves in the dark, and the series of straight lines describe the relationship used in SUBGRO. △, Hofstra and Hesketh, 1969; ● Nevins and Loomis, 1970.

effect of RWC or leaf water potential on sugar beet photosynthesis were not available, so we used Brix's (1962) data for tomato.

The treatment of respiration in productivity models has proved a stumbling block to the verification of their real accuracy. Adjustments at this point can strongly control model predictions (Loomis *et al.* 1967). McCree (1970) demonstrated that respiration in white clover is a function of both photosynthesis (presumably in proportion to rates of growth and storage) and plant weight, but lacking details of a similar relationship in sugar beet, we have used a crude but workable relationship based on photosynthesis alone.

We chose as a reference point the gas-exchange experiments of Thomas and Hill (1949) with sugar beet. For plants with high nutrition, about 25 % of net photosynthesis went to respiration of the total plant. Since McCree and Troughton (1966a, 1966b) showed that plants adjust their respiration rate in about 24 hours when their light environment is changed, we set the respiration rate at 25 °C equal to 25 % of the net photosynthesis on the previous day. This value is then adjusted for the effect of temperature on respiration (Figure 9.6) as measured in sugar beet leaves (Hofstra and Hesketh, 1969; Nevins and Loomis, 1970).

This approach departs from the explanatory mode and is being used only until better opinions are obtained from current research. Although the present model agrees reasonably well with the measurements of Thomas and Hill (1949), it would fail if large plants were placed in environments where photosynthesis is very low or where temperature departed from a favourable range for extended periods. Penning de Vries' (1972) approach with two components of respiration appears to be a superior method. His first component is associated with growth and is computed from the chemical composition of the new material being produced according to the energy exchanges of known biochemical pathways. This is similar to the photosynthetic factor of McCree's (1970) equation. The second component is the temperature-dependent maintenance respiration associated with the weight of tissue already produced. The difference between these two components in their probable response to temperature is a potential source of error in our present simplified model.

The storage rate component of the reserves integral also presents a series of special problems. Sugar accumulation (storage) is considered to occur only in the storage beet (root plus hypocotyl and stem) and is in addition to the portion of the reserves present in the beet. Thus, the total sugar content of the beet is the sum of storage and the reserve fraction and may be expressed as an actual weight or as a percent of total fresh or dry weight. Since the mechanisms for sucrose movement and accumulation in the beet are not known, the storage rate has been modelled from associative information about partitioning ratios between sucrose and non-sucrose dry matter in the root. The partitioning is not influenced by RWC.

Internal water status

The third major physiological theme in SUBGRO relates to the computation of an index of internal water status of the crop (RWC), computed as the relative water content of leaves (Slatyer, 1967). The water deficit of the crop is the integral of the rate of water uptake by the roots minus the rate of transpiration by the canopy. RWC could be estimated directly from the value of the water deficit, but more realistic hourly simulations are reached when it is computed from the rates of water uptake and transpiration in conjunction with the water deficit.

In SUBGRO the uptake of water by roots depends on the effective length of the fibrous root system and the transpirational demand. Effective root length is the length of active roots weighted for the reduction in water uptake capacity caused by suberization. The growth rate in root length is determined from the growth rate in root weight using, at 20 °C, the conversion factor of 290 m g^{-1} (Fick *et al.* 1972). Roots of some plants have a smaller diameter when grown at higher temperature (Brouwer and de Wit, 1969). A temperature effect varying the conversion factor from 200 m g^{-1} for growth occurring at

5 °C to 500 m g^{-1} at 50 °C was included in the model, but experimentally derived factors are needed.

Increments of root length produced during the same time intervals are treated as age-class groups in much the same way that individual leaves are handled separately. The capacity of a particular group of segments to absorb water is determined by the extent of suberization of that group. Suberization depends on temperature (Brouwer and de Wit, 1969) and at 20 °C, it takes a model root segment four days to become fully suberized. At this temperature, suberization progresses at the same rate as growth, so complete suberization would lag behind the growing tip by the length of four days growth. At 35 °C, suberization can progress 1.5 times as fast as growth so the roots would suberize to near the growing apex, as observed by Brouwer and de Wit (1969).

The effect of suberization on water uptake capacity, and hence on effective length of the root system, was derived from work with beans (Brouwer, 1953a; 1953b). Brouwer measured the decreasing capacity of roots to take up water the further they were from the growing apex in the direction of increasing suberization; the same pattern of suberization occurs in sugar beet roots (Fick, 1971).

Based on Brouwer's measurements, 1 m of 'effective' root length was given the potential water uptake capacity of 7×10^{-3} kg day^{-1} (2.91 mm^3 cm^{-1} h^{-1}). Sixteen effective-length weighting factors for the root groups not fully suberized were derived allowing for increased effective length in root-hair zones. The factors increase from 1.2 for the first segment, to 2.95 for the tenth segment and then decrease to 1.5 for the sixteenth segment group. There are an additional 184 groups with a very small weighting factor of 0.05. These are considered to be fully suberized and take up water only slowly (Kramer, 1969). The summation of the 200 effective-length groups multiplied by the root water uptake rate constant of 7×10^{-3} kg m^{-1} day^{-1} gives the possible water uptake rate of the root system.

The possible water uptake rate is realized only when a water deficit is developing or exists in the plant. For the rest of the time the actual water uptake rate will be largely controlled by the rate of transpiration and will be less than the possible rate. Transpiration rate in SUBGRO is calculated with the generalized Penman combination formula (Penman, 1948; Monteith, 1965) that has been demonstrated to be satisfactory for one-hour time intervals, even with high advection (Van Bavel, 1966, 1967). In this formulation, the Brunt equation (Brunt, 1939; Chang, 1968) is used to calculate outgoing long-wave radiation, and the empirical height and wind speed functions of Rijtema (1965) are used to calculate the aerodynamic diffusion resistance.

The surface diffusion resistance of the canopy is also needed to calculate transpiration from this formula. This value depends on the resistance to the movement of water vapour out of the leaf, which is largely determined by stomatal aperture. Stomatal aperture is influenced by the ambient carbon

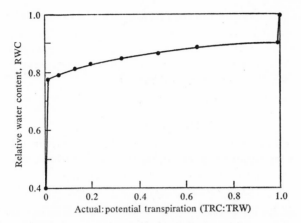

Figure 9.7. The relative water content of leaves (RWC) as a function of the ratio of actual to potential transpiration (TRC:TRW), based on measurements by Brix (1962) and conversion factors from Slatyer (1962).

dioxide concentration, light and the water status of the leaf. However, we found that inclusion of RWC as feedback in the calculation of this factor set up oscillations in predicted transpiration rate that were unrealistic. This was caused by the one-hour delay in response to RWC imposed by the integration time interval. We worked our way around this problem by first ignoring the effect of RWC and calculating a transpiration rate for a crop without a water deficit (TRW). This is an estimate of the possible transpiration rate for the next hour and provides feed forward (as opposed to feedback) control. If the possible water uptake rate of the roots is greater than the rate needed to remove the deficit in the next hour, then the actual transpiration rate of the crop (TRC) will equal the potential rate (TRW). If the possible uptake rate is less than the uptake needed to remove the deficit, TRC will be approximately equal to the possible uptake rate. To allow for the development of a deficit and to set a ceiling on its size, the actual model of TRC is somewhat more complicated than this outline of the essential logic. The model agrees reasonably well with field measurements of sugar beet transpiration (Pruitt and Lourence, 1968). Our simulations for the same environment had the same common range of values with a peak at 8.1 mm day^{-1}.

Since we have not needed a value of RWC to calculate either TRC or TRW, we can use the ratio of TRC:TRW to estimate RWC from a functional relationship of the effect of leaf water potential on the relative transpiration rate of tomato as measured by Brix (1962). This was converted to an RWC scale with factors from Slatyer (1962). When the actual to potential transpiration rate ratio (TRC:TRW) is 1.0, the RWC is 100%. According to the functional relationship (Figure 9.7), RWC drops in a curvilinear pattern so that when the ratio is 0.01, RWC is 77%. This method for estimating RWC reverses the functional relationship by making RWC a function of TRC. The logic of the model could be improved by computing RWC, or leaf water potential, as a function of the water deficit of the plant. However, its present accuracy is sufficient until simulations are run in which the transpiration and plant-water sections are the foci of test hypotheses.

HYPOTHESIS TESTING: PARTITIONING

The completed model represents a theory or hypothesis about plant growth and the simulation results represent predictions made from the hypothesis. We turn to the model when we want to construct hypotheses so complex that we could not in practical ways realize their consequences without the aid of the computer. These systems may incorporate into our explanatory opinions more of the known interacting factors with fewer of the simplifying assumptions, but like predictions from all hypotheses, the model needs to be verified by comparing the predictions with the behaviour of the real system.

As an example, we have developed and tested an hypothesis of the mechanism of dry matter partitioning in the sugar beet (Fick *et al.* 1973). It incorporates into one unified system, the core of SUBGRO, the following series of opinions, and observations drawn from research.

Based on Ulrich's work (1952, 1954, 1955; Loomis and Ulrich, 1959, 1962), the first element of our hypothesis is a hierarchy of priorities among the sinks for the use of reserves. Ulrich concluded that storage beet growth and sugar accumulation required an excess of photosynthates beyond the needs for respiration and for growth of tops and fibrous roots. He demonstrated that environmental conditions that reduced top growth and/or respiration more than photosynthesis, such as low temperature or the early phases of nitrogen shortage, promoted the accumulation of sugar in the beet.

We described earlier this sequence of priorities. It is established in SUBGRO through a series of threshold concentrations of reserves (measured by PRES) at which the sinks may begin activity (Figure 9.4*a*). The highest priority was given to respiration, which must go on as long as the plant is alive, by not assigning a threshold value. The next priorities were given in order to youngest tops and fibrous roots. These sinks were given lower thresholds than the older tops and storage beet, because even at light levels that greatly reduce top growth, new leaves and fibrous roots are formed (Kuiper, 1962; Ulrich, 1952; and Terry, 1968). Lowest priority was given to sugar accumulation. We will examine this last assumption more fully later.

The second element in the hypothesis is that the relative effects of reserves on sink activity differ for various sinks as a function of reserve level (Terry, 1968). This is shown in Figure 9.4*a* by the non-parallel response lines for the different sinks. Intersink competition increases as the reserve supply decreases so the lines are squeezed together toward the threshold values. The general sequence of priorities of tops over fibrous roots over beet is maintained. The physiological basis for this control is not clear-cut but a sensitivity analysis (changes in thresholds and response slopes) showed that the present system reasonably mimics reality. The extent to which sinks really differ in some combinations of transport, substrate dependence, or growth capability not covered by thresholds or estimates of growth rates, remains to be established by further research.

The third element in the hypothesis sets up the actual respiration, growth, and accumulation rates of the sinks. We have already outlined how this was done as functions of temperature and the amount of tissue capable of growth. The final step was the selection of the more limiting factor in the PRES and RWC effects on rate. Other limiting factors were not included in the model, following experiments done by Brouwer that indicated that reserves and water content played the key roles in the balancing of partitioning.

Brouwer (1966) showed that bean plants grown in a controlled environment had a constant top to root ratio, characteristic of the environment. If the environment was changed, the ratio changed. If the ratio was changed by partial defoliation or root pruning while the environment remained the same, the original ratio was restored during the subsequent growth of the plants. This was explained in terms of a 'functional equilibrium' in which the plant regulated the partitioning of reserves between tops and roots so as to maintain a balance between water uptake capacity and transpiration. Thus, with partial defoliation, the supply of reserves through photosynthesis would be reduced while the supply of water would be more than adequate. This combination would shift partitioning in favour of the tops, since they have first priority over the limited supply of reserves, and the original ratio would be restored. With root pruning, a water stress would develop that would limit top growth more than root growth. The reserves would then be channelled more to the roots than to the tops, and again, the original ratio would be restored. We repeated Brouwer's pruning experiments with sugar beet plants at the outset of rapid storage beet development (Fick *et al.* 1971), and found that the functional equilibrium apparently operates in sugar beet (Figure 9.8).

Following these leads, we set out a series of experiments (Fick *et al.* 1973) to test the partitioning hypothesis (Figure 9.4), in which (1) threshold priorities for the use of reserves are set up among the sinks, after Ulrich, (2) the relative competitiveness of the sinks depends on the supply of reserves, after Terry, and (3) the relative rates of growth are adjusted according to the more limiting of the supply of PRES and RWC, after Brouwer.

We shall note especially two of the experiments. To verify first of all that the total model could operate in a reasonable way, we simulated the growth of sugar beets in the field at Davis (Figure 9.9). These growth curves indicated that in general the model made reasonable predictions. However, the

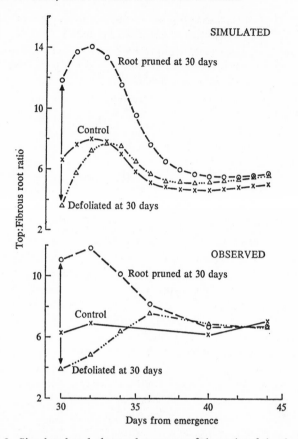

Figure 9.8. Simulated and observed recovery of the ratio of the dry weight of tops (laminae plus petioles) to the dry weight of fibrous roots following defoliation or root pruning. The experiment was conducted in a sunlit controlled environment room; a 'phytotron version I' of SUBGRO produced the simulation.

partitioning within the storage beet between structural material and sugar was not like real plant observations. Since the sum of these two parts of the beet agreed with field observations, only that part of the model that controlled partitioning within the beet appeared to be unsatisfactory. Taken as a whole, reasonable partitioning to the beet was predicted. In a second experiment, we simulated the pruning experiment with sugar beets (Figure 9.8) (Fick *et al.* 1971, 1973), and again found the results to be quite encouraging. Our conclusion was that the partitioning mechanism, which in effect formed the core of SUBGRO, was satisfactory for this stage of model development. The next emphasis should be the refinement of those parts of the model that we knew were not satisfactory.

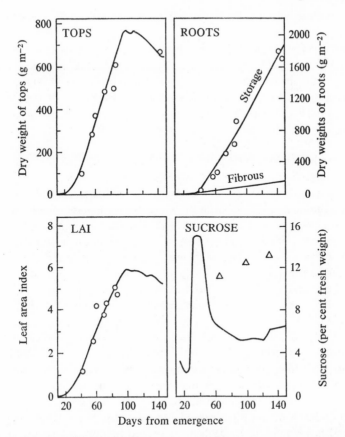

Figure 9.9. Computer simulations (version I, line) and observations (points) of the growth of beet plants in the field at Davis, California. The field experiments were done in 1966 and 1967 at plant populations of approximately 70000 plants ha^{-1}. The simulations used 1967 weather data starting on June 15, 10 days after emergence (Fick *et al.* 1973). Sucrose concentrations (triangles) were observed in other experiments (Loomis *et al.* 1960) and serve only to indicate the normal trend of sucrose concentration at Davis in beets well supplied with nutrients and water.

MODEL REFINEMENT: SUGAR ACCUMULATION

When a section of a model proves unsatisfactory, one rejects the hypothesis that failed and formulates an alternative for further testing (de Wit, 1970). Parameter adjustment could be used to produce a fit of observation and simulation, but by so doing, the model is not improved or refined. Instead, it loses its generality and it becomes a cumbersome and inefficient method of curve fitting. The sugar accumulation mechanism of SUBGRO represented a hypothesis about sugar beet growth to be rejected. Before formulating and testing another alternative, let us take a closer look at the reasons for its failure.

In the first hypothesis (version I, Figure 9.4a) the storage root fraction of the plant was treated as having two parts: (1) structual (and protoplasmic) tissue; and (2) stored sugar plus reserves. The growth rate of each part of the beet was regulated by the reserve level of the crop (PRES). Structural growth started in the beet when PRES reached 6 % (the threshold concentration) and became maximal only when PRES went up to 30 %. The sugar storage rate started at a PRES value of 20 %. Above 20 %, the rate was proportional to the reserve level and to beet growth, based on observations that beet size and sugar storage tend to occur simultaneously (Ulrich, 1954, 1955; Loomis and Ulrich, 1962; Terry, 1968). The growth rate of the structural fraction was further regulated by a maximum value of 3.5 g plant^{-1} day^{-1} based on observed growth rates (Fick *et al.* 1971, 1972). Such a ceiling implied that the amount of potential cambial activity was constant as the beet matured. Since the circumference of the beet increases as the beet grows and the number of active cambia also tends to increase, this implies that the rate of cell division per unit cambial surface must decline. We do not have enough information to model this situation accurately through adjustments in the PRES or RWC controls, so the ceiling was included to control extreme behaviour.

In this hypothesis, structural beet growth is given priority over sugar storage on a concept, outlined by Bouillenne *et al.* (1940), that a storage location needs to be formed before sugar can be stored in it. But this mechanism appears to be the cause of the unrealistic partitioning within the beet since even with a ceiling, structural growth of the beet is such a powerful sink that PRES levels do not often exceed 20 % after the beet reaches the potential for rapid growth. In small beets with limited amounts of tissue capable of growth, PRES levels are above 20 % and simulated sugar storage is rapid (20 to 30 days, Figure 9.9). But when the storage beet is growing rapidly, the PRES levels exceed 20 % enough of the time to maintain a sugar concentration of only about 6 % (fresh wt). Thus, with a one-hour integration interval, the assumption that structure is built before storage occurs results in the structural sink being too competitive with the storage sink. The time lag between production of new cells and storage of sucrose may be quite short (a rapid transition time) since the pathway for substrate transport from sieve tubes to cambial initials lies through the expanding secondary phloem cells.

An alternative hypothesis

In a second hypothesis (version II), we interpreted observations over a longer time scale of parallel growth of structure and storage to mean that the two processes should be given the same potential growth rate. The possible growth rate of beet structure was assumed to be dependent on the amount of reserves in the part of the plant called the beet, the PRES times the total dry weight of the beet divided by 100. A second assumption made with no experimental evidence, for the sake of trial, was that the possible growth rate over the next hour could use up 25 % of this fraction. The possible rate of sugar storage was given the same value.

The effect of PRES on these two rates was also assumed to be identical. The threshold PRES level for both structural beet growth and sugar storage was set at 10 % of total dry weight. This extends into the range of top and fibrous root growth (Figure 9.4 a), so there will still be some beet development at PRES levels that reduce top and fibrous root development (in agreement with Terry, 1968). However, the threshold is high enough that top and fibrous root growth have clear priority for reserves over beet development, thereby conforming to the positive aspects of the original mechanism. Again, the maximum rates are not reached until PRES levels are up to 30 %.

To this point, structural growth and storage in the beet would be equal. A further limit on structural growth restricts beet growth in the new model until the sugar percentage is 4.5 % of fresh weight, approximately as observed by Ulrich (1952), and increases linearly in potential rate until it reaches a maximum at 10 % of fresh weight. This reverses the logic of having structural growth occur before storage; or more accurately, on the time scale of the simulation, it hypothesizes that a certain amount of storage must occur in cells that have already been formed before new ones can be produced. This is compatible with observations of a relatively high sucrose concentration in the vascular cambial rings.

To summarize this alternative hypothesis, the AGR limit on structural growth is replaced by a limiting function based on the amount of reserves in the beet. It uses the same relationships between growth and PRES and RWC as shown in Figure 9.4 a with the exception that the lines for the structural beet and sugar storage are now identical, with a threshold of 10 % PRES and a maximum at 30 %. The consequences of the new hypothesis (II) for field-condition simulations of sugar percentage were much more like real results than those of the original mechanism (I) (Figure 9.10). While the agreement appears to be very good, the observed points were not obtained in the same season. The remainder of the predictions were as good as the original run (Figure 9.9), except for the yield of storage roots which were 16 % higher. Thus, there is a need for further validation and perhaps further refinement.

In this test, version II predicts, as it should, that sucrose concentration is inversely related to minimum air temperatures (Ulrich, 1955). It remains to be seen whether it will do this over series of diurnal temperature patterns and show, for beets in similar environments, an inverse relationship between root size and sucrose concentration (Loomis and Ulrich, 1959, 1962). Sucrose concentration is a critical feature of the simulation and we may need to

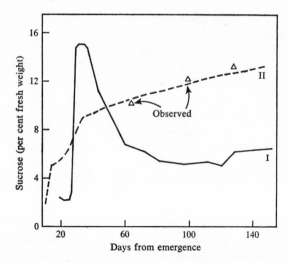

Figure 9.10. Simulations of sucrose concentration in beets under field conditions (lines) and values observed in experiments (triangles) as in Figure 9.9. Version I assumes that structural growth of the beet has priority over sucrose accumulation for photosynthates. Version II assumes equal priorities, and a control on structural growth by sucrose concentration on a fresh basis.

structure yet another version on cell division and enlargement and on cell water relations in order to obtain sufficient precision.

SOME CONCLUSIONS

In a volume devoted to crop physiology we have emphasized computer simulations of crop growth. We have skipped over large areas of important physiological knowledge, and some of the physiology we have included in the model represents rough approximations, and indeed in some places, mere guesses. We have felt free to do this because we believe computer simulation as a new tool for the physiologist's workbench has much to contribute to crop physiology.

What are the contributions to be made? SUBGRO has demonstrated that plant growth modelling is a means of synthesizing what we think we know, and of testing the compatability of various hypotheses in the synthesis. In the case of the partitioning mechanism for the products of photosynthesis, opinions about sugar beet growth from Ulrich, Terry, and Brouwer were integrated into a unified hypothesis. In a complex system that considers as an interacting whole the environment, morphology, and physiology of growth rates, photosynthate supply, and water status of the plant, the hypothesis was shown to give qualitatively correct predictions. It is compatible with that part of the whole system included in the model.

Thus, one of the main contributions that growth modelling can make to the study of crop physiology is to provide a means of using the scientific method to put our knowledge back together again. Experimental research has been by-and-large reductionistic. Synthesis has usually produced more complexity than was scientifically manageable, so what was synthesized was often more philosophical than scientific. The methods of regression analysis provide predictions, but they lack the element of explanation that gives biological hypotheses both their high degree of vulnerability to testing and their power of extrapolation (Hardwick and Wood, 1972). Now the computer makes it possible to make predictions from complex synthetic biological hypotheses that are vulnerable to the scientific tests of their accuracy and realism in mimicry (Fick *et al.* 1973). The requirement for testing means that the crop will not be displaced as the centre of research. The cycle of model formulation and testing starts and ends with the crop.

A second major contribution that simulation makes to crop physiology is that it organizes our knowledge and thinking in a way that reveals what we do not know. A model can be used to estimate the relative importance of the unknowns by sensitivity analysis; but whether this is done or not, the immediate payoff is that the attempt at modelling reveals areas of needed research. Let us take two examples from SUBGRO. We do not yet know how the growth potential of the storage beet, the economic organ, is regulated. Consequently, we cannot accurately predict economic yields with an explanatory hypothesis. Secondly, we know that the nitrogen status of the plant plays a critical role in the determination of economic yield (Hills and Ulrich, 1971). We have not yet tried to model these relationships because we are not sure that we know how to relate the soil supply of nitrogen to the growth of new cells or photosynthesis in ways that will explain the role of nitrogen in sugar beet growth. The attempt to model this mechanism, formulating a new physiological theme analogous to that for water, will answer this question more fully than any other technique we have.

We have emphasized the use of growth models in physiological research. The developed model will also play a powerful role in other fields. The first is in agronomic applications. Of particular importance is the question of ideotype formulation and evaluation. In California there has been a strong trend, for reasons of economy and weed control, towards wide-row (75 cm) culture of sugar beet, and towards the selection of genotypes with vigorous top development to more effectively utilize the greater amount of space per plant available with wide rows. Total sugar production has been improved, but this has been due to a general increase in total primary production since the proportion of total crop weight which occurs as sucrose in the beet root has not increased. In fact, it may be less with the new varieties.

An alternative strategy would be to reduce row spacings and use a plant with less vigorous top growth which partitions a greater proportion of its production to the economic product, sucrose. This strategy can not be tested with real plants because an appropriate set of genotypes does not exist. However, experiments with a validated partitioning model such as SUBGRO would provide quantitative comparisons of the behaviour of the two systems which might justify a major shift in breeding objectives. As Hardwick and Wood (1972) point out, dynamic models may prove to be the only effective means for dealing with complex genotype–environment interactions of this type.

Finally, such models will have a variety of extremely useful applications in crop management. With sugar beet, the model could be used for better co-ordination of the production and processing phases. In California, sugar beets are grown in a wide array of local climates each with a wide range of possible planting and harvest dates. The area to be planted, the dates of planting and harvest, and subsequent transport (over distances of up to 1100 km) must be scheduled according to the processing capabilities of a small number of factories. Harvest and transport schedules are based on yield estimates for each district and it is here that a model like SUBGRO, providing a continually updated estimate of current yield and growth rates as a function of local weather, could be of great value.

But if computer growth-modelling has something to contribute to the development of crop physiology as a science, the basic and applied workers who could use this kind of knowledge also have something to contribute in the opposite direction. This should be clear from the kind of research information and opinion that went into the formulation of SUBGRO. Research that is amenable to quantitative formulation is absolutely essential to model development. Interpretations of research findings that are synthetic in view provide the opinion to be tested. Carefully used, the modelling effort will serve to guide our traditional methods of research and to promote effective co-operation among specialists having a wide diversity of skills and knowledge.

ACKNOWLEDGEMENTS

We are greatly indebted to A. Ulrich and C. T. de Wit for their generous and extensive assistance, and to the National Science Foundation (GB14581 and GB18541) and the California Sugar Beet Industry for financial support.

REFERENCES

Andrews, L. H. (1927). The relation of the sugar beet root system to increased yields. *Through the Leaves* **15**, 16–20.

Artschwager, E. (1926). Anatomy of the vegetative organs of the sugar beet. *J. Agr. Res.* (U.S.) **33**, 143–176.

Belikov, I. F. and Kostetskii, E. Ya. (1964). The distribution of assimilates in sugar beet during the growth period. *Soviet Plant Physiol.* **11**, 508–511.

Bingham, J. (1972). Physiological objectives in breeding for grain yield in wheat. In *The Way Ahead in Plant Breeding*, eds F. G. H. Lupton, G. Jenkins and R. Johnson. Proc. Sixth Congress EUCARPIA, Cambridge, pp. 15–29.

Bouillenne, R., Kronacher, P. G. and de Roubaix, J. (1940). *Etapes morphologique et chimiques dans le cycle vegetatif de la betterave sucriere*. Publ. Inst. Belge Amelior. Betterave. Renaix, 81 pp.

Brix, H. (1962). The effect of water stress on the rates of photosynthesis and respiration in tomato plants and loblolly pine seedlings. *Physiol. Plantar* **15**, 10–20.

Brouwer, R. (1953a). Water absorption by the roots of *Vicia faba* at various transpiration strengths. I. Analysis of the uptake and the factors determining it. *Proc. Kon. Nederl. Akad. Wetensch* **53**C, 106–115.

Brouwer, R. (1953b). Water absorption by the roots of *Vicia faba* at various transpiration strengths. II. Causal relation between suction tension, resistance and uptake. *Proc. Kon. Nederl. Akad. Wetensch.* **53**C, 129–136.

Brouwer, R. (1966). Root growth of grasses and cereals. In *The Growth of Cereals and Grasses*, eds F. L. Milthorpe and J. D. Ivins. Butterworths, London, pp. 153–166.

Brouwer, R. and Wit, C. T. de (1969). A simulation model of plant growth with special attention to root growth and its consequences. In *Root Growth*, ed. W. J. Whittington. Butterworths, London, pp. 224–244.

Brunt, D. (1939). *Physical and Dynamical Meteorology*. Cambridge University Press, London, 428 pp.

Campbell, R. E. and Viets, F. G. Jr. (1967). Yield and sugar production of sugar beets as affected by leaf area variations induced by stand density and nitrogen fertilization. *Agron. J.* **59**, 349–354.

Chang, J. (1968). *Climate and Agriculture*. Aldine Publ. Co., Chicago, 304 pp.

Curry, R. B. (1971). Dynamic simulation of plant growth. Part I. Development of a model. *Trans. Am. Soc. Agr. Eng.* **14**, 946–949, 959.

Curry, R. B. and Chen, L. H. (1971). Dynamic simulation of plant growth. Part II. Incorporation of actual daily weather and partitioning of net photosynthate. *Trans. Am. Soc. Agr. Eng.* **14**, 1170–1174.

Donald, C. M. (1968). The breeding of crop ideotypes. *Euphytica* **17**, 385–403.

Doneen, L. D. (1942). Some soil-moisture conditions in relation to growth and nutrition of the sugar beet plant. *Proc. Am. Soc. Sug. Beet Technol.* **3**, 54–62.

Duncan, W. G., Loomis, R. S., Williams, W. A. and Hanau, R. (1967). A model for simulating photosynthesis in plant communities. *Hilgardia* **38**, 181–205.

Fick, G. W. (1971). Analysis and simulation of the growth of sugar beet (*Beta vulgaris* L.). Ph.D. Thesis, Univ. of Calif., Davis, 242 pp. (*Diss. Abstr.* **32**, 157b).

Fick, G. W., Williams, W. A. and Loomis, R. S. (1971). Recovery from partial defoliation and root pruning in sugar beet. *Crop Sci.* **11**, 718–721.

Fick, G. W., Williams, W. A. and Loomis, R. S. (1973). Computer simulation of dry matter distribution during sugar beet growth. *Crop Sci.* **13**, 413–417.

Fick, G. W., Williams, W. A. and Ulrich, A. (1972). Parameters of the fibrous root system of sugar beet (*Beta vulgaris* L.). *Crop Sci.* **12**, 108–112.

Geiger, D. R. and Christy, A. L. (1971). Effect of sink region anoxia on translocation rate. *Plant Physiol.* **47**, 172–174.

Geiger, D. R., Saunders, M. A. and Cataldo, D. A. (1969). Translocation and accumulation of translocate in the sugar beet petiole. *Plant Physiol.* **44**, 1657–1665.

Goodban, A. E. and McCready, R. M. (1968). The osmolality of sugar beet press juices. *J. Am. Soc. Sug. Beet Technol.* **15**, 120–124.

Gordan, G. (1969). *System Simulation.* Prentice-Hall, Inc., Englewood Cliffs, N.J., 303 pp.

Hall, A. E. (1970). Photosynthetic capabilities of healthy and beet yellow virus infected sugar beets (*Beta vulgaris* L.). Ph.D. Thesis, Univ. Calif., Davis, 220 pp.

Hall, A. E. and Loomis, R. S. (1972). Photosynthesis and respiration by healthy and Beet Yellows Virus-infected sugar beets (*Beta vulgaris* L.). *Crop Sci.* **12**, 566–572.

Hardwick, R. C. and Wood, J. T. (1972). Regression methods for studying genotype-environment interactions. *Heredity* **28**, 209–222.

Hayward, H. E. (1938). *The Structure of Economic Plants.* Macmillan Co., New York, 674 pp.

Heslop-Harrison, J. (1969). Development, differentiation and yield. In *Physiological Aspects of Crop Yield,* eds J. D. Eastin, F. A. Haskins, C. Y. Sullivan and C. H. M. van Bavel. ASA, Madison, Wisconsin, pp. 291–325.

Hills, F. J. and Ulrich, A. (1971). Nitrogen fertilization. In *Advances in Sugarbeet Production: Principles and Practices,* eds R. T. Johnson, J. T. Alexander, G. E. Rush and G. R. Hawkes. Iowa State Univ. Press, Ames, Owa, pp. 111–135.

Hofstra, G. and Hesketh, J. D. (1969). Effects of temperature on the gas exchange of leaves in the light and dark. *Planta* **85**, 228–237.

Hogaboam, G. J. and Snyder, F. W. (1964). Influence of size of fruit and seed on germination of a monogerm sugar beet variety. *J. Am. Soc. Sug. Beet Technol.* **13**, 116–126.

Houba, V. J. G., van Egmond, F. and Wittich, E. M. (1971). Changes in production of organic nitrogen and carboxylates (C–A) in young sugar-beet plants grown in nutrient solutions of different nitrogen composition. *Neth. J. Agr. Sci.* **19**, 39–47.

Humphries, E. C. and French, S. A. W. (1969). Effect of seedling treatment on growth and yield of sugar beet in the field. *Ann. appl. biol.* **64**, 385–393.

IBM (1967). *System/360 continuous system modeling program (360A–CX–16X). Application Description.* Form No. H20–0240–1.

IBM (1969). *System/360 continuous system modeling program (360A–CX–16X). User's Manual.* Form No. H20–0367–3.

Ito, K. (1965). Studies on photosynthesis in the sugar beet plant. II. The difference in photosynthetic and respiratory abilities and response to temperature and light between leaves of different leaf positions. *Proc. Crop Sci. Soc. Jap.* **33**, 487–491.

Jean, F. C. and Weaver, J. E. (1924). *Root behavior and crop yield under irrigation.* Carnegie Inst. Wash., Pub. 357.

Johnson, R. T., Alexander, J. T., Rush, G. E. and Hawkes, G. R. (1971). *Advances in Sugarbeet Production: Principles and Practices.* Iowa State Univ. Press, Ames, Iowa, 470 pp.

Joy, K. W. (1964). Translocation in sugar-beet. I. Assimilation of $^{14}CO_2$ and distribution of materials from leaves. *J. exp. Bot.* **15**, 485–494.

Kelley, J. D. and Ulrich, A. (1966). Distribution of nitrate nitrogen in the blades and petioles of sugar beets grown at deficient and sufficient levels of nitrogen. *J. Am. Soc. Sug. Beet Technol.* **14**, 106–116.

Kramer, P. J. (1969). *Plant and Soil Water Relationships. A Modern Synthesis.* McGraw-Hill Book Co., New York, 482 pp.

Kuiper, P. J. C. (1962). Preliminary observations on the effect of light intensity, temperature, and water on growth and transpiration of young sugar beet plants under controlled conditions. *Meded. LandbHogesch. Wageningen* **62**(7), 1–27.

Loomis, R. S., Brickey, J. H., Broadbent, F. E. and Worker, G. F. Jr. (1960). Comparison of nitrogen source materials for midseason fertilization of sugar beets. *Agron. J.* **52**, 97–101.

Loomis, R. S. and Nevins, D. J. (1963). Interrupted nitrogen nutrition effects on growth, sucrose accumulation and foliar development of the sugar beet plant. *J. Am. Soç. Sug. Beet Technol.* **12**, 309–322.

Loomis, R. S. and Torrey, J. G. (1964). Chemical control of vascular cambium initiation in isolated radish roots. *Proc. Nat. Acad. Sci. USA* **52**, 3–11.

Loomis, R. S. and Ulrich, A. (1959). Response of sugar beets to nitrogen depletion in relation to root size. *J. Am. Soc. Sug. Beet Technol.* **10**, 499–512.

Loomis, R. S. and Ulrich, A. (1962). Responses of sugar beets to nitrogen deficiency as influenced by plant competition. *Crop Sci.* **2**, 37–40.

Loomis, R. S., Ulrich, A. and Terry, N. (1971). Environmental factors. In *Advances in Sugarbeet Production: Principles and Practices*, eds R. T. Johnson, J. T. Alexander, G. E. Rush and G. R. Hawkes. Iowa State Univ. Press, Ames, Iowa, pp. 19–28.

Loomis, R. S., Williams, W. A. and Duncan, W. G. (1967). Community architecture and the productivity of terrestrial plant communities. In *Harvesting the Sun*, eds A. San Pietro, F. A. Greer and T. J. Army. Academic Press, New York, pp. 291–308.

Loomis, R. S. and Worker, G. W. Jr. (1963). Responses of sugar beet to low soil moisture at two levels of nitrogen nutrition. *Agron. J.* **55**, 509–515.

Ludwig, L. J., Saeki, T. and Evans, L. T. (1965). Photosynthesis in artificial communities of cotton plants in relation to leaf area. I. Experiments with progressive defoliation of mature plants. *Aust. J. biol. Sci.* **18**, 1103–1118.

McCree, K. J. (1970). An equation for the rate of respiration of white clover plants grown under controlled conditions. In *Prediction and Measurement of Photosynthetic Productivity*. Proc. IBP/PP Technol. Meeting, Trebon, Czech. Pudoc, Wageningen, The Netherlands, pp. 221–229.

McCree, K. J. and Troughton, J. H. (1966a). Prediction of growth rate at different light levels from measured photosynthesis and respiration rates. *Pl. Physiol.* **41**, 559–566.

McCree, K. J. and Troughton, J. H. (1966b). Non-existence of an optimum leaf area index for the production rate of white clover grown under constant conditions. *Pl. Physiol.* **41**, 1615–1622.

Milford, G. F. and Watson, D. J. (1971). The effect of nitrogen on the growth and sugar content of sugar-beet. *Ann. Bot.* **35**, 287–300.

Monsi, M. and Murata, Y. (1970). Development of photosynthetic systems as influenced by distribution of matter. In *Prediction and Measurement of Photosynthetic Productivity*. Proc. IPB/PP Technical Meeting, Trebon, Czech. Pudoc, Wageningen, The Netherlands, pp. 115–129.

Monteith, J. L. (1965). Light distribution and photosynthesis in field crops. *Ann. Bot. N.S.* **29**, 17–37.

Morton, A. B. and Watson, D. J. (1948). A physiological study of leaf growth. *Ann. Bot. N.S.* **12**, 281–310.

Nevins, D. J. and Loomis, R. S. (1970). A method for determining net photosynthesis and transpiration by plant leaves. *Crop Sci.* **10**, 3–6.

Patefield, W. M. and Austin, R. B. (1971). A model for the simulation of the growth of *Beta vulgaris* L. *Ann. Bot.* **35**, 1227–1250.

Penman, H. L. (1948). Natural evaporation from open water, bare soil and grass. *Proc. Roy. Soc. A.* **193**, 120–145.

Penning de Vries, F. W. T. (1972). Respiration and growth. In *Crop Processes in Controlled Environments*, eds A. R. Rees, K. E. Cockshull, D. W. Hand and R. G. Hurd. Butterworths, London, pp. 327–347.

Pruitt, W. O. and Lourence, F. J. (1968). *Correlation of climatological data with water requirements of crops*. Water Science and Engineering Paper No. 9001. Dept. of Water Science and Engineering, Univ. Calif., Davis, 59 pp.

Radford, P. J. (1967). Growth analysis formulae – their use and abuse. *Crop Sci.* 7, 171–175.

Radke, J. K. and Bauer, R. E. (1969). Growth of sugar beets as affected by root temperatures. Part I: Greenhouse studies. *Agron. J.* 61, 860–863.

Rijtema, P. E. (1965). *An analysis of actual evapotranspiration*. Agr. Res. Rep. 659. Pudoc, Wageningen, The Netherlands, 107 pp.

Savitsky, V. (1950). Monogerm sugar beets in the United States. *Proc. Am. Soc. Sug. Beet Technol.* 6, 156–159.

Savitsky, V. H. and Savitsky, H. (1965). Weight of fruits in self-fertile, male-sterile, and self-sterile diploid and tetraploid monogerm *Beta vulgaris* L. *J. Am. Soc. Sug. Beet Technol.* 13, 621–644.

Slatyer, R. O. (1962). Internal water balance of *Acacia aneura* F. Muell. in relation to environmental conditions. *Arid Zone Res.* 16, 137–146.

Slatyer, R. O. (1967). *Plant-water Relationships*. Academic Press, New York, 366 pp.

Snyder, F. W. (1971). Some agronomic factors affecting processing quality of sugar beets. *J. Am. Soc. Sug. Beet Technol.* 16, 496–507.

Stout, M. (1946). Relation of temperature to reproduction in sugar beets. *J. Agr. Res.* 72, 49–68.

Swaminathan, M. A. (1972). Mutational reconstruction of crop ideotypes. In *Induced Mutations and Plant Improvement*. Int. Atomic Energy Agency, Vienna, pp. 155–171.

Taylor, S. A. and Haddock, J. L. (1956). Soil moisture availability related to power required to remove water. *Soil Sci. Soc. Am. Proc.* 20, 284–288.

Terry, N. (1968). Developmental physiology of sugar beets. I. The influence of light and temperature on growth. *J. exp. Bot.* 19, 795–811.

Terry, N. (1970). Developmental physiology of sugar-beet. II. Effects of temperature and nitrogen supply on the growth, soluble carbohydrate content and nitrogen content of leaves and roots. *J. exp. Bot.* 21, 477–496.

Thomas, M. D. and Hill, G. R. (1949). Photosynthesis under field conditions. In *Photosynthesis in Plants*, eds J. Franck and W. E. Loomis. Iowa State College Press, Ames, Iowa, pp. 19–52.

Thorne, G. N., Ford, M. A. and Watson, D. J. (1967). Effects of temperature variation at different times on growth and yield of sugar beet and barley. *Ann. Bot.* 31, 71–101.

Thornley, J. H. M. (1972*a*). A model to describe the partitioning of photosynthate during vegetative growth. *Ann. Bot.* 36, 419–430.

Thornley, J. H. M. (1972*b*). A balanced quantitative model for root: shoot ratios in vegetative plants. *Ann. Bot.* 36, 431–441.

Ulrich, A. (1952). The influence of temperature and light factors on the growth and development of sugar beets in controlled climatic environments. *Agron. J.* 44, 66–73.

Ulrich, A. (1954). Growth and development of sugar beet plants at two nitrogen levels in a controlled temperature greenhouse. *Proc. Am. Soc. Sug. Beet Technol.* 8(2), 325–338.

Ulrich, A. (1955). Influence of night temperature and nitrogen nutrition on the growth, sucrose accumulation and leaf minerals of sugar beet plants. *Pl. Physiol.* **30**, 250–257.

Ulrich, A. (1956). The influence of antecedent climates upon the subsequent growth and development of the sugar beet plant. *J. Am. Soc. Sug. Beet Technol.* **9**, 97–109.

Ulrich, A. (1961). Variety climate interactions of sugar beet varieties in simulated climates. *J. Am. Soc. Sug. Beet Technol.* **11**, 376–387.

Van Bavel, C. H. M. (1966). Potential evaporation: the combination concept and its experimental verification. *Water Resour. Res.* **2**, 455–467.

Van Bavel, C. H. M. (1967). Changes in canopy resistance to water loss from alfalfa induced by soil water depletion. *Agr. Met.* **4**, 165–176.

Watson, D. J. (1952). The physiological basis of variation in yield. *Adv. Agronomy* **4**, 101–145.

Watson, D. J. and Baptiste, E. C. D. (1938). A comparative physiological study of sugar-beet and mangold with respect to growth and sugar accumulation. I. Growth analysis of the crop in the field. *Ann. Bot.* **2**, 437–480.

Watson, D. J. and Selman, I. W. (1938). A comparative physiological study of sugar-beet and mangold with respect to growth and sugar accumulation. II. Changes in sugar content. *Ann. Bot.* **2**, 827–846.

Watson, D. J. and Witts, K. J. (1959). The net assimilation rates of wild and cultivated beets. *Ann. Bot.* **23**, 431–439.

Winter, H. (1954). Der Einfluss von Wirkstoffen, von Rontgen-und Elektronenstrahlen auf die Cambiumtatigkeit von *Beta vulgaris. Planta* **44**, 636–668.

Wit, C. T. de (1965). *Photosynthesis in leaf canopies.* Agr. Res. Rep. 663. Pudoc, Wageningen, The Netherlands, 57 pp.

Wit, C. T. de (1970). Dynamic concepts in biology. In *Prediction and Measurement of Photosynthetic Productivity.* Proc. IBP/PP Technical Meeting, Trebon, Czech. Pudoc, Wageningen, The Netherlands, pp. 17–23.

Wit, C. T. de, Brouwer, R. and Penning de Vries, F. W. T. (1970). The simulation of photosynthetic systems. In *Prediction and Measurement of Photosynthetic Productivity.* Proc. IBP/PP Technical Meeting, Trebon, Czech. Pudoc, Wageningen, The Netherlands, pp. 47–70.

10

Cotton

J. A. McARTHUR, J. D. HESKETH AND D. N. BAKER

Crop scientists have sought to identify the critical processes in yield development, to find genetic variation in such processes, and to select or develop management practices that will enhance the rates of such processes. In the case of cotton, physiological aspects of the production of seedcoat fibre are of primary concern, although oil and protein composition of the seed are assuming increasing importance.

There are many reviews of the physiology of the cotton plant (Brown, H. 1927; Hector, 1936; Eaton, 1950, 1955; Brown, H. and Ware, 1958; Dastur, 1959; Dastur and Asana, 1960; Carns and Mauney, 1968; and Tollervey, 1970). The symposium on production and utilization of cotton edited by Elliott, Hoover and Porter (1968) included eight chapters devoted to the physiology of cotton, five of which dealt with mineral nutrition. Except for some comments about nitrogen metabolism, mineral nutrition is not considered here. There are many reviews of evolution in the genus *Gossypium* and of cotton genetics (e.g. Hutchinson *et al.* 1947; Saunders, 1961) with histories of the 'Old World' diploid cottons (*G. herbaceum* L. and *G. arboreum* L.) and the 'New World' amphidiploid cottons (*G. hirsutum* L. and *G. barbadense* L.), the New World cottons being of main concern here. There are also several monographs describing the wild and cultivated cottons (e.g. Saunders, 1961).

As mentioned in chapter 1, crop physiology was pioneered by work on the cotton crop in the Sudan (Balls, 1912, 1959). The Empire Cotton Growing Corporation maintained a research station in Trinidad from 1926 to 1944, which was dedicated to basic research on the genus, with emphasis on genetics. Mason, Phillis, Maskell and co-workers at the Trinidad laboratory published many papers on cotton physiology and made an outstanding contribution to the understanding of translocation of assimilates.

This long tradition of fundamental research on *Gossypium* has been continued in numerous laboratories throughout the world, including two new laboratories in the United States Department of Agriculture concerned solely with cotton physiology. A vast store of knowledge has been accumulated in the many disciplines of cotton physiology, and now computer simulation efforts by several groups offer hope of meaningful integration of this knowledge in problem-solving for crop and pest management.

Species evolution in Gossypium

The botany of the genus *Gossypium* was explored intensively at the Trinidad Station. That work, complemented by individual scientists at other locations, is summarized particularly well in two books (Harland, 1939; Hutchinson *et al.* 1947).

Most of the genus has a short-day flowering habit and is sensitive to frost, which restricts it to the tropics and subtropics. Hutchinson *et al.* (1947) proposed an ancient stage of diversification related to the theory of continental drift in the Mesozoic era. Fryxell (1965) suggested several stages of evolution involving adaptation to desert environments, littoral dispersal and spread into the temperate zone with associated loss of the short-day behaviour. Most cottons are found in . . .'xerophytic plant communities in which there is open ground where seedlings can develop without close competition' (Hutchinson *et al.* 1947). The wild amphidiploid cottons '. . . are commonly found in or associated with strand habitats, existing in the open vegetation that characterized the vicinity of marine beaches' (Fryxell, 1965).

The amphidiploid cultivated varieties often escape into xerophytic plant communities or areas where forests are suppressed for some reason. Only one of the thirty or more wild species can survive under shaded conditions. Cotton is one of a few species whose leaves can absorb moisture from humid air (Monteith, 1963; Hoffman *et al.* 1971).

One interesting riddle is how amphidiploid (2AD) genomes arose by hybridization of the South and middle-American D genome with the Asian–African A genome. Hutchinson *et al.* (1947) attribute the hybridization to the introduction of the A genome by man to the Americas, where the D genome was extant. The recent voyage from Africa to the West Indies of Ra II, a boat of an ancient Egyptian type made of reeds, suggested a possible means of transport of the A genome to the Americas (Heyerdahl, 1971). It also has been shown that cottonseed can float for considerable lengths of time on salt water, suggesting another method for dispersal of the species (Stephens, 1958). Hutchinson (1959) and Purseglove (1968) discussed the controversy resulting from this conclusion and summarized other current hypotheses. Fryxell (1965) proposed that the cultivated amphidiploids were obtained from wild types that originated 100000 to one million years ago. Archaeological specimens may help some in establishing the antiquity of cotton (Smith and MacNeish, 1964). Brown, M. (1951) described the spontaneous production of amphidiploids in hybrids between species of the B and D genomes in her garden. One new strain would not cross with its wild parents, theoretically allowing it to become established as a new species. The AD amphidiploid was synthesized earlier from its probable A and D parents (Beasley, J. 1942). The amphidiploid types are now found in Central and South America and throughout the Pacific and Caribbean Islands. At least one of the strains found in the Pacific is a different species (*G. tomentosum*, 'Nutt. ex Seem.').

Comparative physiology

The comparative physiology and biochemistry of the 30 or more wild and cultivated species are important aspects of cotton physiology. Gossypol is harmful to animals if consumed in quantity from cottonseed meal, and considerable variation exists in the genus in seed gossypol content. El-Nockrashy *et al.* (1969) suggest that it can be reduced in seeds of cultivars by appropriate crosses with wild species. Efforts to increase the gossypol content of floral buds and bolls in order to enhance insect resistance have been described by Singh and Weaver (1972).

Modern cultivars, after intensive genetic selection, have larger leaves, larger seeds and more rapid photosynthesis than many of their wild relatives. In the genus, differences occur in responses of floral initiation to environment, in cold tolerance during seedling growth, and in drought tolerance. Thus, a recurring theme of this chapter will be the comparative physiology among the species and races of *Gossypium*. Characters can be transferred from wild species or races to the cultivars. It is very important, then, that remaining races and species in South America, the Caribbean, the Pacific Islands, and Australia be collected, and that their physiological behaviour as well as resistance to cold and drought be characterized.

Seedling growth

Frequent failure to obtain good field stands has resulted in renewed consideration of control mechanisms in seed germination and in early seedling growth (McArthur *et al.* 1971), particularly as the production of cotton fibre is strongly affected by field stand and by seedling vigour (Wilkes and Corley, 1968; Christiansen and Moore, 1959). Cotton seed germination is complicated by water uptake and chilling problems. Seedling emergence and early growth are affected by chilling, soil impedance, mineral nutrition, light, moisture-growth rate balance, and some unknown factors (Powell and Morgan, 1970).

The cotton seed is covered by a dense mat of short and long fibres, by a heavy seedcoat with a chalazal cap-opening mechanism, and by a thin layer of endosperm. Most of the seed mass consists of large, folded cotyledons which surround a hypocotyl–radicle axis (Tharp, 1960). The seed contains about 23 % fat (Brown, H. and Ware, 1958). The glyoxylate cycle (Kornberg and Krebs, 1957) operates in germinating seeds, converting stored fat to carbohydrate (Mohapatra *et al.* 1970).

Early workers found at least two major factors inhibiting cotton seed germination; an impermeable seedcoat mechanism, and impenetrability of seedcoat fibres by water. Fibre removal by concentrated sulphuric acid facilitated water transport toward the seedcoat and resulted in a more permeable seedcoat because of loss of outer seedcoat layers. However, 10–25 % of commercially acid-delinted seeds may remain impervious to water (Christiansen and Moore, 1959). Also, seed stored in exceptionally dry air becomes hardened with time (Toole and Drummond, 1924). Such hard seeds become permeable after exposure to water at 80 °C for about two minutes (Walhood, 1956).

Seed imbibition and splitting of the seedcoat are very complex processes. After a small amount of liquid water reaches the interior of the seedcoat, there is a two-fold increase in size of the seedcoat and opening of the chalazal pore. Large amounts of water then enter the chalazal opening, hydrating the cotyledons, hypocotyl–radicle axis, and endosperm, all of which expand, filling and splitting the seedcoat. Seedcoat expansion and seed imbibition can be separated in time by slow hydration at low water levels. If the sequence of imbibition and splitting of the seedcoat is interrupted by moisture stress, as might be induced by shallow planting (since disturbed topsoil dries rapidly), the seeds revert quickly to the hardened state. However, in the case of such hard seeds, a second wetting usually produces rapid seedcoat expansion and imbibition.

Highly permeable acid-delinted seeds might be inhibited by insufficient aeration under sustained high soil moisture conditions, and fibre impenetrability to water may have been a protective mechanism in the wild state. If such were the case, the chalazal cap pore would be a mechanism for alleviating difficulty in water uptake.

The necessary seed moisture level for germination is about 55 % of the seed fresh weight. If the moisture level of the seeds is raised to 40–45 % of the fresh weight prior to planting, substantial increases in seedling vigour are noted (McArthur and Bowen, unpublished). Pre-germinative hydration results not only in greater seedling vigour but also increases germination percentages and produces a more uniform germination pattern for the seedling population.

Seedling emergence can be advanced by a day or more (emergence time is five days at a planting depth of 2–3 cm) if the humidity surrounding the hypocotyl and cotyledons is increased, which suggests that the rate of seedling emergence in field plantings may, in large part, be determined by the humidity around the upper portion of the seedling (McArthur and Bowen, unpublished). Finally, it should be noted that at least some forms of chilling damage are eliminated by various pre-germinative seed hydration techniques (Christiansen, 1968; Thomas and Christiansen, 1971).

Carbohydrates and amino acids leak out of the radicle during chilling periods or under anaerobic or low pH conditions (Christiansen *et al.* 1970), possibly due to membrane damage during chilling. Chilling may also affect the functioning of a metabolic control enzyme, such as [isocitratase (Mohapatra *et al.* 1970). Some wild species apparently exhibit cold tolerance during germination and early seedling growth (Muramoto *et al.* 1971; Smith *et al.* 1971), which could be transferred to cultivars to improve establishment and yield. Exposure of seedlings to cold early in the season has been reported to result in early flowering and shorter internodes, two characteristics that sometimes result in greater yields (Muramoto *et al.* 1971).

Models for the prediction of time from planting to emergence and of percent emergence, based upon the germination percentage of the seed in laboratory tests, the soil or seed moisture status, and temperature and physical impedance of the soil, have been field tested throughout the American Cotton Belt (Wanjura *et al.* 1971; Wanjura and Buxton, 1972 *a, b*). Errors in the prediction of hours to emergence are typically less than 5 %. Another model being developed for this purpose will include, in addition to the usual factors for seedling emergence, humidity effects, soil oxygen status, impedance increase from crusting (following rain), wind velocity, air temperature and soil moisture tension (cf. Coble and Bowen, 1970).

During early development the seed and seedling have several major control centres: endosperm, cotyledons, hypocotyl, shoot apex (and leaf primordia), and root apex.

The endosperm covering the seed may control germination by preventing the exit of the radicle tip. Preliminary studies have shown that about 30–50 grams of pressure force were necessary to rupture the endosperm. Endosperm strength decreased at advanced stages of hydration and development.

During the first 60–100 hours of germination, the radicle apex is easily damaged by chilling or anaerobic conditions. If the apex is killed, a shallow root system of secondary roots, subject to moisture stress, develops.

Leaf expansion

The optimum temperature for the relative rate of leaf expansion per plant (R_A or $1/A \, dA/dt$) is about 30–33 °C (Hesketh and Low, 1968; Hesketh and Baker, 1969). R_A values are comprised of several components: (*a*) rate of leaf initiation, (*b*) rate of leaf expansion, (*c*) the ultimate size of each leaf, and (*d*) the rate of development of axillary branches (Hesketh, Baker and Duncan, 1972). Components (*a*) and (*b*) are maximum at 30–33 °C (Hesketh, Baker and Duncan, 1972) and component (*c*) is maximum at 18–20 °C (Hesketh and Low, 1968). Axillary branches are best developed at about 20 °C but may be abundant at 30 °C under conditions conducive to rapid supply of photosynthate (intense light, carbon dioxide enriched air, spaced plants), reflecting an effect of photosynthate supply per plant on branching.

Leaf photosynthetic carbon dioxide assimilation

Böhning and Burnside (1956) found that carbon dioxide assimilation by leaves on cotton plants grown in Ohio greenhouses in summer was light-saturated under one-quarter of full sunlight, at a rate of 20 mg CO_2 dm^{-2} h^{-1}. Leaves of plants grown in Arizona in full sunlight, in glasshouses of the Canberra phytotron during summer, and also under artificial light in that phytotron, had carbon dioxide assimilation rates of 45–50 mg CO_2 dm^{-2} h^{-1}, which were not light-saturated (El-Sharkawy *et al.* 1965; Elmore *et al.* 1967; Hesketh, 1968; Troughton, 1969). Reduced carbon dioxide assimilation by plants grown in low light was partly due to stomatal closure (Hesketh, 1968) and partly to reduced Hill reaction activity (Fry, 1970). In many cotton-growing areas, long periods of heavy clouds are not uncommon. The photosynthetic characteristics of the leaf can, therefore, be expected to vary considerably in the field, depending upon available sunlight.

The photosynthetic rates of leaves from plants grown at day temperatures from 18 to 36 °C were compared at 30 °C by Hesketh (1968), who found that leaves of plants grown at a day temperature of 21°C had the highest photosynthetic rate.

Little has been done to describe the effects of mineral nutrition or ageing on photosynthesis in cotton. The photosynthetic mechanism of leaves does deteriorate due both to ageing and to shading (Muramoto *et al.* 1967). De Michele and Sharpe (1972) have developed a comprehensive model of cotton leaf conditions, including the structural characteristics of the guard cell wall, to simulate stomatal action. Guard cell wall thickness and level of enzyme activity were determined by the age and developmental history (presence or absence of drought during development) of the leaf.

Photosynthetic rates among races, cultivars, and species of cotton grown in Arizona fields under irrigation were surveyed (El-Sharkawy *et al.* 1965; Elmore *et al.* 1967). The highest rate (51 mg CO_2 dm^{-2} h^{-1}) was measured in the wild, diploid *G. incanum*, and rates for a number of other wild species were comparable to those of *G. hirsutum*. In another survey, rates varied from 33 to 45 mg CO_2 dm^{-2} h^{-1} among the amphidiploid races and hybrids, with some interaction between varieties and dates of sampling (stage of growth). Varieties with heavy boll loads had low photosynthetic rates. It has been reported that once the boll load builds up, the nitrogen level in leaves often decreases because of boll demand (Tucker, T. and Tucker, B. 1968; Bassett, Anderson, and Werkhoven, 1970). As the sink demand for photosynthate builds up, the leaf is stressed for materials needed for active photosynthesis, which undoubtedly contributes to the lack of correlation of photosynthetic rate with productivity.

When a cotton plant becomes determinate with a load of bolls, no new leaves develop until most of the bolls have matured. Presumably, leaf senescence is delayed in such plants because of the strong boll demand for photosynthate.

Estimation of the various resistances to flow of carbon dioxide and water between leaf and air, as developed by Maskell (1928) and Gaastra (1959), has led to a great deal of work on cotton, which illustrates the potential of mathematical modelling in interaction with experimental measurement for the understanding of a process. The method of sequential resistance analysis is very useful in sorting out stomatal and mesophyll differences among species or varieties, associated with differences in their photosynthetic assimilation (El-Sharkawy and Hesketh, 1965). Mesophyll resistance (r_m) in cotton leaves is greater than in corn (El-Sharkawy and Hesketh, 1965), by about 0.3–0.5 cm s^{-1}, accounting for 10 mg CO_2 dm^{-2} h^{-1}.

When cotton leaves are stressed for water, increased r_s affects photosynthetic carbon dioxide flux into the leaf long before r_m is affected (Troughton, 1969). El-Sharkawy and Hesketh (1964) had suggested that this might be the case, as severely wilted leaves often sustained maximal photosynthetic rates. The issue is not clear, however. Fry (1970), and Vieira da Silva (1972) found that the Hill reaction of chloroplasts was affected below -5 bars leaf water potential, and concluded that there was a direct association between photosynthesis and water stress. A leaf water potential of -5 bars is typical for well-watered plants (Jordan and Ritchie, 1971; Klepper, 1968; Klepper *et al.* 1971). Jarvis and Slatyer (1970) have found a cell wall resistance (r_w) in cotton leaves which appears during rapid transpiration, and which would affect any estimates of carbon dioxide resistance according to the method of Gaastra (1959).

Cyclic rhythms in photosynthesis and transpiration caused by changes in stomatal aperture have been studied using the resistance analytical methods (Barrs, 1968, 1971; Troughton, 1969, Cowan, 1972). Interesting transient depressions in carbon dioxide exchange were observed for leaves showing oscillating transpiration above 37.5 °C (Troughton and Cowan, 1968). By analysis of resistances, this anomalous depression in carbon dioxide exchange, which was reversible, was shown to be independent of r_s.

Estimates of resistances under different environments are useful in computer models simulating leaf photosynthesis and transpiration in cotton (Buxton and Stapleton, 1970; Cowan, 1972), but criticisms of the technique of sequential resistance analysis are accumulating. Leaf temperature and carbon dioxide concentration at the reaction site are critical for accurate partitioning of the resistances (cf. Barrs, 1968; Cowan and Troughton, 1971).

Photosynthesis and the carbon balance

Recent models for plant growth with respiration subsystems require estimates of available gross photosynthate or net photosynthetic rate plus light and dark respiration. There are four methods:

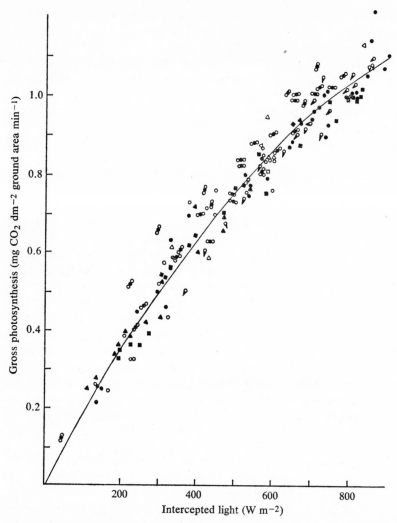

Figure 10.1. The relation between light intensity and gross crop photosynthesis (i.e. net CO_2 assimilation plus respiration adjusted for light intensity) (Baker *et al.* 1972).

(*a*) Combining a leaf model for gross photosynthesis with a model of canopy architecture to estimate canopy photosynthesis. Photosynthesis by other organs, and losses by respiration must then be modelled. The leaf model must account for acclimation and senescence.

(*b*) Developing a model based upon carbon dioxide exchange of canopies enclosed in chambers that can be darkened (Baker *et al.* 1972) (cf. Figure 10.1).

(*c*) Determining carbon dioxide flux to the canopy, with simultaneous determinations of plant and soil respiration in the system. As in (*b*), canopy acclimation and senescence must be described.

(*d*) Dry matter sampling, with analysis into the major components, proteins, carbohydrates, oils, etc., combined with a respiration model (Penning de Vries, 1972; Hesketh, Baker and Duncan, 1972; Thornley and Hesketh, 1972).

Estimation of respiration in the light remains uncertain. During the light period, growth respiration may be slowed because of water stress, but photorespiration may also occur. There is some evidence that synthesis of sugar, fats, and amino acids changes during the light period, greatly complicating instantaneous estimates of light respiration, as carbon dioxide exchange associated with each synthesized product differs.

The only experimental method available for estimating gross photosynthesis is to add net photosynthesis to respiration measured shortly after in the dark, avoiding gushes and gulps of carbon dioxide that occur immediately after the light is shut off. Gas exchange studies have seemed sufficient for the study of rate limiting processes in primary productivity, but it now appears that aspects of respiration and biochemical synthesis can greatly complicate their interpretation.

The various leaf shapes available in cotton by genetic manipulation must also be considered. Okra leaves seem to be as efficient photosynthetically as a canopy of normal leaves, if both are at the same plant density and intercepting the same amount of light (Baker and Myhre, 1969); however, air movement was greater in the okra leaf stand (cf. Andries *et al.* 1969).

Photosynthetic reserve

Leaves store considerable carbohydrate, compared with other parts of the plant. Starch and sugar levels of 2–11 % of leaf dry matter have been reported for large cotton plants (Wadleigh, 1944; Eaton and Rigler, 1945; Eaton and Ergle, 1948; Dastur, 1959; Guinn and Hunter, 1968) and values of up to 47.3 % sugars and starch have been measured in leaves and petioles of squaring plants grown at 1000 ppm CO_2 and 29 °C in greenhouses when the incident light was 650 ly day^{-1} (Hesketh, Fry, Guinn and Mauney, 1972). Leaves of control plants at 300 ppm CO_2 had 16 % starch and sugars on a total dry weight basis. In these experiments, plants at 1000 ppm CO_2 set twice as many bolls as plants at 330 ppm CO_2, and leaves rapidly became chlorotic. Dastur (1959) reported total sugar levels of 8 % in bolls and 6 % for stems, on a dry weight basis.

Translocation

The rate of dry matter accumulation in non-photosynthetic organs is sometimes used to estimate fluxes of carbohydrate. In addition to better estimates of phloem cross-sectional area leading into such an organ, the plant should be growing under optimal conditions for photosynthesis if the capacity of the translocation system is to be estimated. A correction needs to be made for respiratory losses, which may account for 30 to 40% of the photosynthate transported into the growing organ (Baker *et al.* 1972). Canny (1971) recommended that the classic experiments of Mason and Maskell (1928) and Phillis and Maskell (1936) be repeated to generate better experimental data for modelling purposes. Associated calculations may well indicate how much the transport processes limit growth.

Tracer methods have been used with cotton plants to indicate which leaves are interconnected by the vascular network (Tuichibaev and Kruzhilin, 1965; Brown, K. 1968) and to show that bolls closest to photosynthetically-active leaves serve as strong sinks for photosynthate exported from such leaves (Ashley, 1972). However, young bolls are shed despite their closeness to active leaves and older bolls down in the crop get priority for available photosynthate.

Sabbe and Cathey (1969) observed that in plants stressed for water, more of the photosynthate is translocated to the roots. The roots, being stressed the least, would probably be requiring the most photosynthate for growth.

DEVELOPMENTAL PHYSIOLOGY

The cotton stem and root have a cambium which produces wood. Cambial growth is probably controlled by photoperiod and is influenced by activity of the main shoot apical meristem. The shoot apex can become quiescent during growth of bolls, possibly due to competition for photosynthate, but it quickly resumes growth after the bolls mature.

Much of the unique growth behaviour of cotton is controlled by the activity of axillary meristems, as discussed below. Meristematic behaviour, in turn, is controlled by assimilate and nutrient supply and hormones, all interacting with environmental stimuli.

The rate of organ initiation

Mauney (1968) presented a detailed review of cotton morphology. The apical meristem of the main shoot and of non-flowering branches is always vegetative. Flowers, with associated leaves and internodes, are produced on axillary branches. There is a bud in the axil of every leaf on the main shoot as well as in the axil of the associated prophyll, and both buds can produce either vegetative or flowering branches. When the prophyll bud develops into a branch, a third bud develops in the same axil. New buds develop indefinitely in the same axil if the plant is continually pruned to remove older buds. Floral branches consist of a prophyll, a leaf, two internodes, a flower or

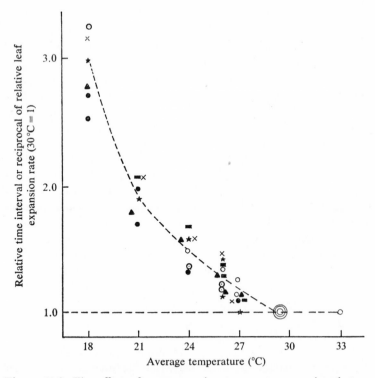

Figure 10.2. The effect of mean growing temperature on time between flowers, squares, and leaves along main shoot nodes or floral branches, on the square and boll periods, and on the reciprocal of the relative rate of leaf area expansion (days) in seedlings. Intervals between: square to flower ⊙, ●; flower to open boll ○; flowers ★; squares ■; leaves ▲. Reciprocal of relative leaf expansion rate × (Hesketh, Baker and Duncan, 1972).

fruit and two axillary buds. Often as many as seven such floral units develop in a series from one main shoot node. With proper pruning and environmental conditions, there could be a large number of floral units on one branch.

In developing a model to simulate the growth of a cotton plant, we found it convenient to resurrect the Zaitzev isophase concept (Hector, 1936) which predicted the effect of temperature on duration of growth periods. Zaitzev's basic unit, the isophase, is the time between expansion of successive leaves. In cotton, the time between the appearance of squares, flowers, or open bolls at equivalent sites on successive nodes on the fruiting branches is dependent upon this same isophase. Isophase units also separate the development of successive morphological events associated with flowering within a branch: expansion of the sympodial leaf or prophyll, appearance of squares or flowers, and the opening of bolls (Figure 10.2). Such a concept neatly describes the gross morphogenesis of the cotton plant at different temperatures.

Organogenesis: carbohydrate requirements

Knowledge of nutritional aspects of organogenesis, together with the iso-phase concept, provides a basis for estimating assimilate demand for any 24 h period at a particular temperature. Maximum rates of dry matter accumulation, and respiratory needs of cotton flower buds, flowers or bolls for plants in controlled environments are now known (Figure 10.3). The average efficiency of boll formation was found to vary from 68 % at 21 °C to 77 % at 27 °C (Baker *et al.* 1972).

Removing every other new boll did not increase the growth rate of the remaining bolls over the first 30 days (Figure 10.4); presumably translocation or boll growth capacity was limiting. The optimum temperature for boll growth was about 27–30 °C (Figure 10.5) for plants growing under intense light and carbon dioxide enriched air. Estimates are available for both the growth and maintenance components of respiration in cotton (Hesketh *et al.* 1971; Thornley and Hesketh, 1972). If the photosynthate produced per day is estimated on a per plant basis, then the carbohydrate deposited daily is estimated as follows:

$$dW/dt = (P - PR_\mathrm{L} - R_0 W)/(1 + G_\mathrm{R}),$$

where W is plant dry weight, P is photosynthate and R_L, R_0 and G_R represent respiratory losses associated with the photosynthetic process, maintenance and growth, respectively (Baker *et al.* 1972). This dW/dt value, plus the content of reserves, represents the available supply of carbohydrate. The demand is obtained by summing the potential growth increments for all meristems in the plant at the particular temperature, taking into account organ age from plots of weight and respiration versus time. If demand is greater than supply, the daily increment of growth equals supply. If not, growth equals demand and the excess is held in reserve with the possibility of feedback inhibition of photosynthesis when the reserves become large. The potential growth increments have been obtained for bolls from experiments in which supply exceeded demand (Hesketh, Fry, Guinn and Mauney, 1972). More informa-tion is needed about the reserve capacity, availability over 24 hours, respira-tory or energy requirements, control systems, and effect of photosynthate accumulation on photosynthetic rates. There are probably correlative interactions between bolls which would be difficult to detect if supply ex-ceeded demand. In the experiments cited, there was no clear-cut evidence that bolls far out on a floral branch were any lighter than those close to the main shoot.

Another missing component is prediction of which meristems will develop into mature organs. We discuss these requirements below.

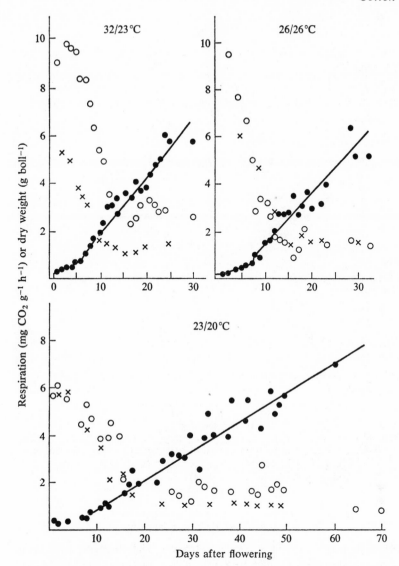

Figure 10.3. Respiration and dry matter accumulation in cotton bolls grown under three temperature regimes. Respiration – day ○; respiration – night × ; dry weight ● (Baker *et al.* 1972).

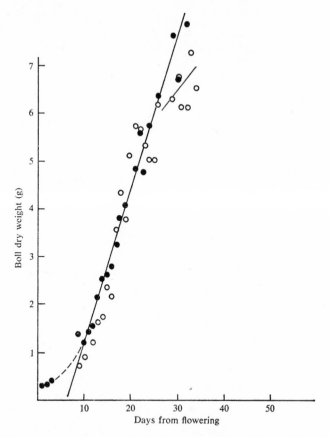

Figure 10.4. Dry matter accumulation in cotton bolls with time from flowering for plants in 32°/23 °C day/night temperatures (○). For some plants, every second flower was removed on the day of flowering (●).

Organogenesis: nitrogen requirements

The cotton boll requires considerable nitrogen. The nitrogen content of leaves typically decreases from 5 to 2.5% of dry weight as the season progresses and nitrogen becomes limiting, being diverted to bolls (Tucker, T. and Tucker, B. 1968; Bassett, Anderson and Werkhoven, 1970).

Jones *et al.* (1973) have developed a nitrogen budget for a cotton plant with rules similar to those for the carbohydrate budget. Maximum and minimum concentrations were set for leaves and shoots. Nitrogen supply depended upon root extension, residual soil nitrogen, nitrogen from breakdown of organic matter, and that added as fertilizer. In the model, demand was generated in much the same way as the carbohydrate demand was

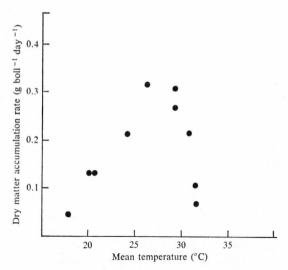

Figure 10.5. Boll growth rates as influenced by temperature. Values at 27 °C and above were determined for plants in air enriched with CO_2, with every other flower removed for some plants as in Figure 10.4.

calculated but without a comparable respiratory component. By varying the relative strictness of rules between the carbohydrate and nitrogen budget subsystems, one or the other could cause the plant to shed floral buds and bolls, and restrict vegetative growth. Particularly impressive was the fact that in initial runs of the computer model, there was considerable restriction on growth and shedding of floral buds and bolls due to the nitrogen subsystem (Jones *et al.* 1973). Better relationships describing the availability of nitrogen from fertilizer (usually about 50 %), and better studies of the effects of nitrogen and photosynthate and their interactions on growth and abscission, are needed.

Growth control mechanisms

(a) *The nutritive theory*

According to the nutritive theory, plant growth is limited mainly by carbohydrate and mineral supply, and the problem is one of priority: which axillary buds develop and which stay dormant? The proximity of a carbohydrate source such as a photosynthesizing leaf may determine priority. Also, the ability of an unfolding axillary leaf to contribute to its own growth requirements by photosynthesis is probably important.

(b) *The nutrient–diversion theory*

Hormones or inhibitors would interact with photosynthate supply to control growth. One suggestion is that auxins from the shoot apex inhibit vascular development to an axillary meristem, forming a resistance to flow of carbo-hydrate and other nutrients to the area (Phillips, 1969). Once the leaf develop-ing there can photosynthesize, the vascular system develops and connects the new leaf to the rest of the plant.

Cytokinins are formed in the root system, and auxins from the shoot apex supposedly divert them to developing organs where they somehow attract nutrients from the vascular system (Fox, 1969). Cotton bolls contain cyto-kinins (Sandstedt, 1970, 1971) with the highest concentrations occurring between the fourth to ninth day after anthesis, i.e. before the linear maxi-mum growth rate of bolls is reached.

(c) *Non-nutritional controls, photo- and thermoperiodism*

Berkley (1931) reported that *G. hirsutum* 'Upland Big Boll' produced squares irrespective of day-length at moderately low temperatures (22 °C), whereas at 32 °C plants in long days did not produce squares, although plants in short days did. Lang (1952) classified several races of *G. hirsutum* as short-day strains and others as day-neutral. Since most strains are tropical, the short days would co-ordinate flowering with the dry winter season when pollen transfer is most effective (cf. Hoffman and Rawlins, 1970). Many of the wild races of *G. hirsutum* and *G. barbadense* collected from Central and South America do not flower in long days (Waddle *et al.* 1961). Mauney and Phillips (1963) screened 23 species of *Gossypium*, and found that short days and cool nights (15 °C) enhanced flowering in most cases.

Following early growth at a cool temperature (21 °C), cotton cultivars initiated floral buds at about the fourth node, independent of day-length (Mauney and Phillips, 1963; Low *et al.* 1969). After a week at 21 °C, exposure to 30 °C did not raise the node of the first floral branch in *G. hirsutum*. Additional exposure to 30 °C did raise the node of the floral branch in *G. barbadense* considerably, and the F_1 interspecific hybrid was intermediate in response to such treatments (Moraghan *et al.* 1968; Low *et al.* 1969).

For some cultivars, temperature has a greater effect than day-length or light intensity on the node of the first floral branch (Mauney, 1966; Low *et al.* 1969). Cultivars of *G. barbadense* are more sensitive to day-length than cultivars of *G. hirsutum* (Low *et al.* 1969), and strains of the former had to be selected for early flowering under longer day-lengths before it could be grown commercially in the United States (Fryxell, 1965).

Fruiting in cotton may be more sensitive to day-length than is flowering. Half of a collection of 600 native or primitive stocks of *G. hirsutum* from Mexico and South America failed to set fruit in College Station, Texas, during the growing season (maximum day-length, 15 h). Lack of fruit set versus normal fruiting at College Station was found to be under multigene control (Waddle *et al.* 1961). Introductions of *G. barbadense* did not flower during the growing season in College Station, Texas, their day-length response being controlled by one gene with the short-day response dominant (Lewis and Richmond, 1960).

The node of the first flower is directly associated with earliness, which in turn is often associated with a heavy crop (Richmond and Radwan, 1962). Strains of *G. barbadense* that differ in node of the first fruiting branch have been selected for commercial production in southwestern United States at different elevations (Feaster and Turcotte, 1965). A chemical for aborting flowers on low-fruiting branches has been reported (Pinkas, 1972).

Numerous fruiting sites per fruiting branch, as well as many well-developed axillary branches with fruiting branches, produce many flowers. When the plant is heavily loaded with bolls, the carbohydrate supply becomes inadequate to meet the demand, growth of vegetative meristems slows, and such plants become effectively determinate. Root growth also slows, reducing the ability to absorb water, nitrogen, and other nutrients, and thereby enhancing the expression of this determinate state. Such a population may become effectively determinate too early in the season, before a fully intercepting canopy is developed, resulting in unused sunlight and lower yields. Such behaviour may, however, be of advantage in some crop rotations, and may facilitate control of pests. Insect problems become worse as the season progresses. When the boll weevil became prevalent in the southeastern United States, earlier varieties were selected which would mature a crop before the insect population increased too greatly.

With rows 1 m apart and 100000 plants ha^{-1}, flower bud initiation occurs while there is considerable space between relatively young plants, whereas during flowering there is considerable interplant competition for light, water, and nutrients. Flower and boll number is thus determined under noncompetitive conditions, with bolls developing under considerable interplant competition. The plant cannot supply all the bolls with photosynthate or other nutrients, and 30 to 70% are shed. If insects destroy floral buds and young bolls, there is a need for the excess of floral sites.

Hormonal diversion of nutrients and control of shedding

Older bolls are usually not shed, and the boll requirement for carbohydrate reaches a stable maximum some 10 days after anthesis (Figure 10.4). Yet older bolls and their leaves may be shaded by newer vegetation. Some sort of control must exist for diverting photosynthate to bolls more than 10 days old. Hormonal effects and shifts in the carbohydrate:nitrogen balance in younger bolls may lead to the formation of an abscission zone and shedding of the boll. Cytokinins and abscisic acid counteract each other in this process (cf. Sandstedt, 1971).

Several abscission-accelerating substances, as well as retardants, have been isolated from young cotton fruit (Carns, 1966; Smith, D. 1969), and much of the early work on abscisic acid was done with cotton leaves and fruit (Carns, 1966; Addicott and Lyon, 1969). Boll weevil larvae produce a proteinaceous shedding factor in an infected boll, which appears to differ from abscisic acid (King and Lane, 1969).

The level of carbohydrate in roots, stems, leaves, and bolls under different levels of supply-demand stress remains stable, leading Eaton (1955) to conclude that limited photosynthate supply was not directly causing all of the observed boll shedding. This conclusion had considerable impact on the direction of subsequent research; and an effort was made to find the controlling hormones and inhibitors (Carns, 1966; Addicott and Lyon, 1969). It might be possible to find excess photosynthate under the following conditions, qualifying Eaton's conclusions:

(a) The plant may eliminate potential meristematic sites during unfavourable conditions, and under more favourable subsequent conditions photosynthate may accumulate for lack of a sink, up to a point where reserve levels may inhibit photosynthesis by feedback.

(b) Within limitations, short-term growth rates depend to some extent on short-term supply, thereby helping to maintain a constant reserve level, which would 'overshoot' under such growth limitations as moisture stress.

It seems likely that growth rates and shedding interact over time to keep the carbohydrate level fairly constant, with drops in carbohydrate level not lasting very long. Presumably, drops in carbohydrate induce abscission in certain organs.

Present growth models for cotton consist mainly of daily or hourly balance sheets of supply and demand for carbohydrate and nitrogen, and there is often not enough photosynthate or nitrogen to mature all the bolls produced by a plant in a normal field population. In fact, according to the SIMCOT model, boll growth is limited by carbohydrate or nutrients under most conditions (Duncan, 1972; Jones *et al.* 1974). Eaton points out that the fresh weight ratio of fruit to plant is fairly constant for a range of environments, although this can change with plant stress or carbon dioxide concentration (Hoffman *et al.* 1971; Hesketh, Fry, Guinn and Mauney, 1972). Fresh weight is an indicator of photosynthetic surface or capacity of the plant to generate photosynthate, and the relatively constant ratio indicates that the plant balances supply and demand. Reported reserve levels (sugars plus starch) are rarely equivalent to more than one or two days' carbohydrate demand. This fact does not give a hormone or inhibitor, independent of carbohydrate supply, much leeway within which to operate. Plants placed in darkness for 48 hours will shed most of their young bolls and flower buds. The peak of shedding occurs two to three days after the treatment. However, plants that had ceased to flower shed no bolls even after eight days of darkness, indicating that the entire plant had probably reached an advanced stable developmental plateau (unpublished data, Hesketh and Baker).

Root physiology

The seedling radicle becomes the main root of the mature plant, which may reach a depth of 2 m or more. Secondary root primordia develop about 12 cm behind the primary root apex, and tertiary roots develop 5 cm behind secondary root apices.

Root extension has been followed by placement of ^{32}P at different points in a volume of well-drained alluvial soil. The taproot extended about 2.5 cm d^{-1} down to 183 cm; lateral roots extended at half this rate (Bassett, Stockton and Dickens, 1970).

Multiple regression models for data from many sources were able to account for 68 % of the variation in root growth over 100 hours with functions of soil temperature, penetrometer resistance, and aluminium activity of pH (Taylor *et al.* 1972). Taylor and Klepper (1971) measured root density, soil moisture, water extraction per unit length of root, and plant water potential during a drying cycle for two-month old cotton plants in transparent containers where root growth could be observed. The root system was extremely dynamic during soil drying, being most active, as expected, in the closest volume of moist soil.

Trouse (1970) found that root growth slowed drastically during the setting and maturing of bolls, the bolls being a stronger sink for photosynthate than the root system.

A computer model simulating root growth in cotton has been developed by M. Huck (personal communication).

Water relations

Cotton is grown under a broad range of atmospheric moisture levels. Daily maximum saturation deficits are commonly of the order of 20 mb in humid areas, and approach 90 mb in many of the irrigated desert regions. Cotton is adapted to large saturation deficits, and the minimum daily leaf water potential can vary from -6 to -28 bars depending mainly on the availability of soil moisture. Namken (1964) ranked the various environmental factors affecting the relative water content of cotton leaves in the following order: soil moisture, air temperature, vapour pressure deficit, pan evaporation and air movement, with the first two factors accounting for 84 % of variation in measurements.

Since commercial plantings of cotton are often stressed for water, water stress physiology is important. Objectives of associated studies have been (*a*) predicting evapotranspiration and available water in soils, (*b*) developing plant indicators of stress levels that depress yields, and (*c*) developing varieties tolerant of stress.

The various physiological processes in cotton appear to respond differently to moisture stress. The method of inducing stress evidently determines stomatal response. El-Sharkawy and Hesketh (1964), Barrs (1968) and Troughton (1969) found near maximum photosynthetic rates in wilted leaves shortly after stress was induced. For cotton canopies stress was induced with the soil moisture near field capacity by imposing a higher than normal radiant heat load on the plants. Under these circumstances both photosynthesis and transpiration proceeded rapidly after complete loss of turgor. In other experiments, where soil moisture tensions ranged from 0.3 to 0.9 bars, no significant change in photosynthesis with leaf water potentials varying from −8 to −20 bars was observed at 20 °C, but at 40 °C a 70 % reduction occurred (Baker, unpublished data). This response was similar to that reported by Boyer (1970) for sunflower. In other experiments, decreases in transpiration with increasing soil suction paralleled, but were more pronounced, than the decreases in photosynthesis.

Jordan and Ritchie (1971) found that stomatal resistance in cotton, estimated from transpiration measurements, was unaffected by water potentials down to −27 bars, although it increased to infinity at a leaf water potential of −16 bars in potted plants. Data from studies with potted plants must be used with reservation in attempting to predict behaviour of plants growing in the field during a prolonged drought.

Namken (1965) reported that at a relative water content of 0.72 (equivalent to −24 bars) leaf wilting was barely visible. Further, he found that this stress, observed in the afternoon, had no effect on vegetative growth but that an afternoon value of 0.64–0.66 significantly reduced growth. Slatyer (1957), Lawlor (1969), and Jordan (1970) reported suppression/reduction of height and leaf growth with leaf water potentials one bar below that at wilting. Similar results have been found for soybeans, sunflowers and corn (Boyer, 1970). Jordan (1970) commented that water potentials must decrease to at least −8 bars during part of the day for growth to be affected.

The diameters of cotton bolls increased consistently for 16 days, even when leaves were severely wilted (Anderson and Kerr, 1943). Meyer *et al.* (1960) suggested that growth in such bolls, as well as in other meristematic regions, was not suppressed by water stress, except when stress was very severe. Clearly photosynthesis, leaf expansion and meristematic activity differ in sensitivity to changes in water stress.

The evolution of the species under arid conditions suggests drought tolerance. Vieira da Silva and co-workers have explored the differential biochemical and physiological basis of drought resistance and avoidance among many cotton species. Some, such as *G. thurberi*, survive drought by shedding leaves, others by developing 'protoplasmic resistance' (e.g. *G. hirsutum* 'marie galante', *G. anomalum*) (Vieira da Silva, 1970; 1969). Characteristics of 'protoplasmic resistance' from *G. anomalum* were transferred successfully into *G. hirsutum* cultivars (Vieira da Silva and Poisson, 1969).

Many theoretical and empirical methods for predicting evapotranspiration have been attempted for field plantings of cotton. Stanhill and Fuchs (1968) summarized the many measurements of albedo and other radiation balance parameters, including heat storage, and were impressed by the variability within and among the many reported values. Fritschen (1966) found little difference among the various components of the radiation balance for cotton growing in wet and dry soil in a desert environment. Stern (1967) and Palmer *et al.* (1964) found evapotranspiration to be fairly constant over a wide range of soil moisture tension values. Rijks (1965) and Fritschen (1966) used the 'Bowen Ratio' method successfully to estimate evapotranspiration of cotton canopies with tests of theoretical and empirical estimates.

Jensen and Haise (1963) published an empirical model for estimating daily evapotranspiration from solar radiation data. It required the development of a coefficient for each crop, but the model has proved to be very accurate in prediction and is in commercial use. A Jensen–Haise crop coefficient has not been published for cotton, but Ritchie (1972) and Richardson and Ritchie (1973) have published models based on net radiation, accounting for light intercepted by the cotton crop, which have been found to be accurate in predicting evapotranspiration.

Lint morphology and physiology

Seed hairs consist of fuzz (approximately 1 mm in length) and fibres (2.5 cm in length, 17 μm in diameter). There is little differentiation between fuzz and fibres in the wild and semi-wild cottons. The fibre of cultivated cottons can be readily detached from the seed, breaking directly above the epidermis in the vicinity of the fibre elbow. Seed hairs receive nutrition from the outer, pigmented layer of the seedcoat underlying the epidermis. This layer is several cells in thickness and is supplied with vascular tissue. The foot of the hair is an absorptive organelle for the fibre (Fryxell, 1963). Lint hairs begin to grow on or near the day the flower opens.

Glucosyl moieties are linked in 1,4 arrangement in cotton fibres as in plant cell walls (Barber and Hassid, 1965). The maximum rate of cellulose deposition occurs 21 days after anthesis (Anderson and Kerr, 1938). Deposition occurs uniformly along the length of the fibre in daily rings (O'Kelly, 1953).

The fibre is covered with a thin waxy cuticle which is significant in germination, and helps maintain individuality during thread and cloth production. The wax is removed during bleaching after the cloth is woven.

Conclusions

We have emphasized the collection and analysis of wild races and species of *Gossypium*. This source of genetic diversity has already contributed insect-, drought- and cold-resistance to cultivars. Cotton has often been used in photosynthetic studies, the results of which have been useful in models simulating primary productivity. More emphasis is needed on the processes controlling leaf expansion.

The morphology of cotton has been described in detail, but cotton has not been used widely in hormonal or photoperiod studies. A general model for floral initiation is needed, although this model might well come from studies with other species. Following the recent success of Beasley, C. (1971) in culturing fibres in vitro, further insight into their physiology should be obtained.

Duncan (1972) and A. B. Hearn (personal communication) have developed growth models which in effect review cotton physiology, and subsystems of these models are being improved by others (cf. Jones *et al.* 1973). The Duncan model is now a part of a general farm management model.

There have been long-term efforts to model many processes: the flow of water in soils; the flow of water, sugars and hormones in plants; gas and energy flow in the soil–plant–atmosphere system; root growth; and photo-synthetic processes in chloroplasts. The plant growth models integrate process models for a pre-determined soil–plant–atmospheric system over a 24 h period, the system being redefined at the end of the time period simulated. This review was completed while information was being collected and in-serted into a cotton model. As inputs about critical processes are improved, the model will better define the factors that limit yield. For example, better measurements of flux rates, concentration gradients, and dimensions of phloem tubes would enable the modeller to evaluate critically the possible limiting role of translocation to production.

Future research should be oriented towards improving the subsystems of these models. Physical, metabolic and hormonal control mechanisms should be sought and translated into quantitative relationships, and communicated for analysis by agronomists and crop physiologists. Such communications may eventually assume the role of general reviews of the literature.

REFERENCES

Addicott, F. T. and Lyon, J. L. (1969). Physiology of abscisic acid and related substances. *Ann. Rev. Plant Physiol.* **20**, 139–164.

Anderson, D. B. and Kerr, T. (1938). Growth and structure of cotton fiber. *Ind. Eng. Chem.* **30**, 48–54.

Anderson, D. B. and Kerr, T. (1943). A note on the growth behavior of cotton bolls. *Plant Physiol., Lancaster* **18**, 261–269.

Andries, J. A., Jones, J. E., Sloane, L. W. and Marshall, J. G. (1969). Effects of okra leaf shape on boll rot, yield, and other important characters of upland cotton, *Gossypium hirsutum* L. *Crop Sci.* **9**, 705–710.

Ashley, D. A. (1972). ^{14}C-Labelled photosynthate translocation and utilization in cotton plants. *Crop Sci.* **12**, 69–74.

Baker, D. N., Hesketh, J. D. and Duncan, W. G. (1972). Simulation of growth and yield in cotton: I. Gross photosynthesis, respiration and growth. *Crop Sci.* **12**, 431–435.

Baker, D. N. and Myhre, D. L. (1969). Effects of leaf shape and boundary layer thickness on photosynthesis in cotton (*Gossypium hirsutum*). *Physiol. Plant.* **22**, 1043–1049.

Balls, W. L. (1912). *The cotton plant in Egypt.* The Macmillan Co., London, 202 pp.

Balls, W. L. (1959). *The yields of a crop, based on an analysis of cotton-growing by irrigation in Egypt.* E. and F. N. Spon, London, 144 pp.

Barber, G. A. and Hassid, W. Z. (1965). Synthesis of cellulose by enzyme preparations from the developing cotton boll. *Nature, Lond.* **207**, 295–296.

Barrs, H. (1968). Effect of cyclic variations in gas exchange under contant environmental conditions on the ratio of transpiration to net photosynthesis. *Physiol. Plant.* **21**, 918–929.

Barrs, H. (1971). Cyclic variations in stomatal aperture, transpiration, and leaf water potential under constant environmental conditions. *Ann. Rev. Plant Physiol.* **22**, 223–236.

Bassett, D. M., Anderson, W. D. and Werkhoven, C. H. E. (1970). Dry matter production and nutrient uptake in irrigated cotton (*Gossypium hirsutum*). *Agron. J.* **62**, 299–303.

Bassett, D. M., Stockton, J. R. and Dickens, W. L. (1970). Root growth of cotton as measured by ^{32}P uptake. *Agron. J.* **62**, 200–203.

Beasley, C. A. (1971). In vitro culture of fertilized cotton ovules. *Bioscience* **21**, 906–907.

Beasley, J. O. (1942). Meiotic chromosome behavior in species, species hybrids, haploids, and induced polyploids of *Gossypium*. *Genetics* **27**, 25–54.

Berkley, E. E. (1931). Studies of the effects of different lengths of day, with variations in temperature, on vegetative growth and reproduction in cotton. *Ann. Mo. bot. Gard.* **18**, 573–601.

Bohning, R. H. and Burnside, C. A. (1956). The effect of light intensity on apparent photosynthesis in leaves of sun and shade plants. *Amer. J. Bot.* **43**, 557–561.

Boyer, J. S. (1970). Leaf enlargement and metabolic rates in corn, soybean and sunflower at various leaf water potentials. *Plant Physiol., Lancaster* **46**, 233–235.

Brown, H. B. (1927). *Cotton.* McGraw-Hill, 517 pp.

Brown, H. B. and Ware, J. O. (1958). *Cotton.* McGraw-Hill, 566 pp.

Brown, K. H. (1968). Translocation of carbohydrate in cotton: movement to the fruiting bodies. *Ann. Bot.* **32**, 703–713.

Brown, M. S. (1951). Spontaneous occurrence of amphidiploidy in species hybrids of *Gossypium*. *Evolution* **5**, 25–41.

Buxton, D. R. and Stapleton, H. N. (1970). Predicted rates of net photosynthesis and transpiration as affected by the microenvironment and size of the cotton leaf. *Beltwide Cotton Prod. Res. Conf.*, pp. 31–34.

Canny, M. J. (1971). Translocation: mechanisms and kinetics. *Ann. Rev. Plant Physiol.* **22**, 237–260.

Carns, H. R. (1966). Abscission and its control. *Ann. Rev. Plant Physiol.* **17**, 295–314.

Carns, H. R. and Mauney, J. R. (1968). Physiology of the cotton plant. In *Advances in production and utilization of quality cotton: Principles and practices*, eds F. C. Elliott, M. Hoover and W. K. Porter. Iowa State Univ. Press, Ames, Iowa, pp. 41–73.

Christiansen, M. N. (1968). Induction and prevention of chilling injury to radicle tips of imbibing cottonseed. *Plant Physiol., Lancaster* **43**, 743–746.

Christiansen, M. N. and Moore, R. P. (1959). Seed coat structural differences that influence water uptake and seed quality in hard seed cotton. *Agron. J.* **51**, 582–584.

Christiansen, M. N., Carns, H. R. and Slyter, D. J. (1970). Stimulation of solute loss from radicles of *Gossypium hirsutum* L. by chilling, anaerobiosis and low pH. *Plant Physiol., Lancaster* **46**, 53–56.

Coble, C. G. and Bowen, H. D. (1970). Physical factors affecting oxygen stress of germinating cotton seeds. *Trans. ASAE* **13**, 162–167.

Cowan, I. R. (1972). Oscillations in stomatal conductance and plant functioning associated with stomatal conductance: Observations and a model. *Planta, Berl.* **106**, 185–219.

Cowan, I. R. and Troughton, J. H. (1971). The relative role of stomata in transpiration and assimilation. *Planta, Berl.* **97**, 325–336.

Dastur, R. H. (1959). *Physiological studies on the cotton crop and their practical applications*. The Indian Central Cotton Committee, Bombay.

Dastur, R. H. and Asana, R. D. (1960). Physiology. In *Cotton in India, a monograph*. The Indian Central Cotton Committee, Bombay, vol. 2, 1–105.

DeMichele, D. W. and Sharpe, P. J. H. (1972). A morphological and physiological model of the stomata. In *Proc. Workshop on Tree Growth Dynamics and Modeling, Duke Univ., Oct. 12, 1971*, eds C. Murphy, J. D. Hesketh and B. R. Strain, pp. 69–85.

Duncan, W. G. (1972). SIMCOT: A simulator of cotton growth and yield. In *Proc. Workshop on Tree Growth Dynamics and Modeling, Duke Univ., Oct. 1971*, ed. C. Murphy *et al.* pp. 115–118.

Eaton, F. M. (1950). Physiology of the cotton plant. *Adv. Agron.* **2**, 11–25.

Eaton, F. M. (1955). Physiology of the cotton plant. *Ann. Rev. Plant Physiol.* **6** 299–328.

Eaton, F. M. and Ergle, D. R. (1948). Carbohydrate accumulation in the cotton plant at low moisture levels. *Plant Physiol., Lancaster* **23**, 169–187.

Eaton, F. M. and Rigler, N. E. (1945). Effect of light intensity, nitrogen supply and fruiting on carbohydrate utilization by the cotton plant. *Plant Physiol., Lancaster* **20**, 380–411.

Elliott, F. C., Hoover, M. and Porter, W. K. Jr. (1968). *Advances in production and utilization of quality cotton: Principles and practices*. Iowa State Univ. Press, Ames, Iowa, 532 pp.

Elmore, C. D., Hesketh, J. D. and Muramoto, H. (1967). A survey of rates of leaf growth, leaf aging and leaf photosynthetic rates among and within species. *J. Ariz. Acad. Sci.* **4**, 215–219.

El-Nockrashy, A. S., Simmons, T. G. and Frampton, V. L. (1969). A chemical survey of seeds of the genus *Gossypium*. *Phytochemistry* **8**, 1948–1958.

El-Sharkawy, M. and Hesketh, J. D. (1964). Effects of temperature and water deficit on leaf photosynthetic rates of different species. *Crop Sci.* **4**, 514–518.

El-Sharkawy, M. and Hesketh, J. D. (1965). Photosynthesis among species in relation to characteristics of leaf anatomy and CO_2 diffusion resistances. *Crop Sci.* **5**, 517–521.

El-Sharkawy, M. Hesketh, J. D. and Muramoto, H. (1965). Leaf photosynthetic rates and other growth characteristics among 26 species of *Gossypium. Crop Sci.* **5**, 173–175.

Feaster, C. V. and Turcotte, E. L. (1965). Fruiting height response: a consideration in varietal improvement of Pima cotton, *Gossypium barbadense* L. *Crop Sci.* **5**, 460–464.

Fox, J. E. (1969). The cytokinins. In *Physiology of plant growth and development*, ed. M. B. Wilkins. McGraw-Hill, New York, pp. 83–123.

Fritschen, L. J. (1966). Evapotranspiration rates of field crops determined by the Bowen ratio method. *Agron. J.* **58**, 339–342.

Fry, K. E. (1970). Some factors affecting the Hill reaction activity in cotton chloroplasts. *Plant Physiol., Lancaster* **45**, 465–469.

Fryxell, P. A. (1963). Morphology of the base of seed hairs of *Gossypium*. I. Gross morphology. *Bot. Gaz.* **124**, 196–199.

Fryxell, P. A. (1965). Stages in the evolution of *Gossypium* L. *Adv. Front. Plant Sci.* **10**, 31–35.

Gaastra, P. (1959). Photosynthesis of crop plants as influenced by light, carbon dioxide, temperature and stomatal diffusion resistance. *Meded. Landbouwhogesch. Wageningen* **59**, 1–68.

Guinn, G. and Hunter, R. E. (1968). Root temperature and carbohydrate status of young cotton plants. *Crop Sci.* **8**, 67–70.

Harland, S. C. (1939). *The genetics of cotton.* Jonathan Cape, London, 193 pp.

Hector, J. M. (1936). *Introduction to the botany of field crops*, vol. II, *Non-cereals*. Central News Agency, Ltd. Johannesburg, South Africa, 1127 pp.

Hesketh, J. D. (1968). Effects of light and temperature during plant growth on subsequent leaf CO_2 assimilation rates under standard conditions. *Aust. J. biol. Sci.* **21**, 235–241.

Hesketh, J. D. and Baker, D. N. (1969). Relative rates of leaf expansion in seedlings of species with differing photosynthetic rates. *J. Ariz. Acad. Sci.* **5**, 216–221.

Hesketh, J. D., Baker, D. N. and Duncan, W. G. (1971). Simulation of growth and yield in cotton: Respiration and the carbon balance. *Crop Sci.* **11**, 394–398.

Hesketh, J. D., Baker, D. N. and Duncan, W. G. (1972). Simulation of growth and yield in cotton: II. Environmental control of morphogenesis. *Crop Sci.* **12**, 436–439.

Hesketh, J. D., Fry, K. C., Guinn, G. and Mauney, J. R. (1972). Experimental aspects of growth modeling: Potential carbohydrate requirement of cotton bolls. In *Proceedings of a Workshop on Tree Growth Dynamics and Modeling, Duke University, Oct. 12, 1971*, ed. C. Murphy *et al.* pp. 123–127.

Hesketh, J. D. and Low, A. (1968). Effect of temperature on components of yield and fibre quality of cotton varieties of diverse origin. *Cott. Gr. Rev.* **45**, 243–257.

Heyerdahl, T. (1971). The voyage of Ra II. *Nat'l Geographic* **139**, 44–71.

Hoffman, G. and Rawlins, S. L. (1970). Infertility of cotton flowers at both high and low relative humidities. *Crop Sci.* **10**, 721–723.

Hoffman, G. J., Rawlins, S. L., Garber, M. J. and Cullen, E. M. (1971). Water relations and growth of cotton as influenced by salinity and relative humidity. *Agron. J.* **63**, 822–826.

Hutchinson, J. B. (1959). *The application of genetics to cotton improvement.* Cambridge University Press, 86 pp.

Hutchinson, J. B., Silow, R. A. and Stephens, S. G. (1947). *The evolution of Gossypium and the differentiation of the cultivated cottons.* Oxford University Press, London, Toronto, N.Y., 186 pp.

Jarvis, P. G. and Slatyer, R. O. (1970). The role of the mesophyll cell wall in leaf transpiration. *Planta, Berl.* **90**, 303–322.

Jensen, M. E. and Haise, H. R. (1963). Estimating evapotranspiration from solar radiation. *Amer. Soc. Civil Engin. Irrig. Drain. Div.*, J. 89 (IR4) 15–41.

Jones, J. W., Hesketh, J. D., Kramprath, E. J. and Bowen, H. D. (1974). Development of a nitrogen balance for cotton growth models – a first approximation. *Crop Sci.* **14**, (in press).

Jordan, W R. (1970). Growth of cotton seedlings in relation to maximum daily plant-water potential. *Agron. J.* **62**, 699–701.

Jordan, W. R. and Ritchie, J. T. (1971). Influence of soil water stress on evaporation, root absorption, and internal water status of cotton. *Plant Physiol., Lancaster* **48**, 783–788.

King, E. E. and Lane, H. C. (1969). Abscission of cotton flower buds and petioles caused by protein from boll weevil larvae. *Plant Physiol., Lancaster* **44**, 903–906.

Klepper, B. (1968). Diurnal pattern of water potential in woody plants. *Plant Physiol., Lancaster* **43**, 1931–1934.

Klepper, B., Douglas Browning, V. and Taylor, H. M. (1971). Stem diameter in relation to plant water status. *Plant Physiol., Lancaster* **48**, 683–685.

Kornberg, H. L. and Krebs, H. A. (1957). Synthesis of cell constituents from C_2-units by a modified tricarboxylic acid cycle. *Nature, Lond.* **179**, 988–991.

Lang, A. (1952). Physiology of flowering. *Ann. Rev. Plant Physiol.* **3**, 265–306.

Lawlor, D. W. (1969). Plant growth in polyethylene glycol solutions in relation to the osmotic potential of the root medium and the leaf water potential. *J. Expt. Bot.* **20**, 895–911.

Lewis, C. F. and Richmond, T. R. (1960). The genetics of flowering response in cotton. II. Inheritance of flowering response in a *Gossypium barbadense* cross. *Genetics* **45**, 79–85.

Low, A., Hesketh, J. D. and Muramoto, H. (1969). Some environmental effects on the varietal node number of the first fruiting branch. *Cott. Gr. Rev.* **46**, 181–188.

McArthur, J. A., Butler, W. L. and Mauney, J. R. (1971). *In vivo* spectrophotometric examinations of cotton embryos and seedlings. *Proc. Cotton Physiology Defoliation Conference*, Atlanta, Ga., p. 45 (abstract).

Mason, T. G. and Maskell, E. J. (1928). Studies on the transport of carbohydrates in the cotton plant. I. A study of diurnal variation in the carbohydrates of leaf, bark and wood, and of the effects of ringing. *Ann. Bot.* **420**, 189–253.

Maskell, E. J. (1928). Experimental researches on vegetable assimilation and respiration. XVIII. The relation between stomatal opening and assimilation. *Proc. Roy. Soc. B.* **102**, 488–533.

Mauney, J. R. (1966). Floral initiation of upland cotton *Gossypium hirsutum* L. in response to temperatures. *J. expt. Bot.* **17**, 452–459.

Mauney, J. R. (1968). Morphology of the cotton plant. In *Advances in Production and Utilization of Quality Cotton: Principles and Practices*, ed. F. C. Elliott *et al.* Iowa Univ. Press, Ames, Iowa, pp. 23–40.

Mauney, J. R. and Phillips, L. L. (1963). Influence of daylength and night temperature on flowering of *Gossypium*. *Bot. Gaz.* **124**, 278–283.

Meyer, B. S., Anderson, D. B. and Bohning, R. H. (1960). *Introduction to plant physiology*. Van Nostrand Co., New York and London, pp. 159–160.

Mohapatra, N., Smith, E. W., Fites, R. C. and Noggle, G. R. (1970). Chilling temperature depression of isocitratase activity from cotyledons of germinating cotton. *Biochem. Biophy. Res. Comm.* **40**, 1253–1258.

Monteith, J. L. (1963). Dew: Facts and fallacies. In *The water relations of plants*, eds A. H. Rutter and F. H. Whitehead. Blackwell Scientific Publications, London, pp. 37–56.

Moraghan, B. J., Hesketh, J. D. and Low, A. (1968). Effects of temperature and photoperiod on floral initiation among strains of cotton. *Cott. Gr. Rev.* **45**, 91–100.

Muramoto, H., Hesketh, J. D. and Baker, D. N. (1971). Cold tolerance in a hexaploid cotton. *Crop Sci.* **11**, 589–591.

Muramoto, H., Hesketh, J. D. and Elmore, C. D. (1967). Leaf growth, leaf aging, and leaf photosynthetic rates of cotton plants. *Proc. Cotton Physiology Defoliation Conf.* pp. 161–165.

Namken, L. N. (1964). The influence of crop environment on the internal water balance of cotton. *Soil Sci. Soc. Amer. Proc.* **28**, 12–15.

Namken, L. N. (1965). Relative turgidity technique for scheduling cotton (*Gossypium hirsutum*) irrigation. *Agron. J.* **57**, 38–41.

O'Kelly, J. C. (1953). The use of ^{14}C in locating growth regions in cell walls of elongating cotton fibres. *Plant Physiol., Lancaster* **28**, 281–286.

Palmer, J. H., Trickett, E. S. and Linacre, E. T. (1964). Transpiration response of *Atriplex nummularia* Lindl. and Upland cotton vegetation to soil-moisture stress. *Agr. Meteorol.* **1**, 282–293.

Penning de Vries, R. W. T. (1972). Respiration and growth. In *Crop processes in controlled environments*, eds A. R. Rees, K. E. Cockshull, D. W. Hand and R. G. Hurd. Academic Press, London, pp. 327–346.

Phillips, I. D. J. (1969). Apical dominance. In *Physiology of plant growth and development*, ed. M. B. Wilkins. McGraw-Hill, New York, pp. 163–202.

Phillis, E. and Maskell, T. G. (1936). Further studies on transport in the cotton plant. IV. On the simultaneous movement of solutes in opposite directions through the phloem. *Ann. Bot.* **50**, 161–174.

Pinkas, L. L. H. (1972). Modification of flowering in Pima cotton with ethephon. *Crop Sci.* **12**, 465–466.

Powell, R. D. and Morgan, P. W. (1970). Factors involved in the opening of the hypocotyl hook of cotton and beans. *Plant Physiol., Lancaster* **54**, 548–552.

Purseglove, J. W. (1968). *Tropical crops: Dicotyledons*. John Wiley and Sons, New York, 719 pp.

Richardson, C. W. and Ritchie, J. T. (1973). Soil water balance of small watersheds. *Trans. ASAE* **16**, 72–77.

Richmond, T. R. and Radwan, S. R. H. (1962). A comparative study of seven methods of measuring earliness of crop maturity in cotton. *Crop Sci.* **2**, 397–400.

Ritchie, J. T. (1972). Model for predicting evaporation from a row crop with incomplete cover. *Water Resources Res.* **8**, 1204–1213.

Rijks, D. A. (1965). The use of water by cotton crops in Abyan, South Arabia. *J. Appl. Ecol.* **2**, 317–343.

Sabbe, W. E. and Cathey, G. W. (1969). Translocation of labelled sucrose from selected cotton leaves. *Agron. J.* **61**, 436–438.

Sandstedt, R. (1970). Partial purification of cytokinin from immature cotton fruit. *1970 Proc. Beltwide Cotton Prod. Res. Conf.*, pp. 29–31.

Sandstedt, R. (1971). Cytokinin activity during development of cotton fruit. *Physiol. Plant.* **24**, 408–410.

Saunders, J. H. (1961). *The wild species of Gossypium and their evolutionary history* Oxford University Press, London, 62 pp.

Singh, I. B. and Weaver, J. B. (1972). Effect of gossypol on the incidence of boll rot and agronomic characters of cotton. *Cott. Gr. Rev.* **49**, 242–249.

Slatyer, R. O. (1957). The influence of progressive increases in total soil moisture stress on transpiration, growth and internal water relationships of plants. *Aust. J. biol. Sci.* **10**, 320–336.

Smith, C. E. and MacNeish, R. C. (1964). Antiquity of American polyploid cotton. *Science* **143**, 675–676.

Smith, D. E. (1969). Changes in abscission-accelerating substances with developing cotton bolls. *New Phytol.* **68**, 311–322.

Smith, E. W., Fites, R. C. and Noggle, G. R. (1971). Effects of chilling temperature on isocitratase and malate synthetase levels during cotton seed germination. *1971 Cotton Defoliation–physiology Conference*, Atlanta, Ga., pp. 47–54.

Stanhill, G. and Fuchs, M. (1968). The climate of the cotton crop: Physical characteristics and microclimate relationships. *Agr. Meteorol.* **5**, 183–202.

Stephens, S. G. (1958). Salt water tolerance of seeds of *Gossypium* species as a possible factor in seed dispersal. *Amer. Nat.* **92**, 83–92.

Stern, W. R. (1967). Seasonal evapotranspiration of irrigated cotton in a low-latitude environment. *Aust. J. biol. Sci.* **18**, 259–269.

Taylor, H. M., Huck, M. G. and Klepper, B. (1972). Root development in relation to soil physical conditions. In *Optimizing the soil physical environment towards greater crop yields*, ed. D. Hillel. Academic Press, New York and London, pp. 57–77.

Taylor, H. M. and Klepper, B. (1971). Water uptake by cotton roots during an irrigation cycle. *Aust. J. biol. Sci.* **24**, 853–859.

Tharp, W. H. (1960). *The cotton plant. How it grows.* U.S.D.A. Handbook No. 178.

Thomas, R. O. and Christiansen, M. N. (1971). Seed hydration – chilling treatment effects on germination and subsequent growth and fruiting of cotton. *Crop Sci.* **11**, 454–456.

Thornley, J. H. M. and Hesketh, J. D. (1972). Growth and respiration in cotton bolls. *J. appl. Ecol.* **9**, 315–317.

Tollervey, P. E. (1970). Physiology of the cotton plant. *Cott. Gr. Rev.* **47**, 245–256.

Toole, E. H. and Drummond, P. L. (1924). The germination of cotton seed. *J. agr. Res.* **38**, 287–296.

Troughton, J. H. (1969). Plant water status and carbon dioxide exchange of cotton leaves. *Aust. J. biol. Sci.* **22**, 289–302.

Troughton, J. H. and Cowan, I. R. (1968). Carbon dioxide exchange in cotton: Some anomalous fluctuations. *Science* **161**, 281–283.

Trouse, A. (1970). The dynamic nature of development of the root system of the cotton plant. *Assoc. Southern Agric. Workers, 67th Ann. Proc.* p. 51 (abstract).

Tucker, T. C. and Tucker, B. B. (1968). Nitrogen nutrition. In *Advances in Production and Utilization of Quality Cotton: Principles and Practices*, ed. F. C. Elliott *et al.* Iowa Univ. Press, Ames, Iowa, pp. 183–211.

Tuichibaev, M. and Kruzhilin, A. S. (1965). Translocation of labeled assimilates from individual cotton leaves. *Soviet Plant Physiology* **12**, 919–922.

Vieira da Silva, J. (1969). Comparaison entre cinq espèces de *Gossypium* quant à l'activité de la phosphatase acide après un traitement osmotique. Etude de la vitesse de solubilisation et de formation de l'enzyme. *Z. Pflanzenphysiol.* **60**, 385–387.

Vieira da Silva, J. (1970). Thesis ORSAY. CNRS (Paris) no. A.O. 4685.

Vieira da Silva, J. (1972). Influence de la sécheresse sur la photosynthèse et la croissance du cotonnier. *Physiol. Plant.* **25**, 213–220.

Vieira da Silva, J. and Poisson, Ch. (1969). Solubilisation d'enzymes hydrolytiques chez *Gossypium hirsutum*, *G. anomalum* et des dérivés de l'hybridation entre ces deux espèces. *Canad. J. Genet. Cyt.* **11**, 582–596.

Waddle, B. M., Lewis, C. F. and Richmond, T. R. (1961). The genetics of flowering response in cotton. III. Fruiting behavior of *Gossypium hirsutum* race *latifolium* in a cross with a variety of cultivated American Upland cotton. *Genetics* **46**, 427–437.

Wadleigh, C. H. (1944). Growth status of the cotton plant as influenced by the supply of nitrogen. *Arkansas Ag. Expt. Sta. Bul.* **446**.

Walhood, V. T. (1956). A method of reducing the hard seed problem in cotton. *Agron. J.* **48**, 141–142.

Wanjura, D. F. and Buxton, D. R. (1972*a*). Hypocotyl and radicle elongation of cotton as affected by soil environment. *Agron. J.* **64**, 431–434.

Wanjura, D. F. and Buxton, D. R. (1972*b*). Water uptake and radicle emergence of cottonseed as affected by soil moisture and temperature. *Agron. J.* **64**, 427–431.

Wanjura, D. F., Buxton, D. R. and Stapleton, H. N. (1971). A model for describing cotton growth during emergence. Presented at 1971 ASAE summer meeting, Pullman, Washington.

Wilkes, L. H. and Corley, T. E. (1968). Cotton: planting and cultivation. In *Advances in Production and Utilization of Quality Cotton: Principles and Practices*, ed. F. C. Elliott *et al.* Iowa State Univ. Press, Ames, Iowa, pp. 117–149.

11

The physiological basis of crop yield

L. T. EVANS

Perhaps the overriding impression from the preceding case histories is the great physiological diversity evident among the major crop plants. Some store sugar, some starch; others are valued for their stored proteins or oils, cotton for its cellulosic lint. Some fix nitrogen symbiotically, most do not. Photosynthesis may be by the Calvin cycle, or by the C_4-dicarboxylic acid pathway. The storage organ harvested by man is often the inflorescence, but in cane it is the stem, in potatoes the underground stem or tuber, and in sugar beet it is the root. Flowering is crucial to yield in many crops, but in the potato it is irrelevant, and in both sugar cane and beet it is unwanted, except for plant breeding purposes. When the inflorescence is the storage organ it may be terminal, as in wheat, sorghum and rice, or axillary as in maize with one cob, or with many sequentially developed fruits as in soybeans and peas. Leaves may be erect or mainly of horizontal inclination. Some crops may be tall, like maize or cane, whereas others form a rosette, like sugar beet. Branching or tillering may be important, as in potatoes and rice, or largely suppressed as in maize. The list could be continued, but its point is clear, that there is no one constellation of characteristics and no one path to high yield and success as a crop plant.

Selection for yield seems in many cases to have been accompanied by a shortening of the life cycle, as in cotton, sugar beet, rice and wheat. Net production of dry matter tends to be greater in perennial crops because of their more extended interception of light throughout the year, but their greater biomass increases respiratory losses, and perennials tend to allocate a smaller proportion of their biomass to reproduction. In fact, selection for heavy investment by the plant in its reproductive organs at the expense of its roots and branches may have led to the shortening of the life cycle.

THE BASIS OF YIELD ASSESSMENT

In Biblical times yield was assessed in terms of the number of grains harvested for each grain sown, and thus foregone as food, even in years of famine. As agriculture became more settled and good arable land more scarce, yield per unit ground area became of prime importance and, with increasing pressure of population, remains so. We are reaching the stage, however, when other bases of yield estimation must also be considered, such as yield per unit of water or phosphorus or energy consumed or pesticide discharged into the environment, which may lead to a reversal of some past trends.

Fresh water, although a renewable resource, is increasingly the subject of competition between agricultural, urban and industrial requirements. As urban needs increase, less water may become available for irrigation, giving less scope for increase in yield through intensive agriculture since response to other inputs may be restricted by lack of water. Integrated closed systems to conserve desalinated water in vegetable crop production have already been developed at Puerto Penasco and Abu Dhabi (Hodges and Hodge, 1971), and several Oak Ridge National Laboratory study groups have considered the design of much larger integrated nuclear agro-industrial complexes in which desalinated water is used for open irrigation (e.g. ORNL–4291 UC80, 1968). On the assumption that efficient cereal food production requires only 0.5 m^3 of water per day, about the same as present daily water use by urban man, to produce the 1×10^7 J of food per day needed by each adult, they believe that intensive cereal production by irrigation with such water is feasible. More acceptable and balanced diets would require far more water than that, however, and efficiency of water use will undoubtedly become an increasingly important criterion in assessing crop productivity. Plants with the C_4-dicarboxylic acid pathway of photosynthesis use considerably less water per unit of additional dry weight than do Calvin cycle plants (50–100 compared with 150–250), and plants with Crassulacean acid metabolism (CAM) also have low transpiration ratios, 30–116 (Neales *et al.* 1968; Evans, 1971). Remarkably few CAM plants have been developed as crops (pineapple, sisal and opuntia), and the use of others should be explored in this context.

The efficiency of phosphorus use by crops may also assume increasing importance. World use of phosphatic fertilizers has trebled in the last twenty years, increasing exponentially at a faster rate than food production. Reserves of the mineral are large, and phosphorus is not mentioned by Meadows *et al.* (1972) in their *Limits to Growth*, but if its use continues to increase exponentially, presently workable deposits will be exhausted in 100 years or so. Phosphorus is the backbone of life, of nucleic acids on the one hand and of respiratory energy transduction on the other, and there can be no substitution technology for it. Yet present use is completely dissipative, most of the phosphorus in human food ending up in the sewer and the sea. It is the element for which the development of control and recovery measures is most urgent (Goeller, 1972).

EFFICIENCY OF ENERGY USE IN AGRICULTURE

Crop production is a system for harvesting energy from the sun, as food, feed and fibre, and Loomis *et al.* (1971) have suggested that its efficiency should be

Table 11.1. *Ratio of consumable output energy to consumed human, draught animal and fossil fuel input energy in several cropping systems*

Fishing		
Pacific atoll	6	(Alkire, 1965)
Food gathering		
Wild wheats, Turkey	40–50	(estimated from Harlan, 1967)
Hand cultivation		
Thirteen examples	17 (3–34)	(Black, 1971)
Sweet potatoes, New Guinea	16	(Rappaport, 1967)
Draught animals		
Two examples	10 (6–14)	(Black, 1971)
Mechanized agriculture		
(*a*) Primary inputs only	22 (9–34)	(Black, 1971)
(*b*) Including secondary inputs		
Australia	3	(Gifford and Millington, 1974)
USA	0.2	(Perelman, 1972)

assessed in terms of the conversion of solar radiation into useful end products. Higher yields per unit ground area per year may mean more efficient conversion of contemporary radiation from the sun, but to an increasing degree they are achieved because of substantial inputs of energy from fossil fuels and prehistoric photosynthesis. Consider the entries in Table 11.1.

The estimate from data given by Harlan (1967) on the rate at which he harvested grain from stands of wild cereals in the Middle East indicates that food gathering under favourable conditions could yield a far more abundant harvest of energy per unit of expended energy than does fishing, and probably hunting too. Moreover, the harvested grain could readily be stored. In the thirteen examples of simple hand cultivation analysed by Black (1971) the average ratio of output:input energy was 17, close to the figure derived by Rappaport (1967) from a thorough analysis of the swidden cropping of the Tsembaga people in New Guinea. In the cases analysed by Black (1971) the output:input energy ratios in cultivation with draught animals and in mechanized agriculture were of similar magnitude. However, Black made no allowance for secondary inputs such as the energy used in the manufacture and distribution of fertilizers, pesticides, implements, etc., or for tertiary inputs. An estimate by Odum (1971) for agriculture in the USA, indicates a return of 7:1 when secondary inputs are included, and an overall return of 3:1 at the farm gate has been estimated for Australian agriculture by Gifford and Millington (1974). Where much of the harvested crop is used to feed animals, agriculture itself may be an energy sink (cf. Perelman, 1972).

Clearly, our more efficient harvesting of contemporary sunshine with higher crop yields depends on substantial inputs of other energy, mostly

Table 11.2. *Global power* (kW)

1. Solar power at the earth's surface	7×10^{13}
2. Fixed in photosynthesis – terrestrial	9.2×10^{10}
– oceanic	3.8×10^{10}
3. Harvested from crops and animals	7.8×10^{8}

from earlier photosynthesis. As epitomized by Odum (1971), our potatoes are now made partly of oil, and the law of diminishing returns leads us to expect that further yield increases will require them to be more so, and we may soon reach the point where, in terms of energy expenditure, chemical synthesis may be as efficient as agriculture in the production of food.

The fuel for mechanical cultivation and harvest replaces the energy of draught animal or man, but the additional energy input to make and distribute fertilizers, herbicides, insecticides, fungicides, and irrigation water increases yield partly by increasing the duration and rate of crop photosynthesis and partly by increasing the proportion of dry weight harvested. The maintenance of optimum levels of water and nutrients requires proportionally less investment in root growth to seek them out. Control of weeds eliminates the need for taller stems and large, densely shading leaves to compete more effectively with them. Control of pests and diseases frees the substrate used in their support for storage in the harvested organs. Given this supporting energy input, the crop plant can be selected for investment of a still greater proportion of its accumulated reserves into the products required by man. Without such support these high yielding varieties are more vulnerable to adverse environments than more competitive, lower yielding varieties.

In terms of maximum daily rates of photosynthesis and solar energy capture, well developed crops and pastures appear to be more efficient than native plant communities, but over longer periods of time this advantage disappears (Loomis *et al.* 1971). Most crops cover the ground and intercept a large proportion of the incident radiation for only a few months of each year. Consequently, although arable land constitutes 11% of the world's land surface, the harvested products of agriculture (given in Table 1.1) have an energy content equivalent to less than 1% of terrestrial photosynthesis, as may be seen from Table 11.2. Total crop photosynthesis will be two to three times as great, but it is clear that crop production is not exceptionally efficient in harvesting the sun. It may be no more so than grazed pastures.

LIMITING PROCESSES – SOURCE OR SINK?

F. F. Blackman's law of limiting factors or processes was a cornerstone in the foundation of crop physiology, and continues to play an important role.

In modelling crop growth, for example, it may be invoked to deal with the effects of several stresses whose interactions are not known (e.g. p. 274). It is also the rationale behind much contemporary discussion of whether the supply of assimilates (source) or the capacity for their storage (sink) limits crop yields.

'Source or sink?' is too polarized a question. There may well be situations where neither source nor sink is limiting, but rather the capacity to translocate assimilates from the one to the other. Or other processes, such as water or nutrient uptake or transport, may be limiting. Moreover, even if the source or the sink were the major limitation to yield in a particular case, it is doubtful whether that conclusion would apply to other crops in the same environment, or to the same crop in other environments. The seasonal sequence of conditions plays a major role in determining whether source or sink is more limiting. In the cereals, for example, grain filling is largely dependent on photosynthesis and environmenal conditions after flowering, but the capacity for storage is determined by conditions before flowering, and these may have the dominant influence on yield, as in the Japanese IBP experiments on rice (p. 90). In plants with sequential axillary flowering, the period when storage capacity is determined overlaps the period of storage itself, and mutual adjustment of the two components may be more readily achieved. This is true also of sugar beet and potatoes in which the roots and tubers continue cell division and growth throughout storage.

Whether source or sink is limiting is difficult to unravel in many cases, because the demand for assimilates for storage can have a pronounced feedback effect on the rate of photosynthesis, with the result that spare photosynthetic capacity may be present but not evident unless storage capacity is increased. Several examples of such feedback effects have been given in the preceding chapters. Removal of the ear in wheat halved the photosynthetic rate of the flag leaf, but the original rate was restored when the flag became the only support for the roots and young tillers (p. 110). Similarly, cooling or removal of tubers led to lower photosynthetic rates in potato leaves (p. 245). Prevention of tuber growth in sweet potatoes by exposing the tubers to light led to accumulation of carbohydrate in the leaves and a depression of photosynthesis (Tsuno and Fujise, 1965). In cross grafts between tops and roots of sugar beet and spinach beet, tops grafted to sugar beet roots had a higher net assimilation rate, regardless of their origin, than those grafted on spinach beet roots (Thorne and Evans, 1964).

Ontogenetic drifts in leaf photosynthetic rate may reflect changes in the demand for assimilates from them. This is evident in many wheat varieties, in which the photosynthetic rate of flag leaves under controlled environmental conditions falls substantially during the period between the end of stem growth and the beginning of grain growth, but rises again as export of assimilates to the grain increases (p. 110). In potato crops the photosynthetic rate of the leaves may rise severalfold, and the proportion of assimilates exported may increase, as soon as tubers are initiated (p. 239). Photosynthetic rate of the early leaves on pea plants reaches a sharp peak and then declines rapidly, whereas in leaves subtending pods the rate remains high over a long period (p. 199). Rates of leaf photosynthesis in soybeans are much higher during pod filling than during flowering, even under water stress, presumably due to increased demand at the later stage (p. 167). On apple trees, leaves subtending fruit may have a photosynthetic rate more than 50 % higher than matched leaves without fruit (Hansen, 1970; Kazaryan *et al.* 1965). In cotton, on the other hand, the leaves supporting heavy bolls may have low rates of photosynthesis due to the withdrawal of nitrogen from them by the boll (p. 302). High demand, leading to mobilization of protein from leaves to fruit, may cause leaf senescence, as may very low demand. Leaves of cotton plants grown under high carbon dioxide concentrations may accumulate starch and sugars up to almost half their total dry weight, but this leads to their rapidly becoming chlorotic (p. 305). Similarly, removal of ears or prevention of pollination in maize leads to increased starch and sugar content in the leaves, and their premature senescence (Allison and Weinmann, 1970). Clearly, therefore, the rates of both photosynthesis and leaf senescence are controlled by the demand for assimilates, and photosynthesis and storage may appear to be in balance although the storage processes may actually be limiting. Consequently, parallel rankings for photosynthetic rate and yield need not imply that yield is limited by photosynthetic rate.

Consider also the nature of plant breeding for yield. In the early stages of domestication of a wild plant it is the capacity for storage that most probably limits yield. However, the components of storage capacity are readily observable and subject to selection by the plant breeder, whereas photosynthesis is not, and storage capacity and yield will tend to increase in parallel until they reach the limit set by photosynthetic capacity. Further progress will then require both photosynthetic and storage capacity to increase in a more or less co-ordinated way. More assimilate without more storage capacity would not increase yield, whereas more storage capacity without more assimilate would merely result in fruits that fail or are shrivelled. Yields in the former cases are likely to be relatively more stable across sites and years, whereas in the latter they will be highly responsive to environmental conditions. Most rapid progress in breeding for yield may therefore be made by crossing these two types, the stable with the responsive, as Finlay (1970) has suggested. In the present context, however, the relevant conclusion is that modern cultivars of our major crop plants are likely to have their photosynthetic capacity and their storage capacity fairly closely balanced in the environments to which they are adapted.

Nevertheless, there has been much work in recent years based on the assumption that yield is limited by the photosynthetic rate of crop stands, such as the programmes of selection for plants with more erect leaves or higher photosynthetic rates, or for Calvin cycle plants lacking photorespiration, on the assumption that these characteristics will result in increased yields. The kinds of evidence which suggest that source or sink may be more limiting in particular instances should therefore be considered.

(i) Detailed balance sheets of the supply of and requirements for photosynthetic assimilates under defined conditions during the storage phase have not often been compiled. In one example with wheat there was more than enough assimilate at all stages of grain filling (p. 131). Hackett (1973) has simulated the carbon budget of tobacco plants in an early growth analysis experiment and concluded that for much of their life the plants had far more assimilating capacity than was required. With cotton, on the other hand, although assimilate supply is ample at low temperatures, at which growth is slow, boll growth at higher temperatures is probably limited by carbohydrate supply in most conditions (p. 314).

(ii) Substantial reserves of mobilizable carbohydrate may be present in crop plants even to the end of their life cycle. In maize plants grown at low latitudes, up to 40 % of stem weight may be sucrose not called on for storage in the grain. At higher latitudes these reserves may be fully mobilized. According to Johnson and Tanner (1972) the rate of grain filling is faster in hybrids than in inbreds, whereas inbreds have a higher sugar content in their stems, suggesting that the rate of storage is more limiting to yield than is the rate of photosynthesis.

(iii) Shading or reduction of leaf area during the period of storage sometimes reduces yield substantially, sometimes slightly. In the experiments of Willey and Holliday (1971 a, b), for example, shading had no effect on grain filling in barley but greatly reduced yield in wheat. Substantial reductions do not necessarily imply that, at full light or leaf area, yield was source-limited, only that it was so when photosynthesis was reduced. Lack of yield reduction does, however, imply sink-limitation, except where partial defoliation still leaves sufficient leaf area for full light interception and canopy photosynthesis. In other cases, increased mobilization of reserves or a greater rate of photosynthesis in the remaining leaves may compensate for leaves removed. Hicks and Pendleton (1969) and McAlister and Krober (1958) found that up to one third of the pods in a soybean crop could be removed without reducing yield, whereas Begum and Eden (1965) removed up to two-thirds of the leaves without reducing grain yield. Such techniques, on their own, are clearly inadequate to resolve the extent of source limitation on yield.

(iv) For many crops there is a phase when the storage of assimilates continues at a fairly constant rate regardless of variation in incident radiation or leaf area, as in maize (Duncan *et al.* 1965), wheat (Sofield *et al.* 1974), and rice (Murayama *et al.* 1955). Radley (1963) found the tuber bulking rate per unit land area to be remarkably constant for potato crops over a range of leaf area index from 1 to 4, and independent of the weather. With soybeans, Hanway and Weber (1971) found many cultivars with the same rate of grain storage, 9.9 g m^{-2} day^{-1}, although crop growth rates at that time varied from 8.8 to 14.9 g m^{-2} day^{-1} in different cultivars. Removal of half the flowers in cotton had no effect on the growth rate of the remaining bolls (Figure 10.4). Such results imply that photosynthetic rate is not limiting storage rate, although the rate of translocation may be.

(v) If photosynthesis limited storage and yield we might expect higher yielding varieties to display higher photosynthetic rates while the leaves are young. This last proviso is important because photosynthetic rate may parallel yield late in the storage phase, reflecting differences between varieties in their *duration* of photosynthesis and storage rather than in photosynthetic capacity. The relation between yield and photosynthetic rate is discussed more fully below, but there is little evidence of any positive relation between them, nor any instance where selection for a greater rate of photosynthesis has led to increase in yield.

(vi) Where there are a number of components to yield, negative correlations among them are frequently observed. In rice, for example, the more panicles per unit area there are, the fewer are the spikelets per panicle (cf p. 87); in maize, the more ears per plant, the fewer kernels per ear (p. 35). Such negative correlations between yield components are often interpreted as indicating that yield is limited by the supply of assimilates, but Adams (1967) has pointed out that other explanations may hold. In his experiments with field beans the negative correlations were lowest in the highest yielding lines, and had little to do with establishing actual yield levels.

(vii) Treatments which primarily increase storage capacity may lead to increased yields. Grafting experiments with several potato varieties by Börger *et al.* (1956) suggest that tuber size and starch content are largely independent of the characteristics of the above ground parts. Thus, yield must be largely sink-determined, as also in the case of reciprocal grafts between sugar and spinach beets (Thorne and Evans, 1964), and between high and low yielding sweet potatoes (Hozyo and Park, 1971; Kato and Hozyo, 1972). With sugar beet, Humphries and French (1969) obtained higher yields when seedlings were grown before transplanting in continuous light rather than in the field, although the two crops had similar maximum LAI and duration. Such seedlings had a larger root:shoot ratio at transplanting and presumably retained a more active sink. Similarly, treatment of sugar beet plants with growth substances which could increase cambial activity of the root may also increase yield (Das Gupta, 1972), and application of gibberellic acid to young sugar cane plants may increase dry matter production considerably (p. 64). In wheat, treatments which increased the number of grains set, without affecting the photosynthetic structure, may increase yields substantially (p. 118).

Increase in the concentration of carbon dioxide during inflorescence development in cereals may increase storage capacity and yield. The number of ear-bearing tillers was particularly increased by such treatment in barley (Gifford *et al.* 1973), whereas in rice both grain number and grain size were increased (Yoshida, S. *et al.* 1972). Carbon dioxide enrichment during grain filling may also increase yields, even with plants not enriched before flowering, as found by both Gifford *et al.* (1973) and Yoshida, S. *et al.* (1972). With soybeans in the field, carbon dioxide enrichment at flowering increased the number of pods but seed size was reduced, whereas enrichment during grain filling increased both seed size and yield (Hardman and Brun, 1971). Carbon dioxide enrichment may also increase the yield of sugar beet and kale (Ford and Thorne, 1967), as well as of many glasshouse crops (cf. Wittwer, 1970). The carbon dioxide enrichment experiments with cereals are important because they suggest that increase in either source or sink can lead to quite substantial increases in yield, i.e. that both source and sink could be simultaneously limiting, and Gifford *et al.* (1973) have developed a method for quantifying the relative limitation on yield by source and sink in particular crops.

Photosynthesis during the storage phase can be an important determinant of yield, but photosynthesis prior to that contributes to the determination of storage capacity and generates reserves that may be mobilized during the storage phase. These reserves were once thought to be a major component of yield, and may well have been so in earlier times when, with less nitrogen fertilizer applied and no late applications, nitrogen mobilization from the leaves may have been early and extensive, reducing the rate and duration of photosynthesis during storage, thereby increasing the dependence on reserves. For most crops such reserves now appear to contribute a relatively small proportion to final yield, 5–10 % in wheat not under stress (p. 109), about 10 % in potatoes (p. 240), 20 % in maize (p. 39) and barley (Archbold and Mukerjee, 1942), and up to 40 % in rice (p. 80). Most fruit growth in apples is based on current photosynthesis (Hansen, 1971), and mobilized reserves contribute little to seed weight in soybean, but much to the nitrogen content of the seed (Figure 6.2). Under present agronomic conditions, therefore, yield is largely derived from photosynthesis during the storage phase.

CANOPY STRUCTURE

The rate of crop photosynthesis depends on the leaf area index (LAI) and structure of the canopy and on photosynthetic rate per unit leaf area. Initial canopy development tends to be slow, with the result that light interception during early crop growth may be slight, even when growing conditions are favourable, and Watson (1971) believes this to be a major source of inefficiency in crop production systems. Selection for large leaves, their rapid expansion, a marked increase in the size of successive leaves, and early branching, may reduce the initial lag in LAI increase, as does closer planting. Canopy structure can be controlled by planting density in a few crop plants, e.g. maize (p. 41), but for many of them the extent of branching and leaf area per plant eventually compensates for changes in density of planting.

After the initial lag, there is usually a rapid increase in LAI, followed by a fall which may also be quite rapid. Light interception usually approaches a high proportion only when LAI exceeds 3–4, which may be the case for only two to three months in many annual crops, and not always at the times of highest radiation or most favourable growing conditions. Some years ago there was concern that the peak LAI values attained by crops could be super-optimal, net photosynthesis being reduced because respiration of the lower shaded leaves might exceed their photosynthesis. It is now known that such older leaves are not parasitic on the plant, have low respiration rates, and soon die if these exceed their photosynthetic rate, with the result that net photosynthesis tends to plateau at high LAI. Many examples of such behaviour have been reviewed by Yoshida, S. (1972).

At high LAI, canopies with more vertically inclined leaves have a higher photosynthetic rate than those with horizontal leaves, at least under clear skies with the sun at high elevations, because of reduced light saturation of the upper leaves and more uniform distribution of light throughout the canopy. More vertical leaves may therefore be of significant advantage to crops, and Watson and Witts (1959) suggested that more inclined leaves constitute one of the important differences between wild and cultivated sugar beet. Reducing the inclination of rice leaves with paper clips reduced crop photosynthesis (p. 77), and productive modern rice varieties tend to have more erect leaves than do older varieties (Tanaka *et al.* 1964). Tanner *et al.* (1966) have likewise found a strong association between high yield and erect leaves among barley, wheat and oat varieties. Comparing an erect with a lax leafed variety of barley, Angus *et al.* (1972) found crop photosynthesis to be greater with erect leaves, as was yield at high sowing density. However, at a lower sowing density the lax leafed variety had the highest yield of all stands. Increased photosynthesis was not necessarily associated with increased yield, therefore. For maize, Duncan (1971) estimated that erect leaves could double crop photosynthesis at high LAI, but Sinclair (quoted by Wallace *et al.* 1972) did not find greater photosynthesis in more erect leafed stands, and Russell (1972) found yield to be higher with less erect than with more erect leaves in comparisons at several plant spacings.

Such paradoxical behaviour may be due to the fact that leaf inclination has been considered almost entirely in terms of light distribution in the canopy, whereas it may have other, counteracting, effects. For example, more erect leaves may adversely affect the influx of carbon dioxide into the canopy, or its diffusivity within it. Moreover, only the uppermost leaves are of importance in supplying the grain with assimilates in crops such as wheat, rice and barley, and greater photosynthesis by the lower leaves at the expense of the upper leaves may be of no advantage. More erect leaves would seem to be of greatest advantage for crops with axillary inflorescences, particularly those like soybean, pea and cotton which bear fruit at many nodes. There is, however, at least one reason why better penetration of light to the lower leaves may be advantageous even for cereals. The lower leaves export a higher proportion of their assimilates to the roots than do upper leaves, and it is possible that more erect leaves could be associated with more active and prolonged root growth, and therefore with more prolonged water and nutrient uptake and, possibly, export of cytokinins to the shoot.

VARIATION IN PHOTOSYNTHETIC RATE WITHIN SPECIES

Considerable differences between varieties in their photosynthetic rate per unit leaf area have been found in many crop plants, such as sugar cane (Irvine, 1967), beans (Izhar and Wallace, 1967), cotton (Muramoto *et al.* 1965), barley (Apel and Lehmann, 1969) and lucerne (Pearce *et al.* 1969), in none of which were they closely related to differences in productivity. Soybean varieties also differ significantly in their photosynthetic rate, as shown by Ojima and Kawashima (1968), Curtis *et al.* (1969), Dreger *et al.* (1969) and Dornhoff and Shibles (1970); in two of these comparisons there was a trend towards higher photosynthetic rates in the higher yielding varieties, but with striking exceptions. In rice there are also differences between varieties (e.g. Osada, 1966), but Murata (1965) found little correlation between photosynthetic rate and grain yield. Among strains of ryegrass, crop growth rate was inversely related to photosynthetic rate (Rhodes, 1972). In wheat, photosynthetic rate has fallen in the course of evolution as leaf and grain size has increased: the relative growth rates of wheats from all stages of evolution of the crop are similar (Evans and Dunstone, 1970), just as modern highly productive cultivars of tomatoes have relative growth rates no higher than the wild progenitors (Warren Wilson, 1972). The highest photosynthetic rate among cottons is for a wild species (p. 302). With maize, Duncan and Hesketh (1968) found varietal differences in dry matter production to be more dependent on the differences in leaf growth than on those in photosynthetic rate. Much larger varietal differences in the light-saturated net photosynthetic rate of maize have been reported by Heichel and Musgrave (1969), and five generations of selection for high and low photosynthetic rate have been carried out. Heritability is high, but effects on yield have not been established (see Moss and Musgrave, 1971).

Thus, although increase in photosynthetic rate by carbon dioxide enrichment during the storage phase leads to increased yield in several crops, there is no clear evidence that increases in crop productivity have been accompanied by a rise in leaf photosynthetic rate, and in some cases at least the rate has fallen. To reconcile these findings we must assume that there are counter-productive associations with high photosynthetic rate, and it is one of the urgent tasks of crop physiology to identify these. One possible explanation is that there is extensive mobilization of nitrogen from the leaves in high yielding crops, lowering their photosynthetic rate (cf. p. 302). In the wheat comparisons, however, the pronounced differences in photosynthetic rate were evident even before grain set, and some other explanation is required. Another contributing factor could be the frequently negative relation between leaf size and photosynthetic rate, as in wheat, rice, and maize (Hanson, 1971), although Berdahl *et al.* (1972) could detect no difference in photosynthetic rate between leaves of large- and small-leaved lines of barley. Wilson and Cooper (1969) found a negative relation between mesophyll cell size and photosynthetic rate in ryegrass, which they associated with a fall in surface: volume ratio of the cells, and a rise in the resistance to carbon dioxide exchange, as cell size increased, but among the species of wheat there is no clear relation between cell size and leaf or grain size, and only a weak negative correlation between cell size and photosynthetic rate (Dunstone and Evans, 1973). Chloroplasts usually position themselves with their flat surface close to the cell wall, so that the surface:volume ratio of the cells is unlikely to influence the path length for diffusion of carbon dioxide from outside the cell to the site of fixation. At the moment, therefore, we lack a satisfactory explanation of why photosynthetic rate and leaf size may be negatively associated and why photosynthetic rate and yield potential so rarely show a positive relationship, unless it is that yield is not often limited by the supply of assimilate.

Until more progress is made on these questions there is little point in selecting crop plants for high photosynthetic rate, particularly in view of the many difficulties in the way of doing this effectively. For example, conditions prior to measurement, particularly the light environment, may have a profound effect on the light-saturated photosynthetic rate, as in soybeans (Bowes *et al.* 1972) and wheat (Dunstone, Gifford and Evans, 1973). So also do the ontogenetic rank and position of the leaves, leaf age and the internal demand for assimilates. Given all these sources of variation, as well as the need to compare rates under standard conditions of illumination, atmospheric carbon dioxide level, leaf temperature and water potential, it is apparent that selection for photosynthetic rate in the field from a wide range of genotypes presents very great problems with little surety of return.

PHOTOSYNTHETIC PATHWAY AND CROP PRODUCTION

Selection for the absence of photorespiration, usually associated with photosynthesis by the C_4 pathway, can be done effectively, but Moss and his colleagues (Cannell *et al.* 1969; Moss and Musgrave, 1971) have screened about 2500 soybean genotypes, 10000 oat seedlings and 50000

Table 11.3. *Maximum recorded short term crop growth rates* (*CGR*)

Species	Photosynthetic pathway	CGR (g m^{-2} ground day^{-1})	Reference
Helianthus annuus	C$_3$	68	Hiroi and Monsi, 1966
Pennisetum purpureum	C$_4$	60	Arias and Butterworth, 1965
Agrostemma githago	C$_3$	57	Blackman, 1962
Phragmites communis	C$_3$	57	Dykyjova, 1971
Oryza sativa	C$_3$	55	Tanaka *et al.* 1966
Pennisetum typhoides	C$_4$	54	Begg, 1965
Typha latifolia	C$_3$	53	Williams *et al.* 1965
Zea mays	C$_4$	52	Williams *et al.* 1965
Sorghum vulgare	C$_4$	51	Loomis and Williams, 1963
Saccharum officinarum	C$_4$	38	Borden, 1942
Solanum tuberosum	C$_3$	37	Lorenz, 1944
Beta maritima	C$_3$	31	Watson, 1958
Ananas comosus	CAM	28	Eckern, 1964
Glycine max	C$_3$	17	(p. 159)

wheat seedlings without finding a single plant with a low carbon dioxide compensation point, nor have any been found among species of *Beta* (Hall, 1972) or *Triticum* (Dvorak and Nátr, 1971). Disappointing as this may be we should, in the light of the discussion above, reconsider the objective of such selection. In essence it is that plants lacking photorespiration, like those with the C$_4$-dicarboxylic acid pathway, will have higher photosynthetic rates and therefore higher yields (cf. Zelitch, 1971).

The rate of photosynthesis per unit leaf area by single leaves under high light intensity at atmospheric carbon dioxide levels is generally greater among C$_4$ plants than in most Calvin cycle plants, although the highest rates have been approached by leaves of the primitive wheats (Evans and Dunstone, 1970) and of the bulrush, *Typha* (McNaughton and Fullem, 1970). If the atmospheric concentration of carbon dioxide continues to increase in future, the relative advantage of the C$_4$-dicarboxylic acid pathway at higher light intensities will decline. At low levels of irradiation C$_3$ crops may have an advantage because of the greater stomatal resistance in C$_4$ plants (Gifford, 1974).

For crop photosynthesis, the relative advantage of C$_4$ over C$_3$ plants is less than at the level of the single leaf due to shading of the lower leaves and to the increasing importance of the aerodynamic compared with the leaf resistances to carbon dioxide exchange (Slatyer, 1970). The highest rates of photosynthesis measured on wheat crops by aerodynamic methods (Denmead, 1970) are only slightly less than those for maize crops (Lemon, 1967; Inoue *et al.* 1968).

Similarly, the highest short term crop growth rates (CGR) recorded for several C_3 plants are quite comparable with the highest values for C_4 plants (Table 11.3). During the intervals of a week or more over which crop growth rates are recorded there will be periods of low irradiation when any photosynthetic advantage of C_4 crops is minimized, and prolonged periods of night-time respiration which further mute their net photosynthetic advantage. Moreover, differences in the extent of investment in leaf growth may be far more important than differences in photosynthetic rate in the determination of crop growth rate (e.g. Slatyer, 1970). Crop yields do not lend themselves to satisfactory comparisons because the duration of the storage phase is so important a determinant of yield, and depends on the climatic conditions in which the plants are grown. Moreover, most record yields are inadequately documented. Of the record yields noted in the preceding chapters, the highest – in terms of useable product – are for sugar cane (p. 67), sugar beet (p. 265) and potato (p. 250), two of them Calvin cycle plants.

The very great photosynthetic advantage of C_4 compared with C_3 plants at the level of carboxylation is attenuated at the whole leaf level by combination with stomatal and mesophyll resistances (cf. Gifford, 1974), and still more so at the level of crop photosynthesis by shading, periods of low light, and respiration, with the result that no consistent advantage of the C_4 pathway is evident in the maximum crop growth rates and crop yields. The real value of the C_4 pathway probably lies elsewhere, in its better adaptation to high temperature–high insolation conditions, provided the nights are not cold, just as C_3 plants perform better under cool conditions with only moderate insolation. Since the conditions for optimum growth of C_4 and C_3 plants are quite different, valid comparisons of their performance can hardly be made.

CROP RESPIRATION

The evidence that respiration is usually coupled tightly to cellular work has been reviewed by Beevers (1970). In the few instances where natural uncoupling occurs, advantage may accrue to the plant from the heat or carbon dioxide generated, as in ripening fruits, flowers at anthesis, or spadices thrusting through snow. Because of the usual tight coupling of common synthetic pathways, marked differences in respiratory efficiency between cultivars and between species are unlikely to be found, although there may well be significant differences in the extent to which carbon dioxide from respiration is recaptured in photosynthesis, within the leaf, or by glumes and husks around the storage organs, or by the canopy from root and stem respiration. Effective coupling of respiration also provides a rationale for the estimation of respiration in crop modelling, which would be difficult if uncoupling were extensive.

Several early models of plant communities assumed that crop respiration was proportional to accumulated dry weight or LAI, although great differences between organs in their respiration rate had long been known. It was then shown that crop respiration, like crop photosynthesis, approaches a plateau as LAI increases (Ludwig *et al.* 1965; McCree and Troughton, 1966; Figure 5.4). Respiration is thus not proportional to dry weight, but roughly proportional to gross photosynthesis, and is estimated on this basis in some models. A more satisfactory approach is that introduced by McCree (1970) in which crop respiration is estimated as the sum of two terms, one for maintenance and the other for growth, although the distinction between these is operational rather than biochemical. Maintenance consists of the processes which replace structures undergoing degradation, and McCree estimates it as proportional to the accumulated dry weight. On this basis it becomes the dominant term for respiration of crops towards the end of their life cycle, but it would be more satisfactory to base its estimation on the accumulation of non-storage proteins and membrane components which turn over far more rapidly than cell wall constituents and storage polysaccharides, proteins and oils. For white clover at 20 °C, McCree's (1970) estimate of maintenance respiration was 0.015 g g^{-1} day^{-1}, and Penning de Vries (1972) has tabulated many rates ranging from 0.008 to 0.022 g g^{-1} day^{-1} for maize and bean seedlings at 15–25 °C. For cotton leaves at 30 °C Hesketh *et al.* (1971) estimated a rate of 0.026 g g^{-1} day^{-1}, whereas the rate for the cotton boll during linear growth was only 0.003–0.006 g g^{-1} day^{-1} (cf. Thornley and Hesketh, 1972). Towards the end of the crop life cycle when structural and storage compounds comprise most of the accumulated dry weight, maintenance respiration is likely to be a far smaller proportion of dry weight than in the seedling stage.

McCree relates the other respiration term to photosynthesis: it represents the metabolic cost of converting the translocated products of photosynthesis to structural, cytoplasmic or storage compounds. Penning de Vries (1972, 1974) has calculated the theoretical efficiency of the various conversions, and has shown them to be unaffected by temperature. His review of earlier experiments on the relative yield of plant tissue from grains of several cereals gives values of 0.51–0.78 g g^{-1} with comparable values for legume seeds, and values of 0.88–0.96 g g^{-1} for oilseeds. McCree derived a value of 0.75 g g^{-1} for the growth of white clover, and a similar value was estimated by Thornley and Hesketh (1972) for the growth of cotton bolls. However in view of the considerable variation in the magnitude of both components of respiration, more work is required before respiration losses by crops can be estimated with confidence.

THE CAPACITY FOR ASSIMILATE TRANSLOCATION

As noted above, there may well be situations where neither source nor sink limits the rate of storage, and where the capacity of the translocation system may be limiting at some point between the source and the sink.

Near the source, for example, diffusion of sugars to the veins in the leaf, their loading into the phloem, or their movement through the petiole, leaf sheath or stem may impose a bottleneck. In many systems, however, the rate of export seems to vary in parallel with the rate of photosynthesis. Comparing nine different crop plants, Hofstra and Nelson (1969) found the proportion of assimilate exported within six hours to increase in proportion to the photosynthetic rate, export being most rapid in plants with C_4 photosynthesis. Similarly, the proportion of photosynthate exported by leaves of a plant may remain constant over a wide range of light intensities and photosynthetic rates, as in sugar cane (Hartt, 1965), sugar beet (Habeshaw, 1969), and several tropical grasses up to an intensity beyond full sunlight (Lush and Evans, 1974). It appears, therefore, that the accumulation of assimilates in leaves need not imply an insufficient capacity for their loading and translocation, because a rise in light intensity and photosynthesis, or increased demand for assimilates elsewhere in the plant (King et al. 1967), can lead to increase in the rate of assimilate export from leaves.

Milthorpe and Moorby (1969) concluded that the transport systems of plants are more than adequate to meet the demands on them except possibly for the supply of sucrose to apical meristems and young primordia. In their chapter on the potato, however, they point out that whereas translocation is unlikely to limit tuber growth rate to 25 g m^{-2} day^{-1} when all tubers grow simultaneously, it could do so when only one third of the tubers grow at any one time (p. 240), as is often the case. The fact that overall tuber growth rate is little affected by LAI or light intensity on the one hand, or by the number of tubers growing at any one time on the other, could suggest a limitation by the transport system. Murata and Matsushima suggest (p. 82) that the slow rate of grain filling in rice at low temperatures may be due to a limitation by translocation, but this seems unlikely since the rate of translocation soon recovers even when temporarily depressed by low temperatures, at least in sugar beet (Geiger and Sovonick, 1970). Among wheats from all stages in the evolution of the crop the cross-sectional area of phloem in the peduncle has increased in parallel with the peak rate of storage in the ear (Evans et al. 1970). Likewise, there is a close proportionality in grapes between bunch growth and phloem cross-section (Singh and Sharma, 1972). Geiger et al. (1969) also found a high correlation between phloem cross-section and the rate of sugar movement through petioles of sugar beet, and argued that this, together with the similarity of mass transfer per unit area of phloem in many plants (Canny, 1960), indicates a limitation by phloem on the rate of translocation.

So long as the mechanism of translocation remains unknown, it will be difficult to come to a firm conclusion. Even with a given amount of phloem tissue, the rate of translocation may be increased by a rise in the speed of movement (which can vary according to demand as shown by Wardlaw (1965) with wheat), or in the concentration of sucrose if mass flow is involved. Moreover, much higher rates of mass transfer per unit phloem area than those reviewed by Canny (1960) have been found recently, in soybean petioles (Fisher, 1970), in leaves of several C_4 grasses (Lush and Evans, 1974), and in wheat roots (Passioura and Ashford, 1974). The phloem, therefore, may have a capacity for translocation well beyond that measured in most systems.

THE PARTITIONING OF ASSIMILATES

The pattern of assimilate distribution is determined by that of photosynthesis on the one hand, and on the other by the strength and proximity of the various sinks, modified to some extent by the pattern of vascular connections (cf. Wardlaw, 1968), and by environmental conditions. In sugar beet, for example, a limiting supply of assimilate usually is distributed preferentially to the young leaves, but goes to the roots instead in plants under water stress (cf. Figure 9.4). The pattern of distribution changes as plants grow and develop new leaves and new sinks, as described in wheat by Rawson and Hofstra (1969). During the period of grain growth, the flag and penultimate leaves are the main suppliers to the ear, while the lower leaves support the roots and new tillers, but the pattern is flexible: removal of the flag leaf causes the leaves below it to supply more to the ear, while shading of the lower leaves may cause the flag leaf to support the roots (King *et al.* 1967). In crops with many axillary inflorescences, such as pea, soybean or cotton, each inflorescence is usually supported mainly by its subtending leaf (Wardlaw, 1968). However, a well established cotton boll whose supply leaf is shaded may gain preference in supply from a younger leaf in the light, causing failure of the young boll it subtends (p. 306). With root and tuber crops even the uppermost leaves supply the underground storage organs. Thus, it is the relative strength of the sinks that largely determines the pattern of movement, and from this point of view there is no preferred position for storage organs, in spite of Mangelsdorf's (1966) belief in the advantage of a centrally placed corn cob.

The pattern of vascular connections can be restrictive, in spite of dominance by the sink, because many crops show little lateral movement of assimilate out of the phloem. Older tobacco leaves preferentially support younger ones three, five and eight nodes above them (Jones *et al.* 1959; Shiroya *et al.* 1961). Each sugar beet leaf may supply a particular part of the root system (p. 263) and each sunflower leaf may supply seeds in only certain parts of the inflorescence (Prokofyev *et al.* 1957). Soybean leaves mainly support the pods in their axil, but also those two leaves above and two leaves below (p. 171). These phyllotactic patterns of distribution may, however, be modified, as in partially defoliated sugar beet (p. 263).

Sink strength is an important determinant of translocation patterns, of the partitioning of dry matter, and therefore of yield, but we have little understanding of what determines it or how it governs translocation, and it is one of the important tasks before crop physiology to analyse the basis of the attractive and competing power of an organ for assimilates, because it is on empirical selection for this that past increases in yield have largely been based. The number, size, proximity, synchrony and potential growth rate of storage organs are undoubtedly important determinants of sink strength, but whether this latter depends on the activation of local translocation, unloading, enzymic conversion or storage processes, with or without the aid of hormones produced by the sink, is not clear.

In situations where several sinks are competing for a limited supply of assimilates, the relative magnitude of the sinks may be of overriding importance in partitioning, with a pronounced bias in favour of the largest sink. This is shown in an experiment by Peel and Ho (1970) in which aphid colonies of four to five individuals competed with colonies of 15–20 aphids for labelled sugars from a symmetrically placed source. In all experiments the larger colony received far more than its proportional share of the sugar, there being up to a thousandfold bias in favour of the larger colony. Consider primitive and modern wheats during grain filling in the light of this experiment. In the primitive wheats there are several sinks of comparable magnitude, for root growth, tiller establishment, stem growth and grain filling, and assimilates are partitioned more or less evenly among all of them. In the modern wheats at this stage, however, nearly all the assimilate goes to the grain. This 'winner takes all' bias in distribution, which has undoubtedly played a major role in crop evolution, can be seen in several chapters of this book. In pea plants with two equal shoots, for example, any setback to one shoot results in the more vigorous shoot draining its sister of assimilates (p. 194). Similarly, few of the many sprouts per tuber in potatoes eventually become dominant and receive all the assimilate (p. 228), which is advantageous to yield, as in cassava (Enyi, 1972). Although several ears may differentiate on a maize plant, only the uppermost one or two develop in modern cultivars, whereas primitive maize had many small ears.

Such bias in favour of the storage sink may increase yield up to a point, by increasing the proportion of assimilates stored in the harvested organs. It will also operate to increase the synchrony of storage and ripening, since later formed sinks will tend to fail. In the potato, for example, only tubers initiated over a relatively brief period grow significantly (p. 239). The development of a substantial storage bias is made possible by the separation in time of the growth and storage phases of a crop, and in turn reinforces it. Such separation is perhaps clearest in the small-grain cereals, and least evident in axillary flowering plants such as soybean and cotton in which continuing vegetative growth competes with fruit development at all stages, and in sugar cane and beet in which sugar storage competes with continued vegetative growth.

In the longer term translocation bias may be detrimental to further yield increase in that, by cutting short root and leaf renewal, it may limit the duration of storage, a powerful determinant of yield in many crops. In fact, the various mechanisms within the plant which are involved in the integration of its growth and in the maintenance of effective balance between organs may limit the operation of sink bias. Following partial defoliation or root pruning, for example, the relative investment in new leaf or root growth respectively increases transiently until the equilibrium root:shoot ratio characteristic of the particular environment is restored (e.g. Figure 9.8). This ratio does, of course, change during crop development, or with change in environmental conditions, but tends to do so relatively smoothly and progressively, as in growth of beet (Patefield and Austin, 1971). Assimilate supply may be adequate in such cases, with little competition between sinks and therefore less tendency for an extreme bias in assimilate distribution to develop. Under such conditions, the partitioning of assimilates can be empirically modelled in crops by the use of smoothly changing allometric ratios, but explanatory models, such as that attempted for sugar beet in chapter 9, will eventually give us more insight into the processes which control partitioning and yield.

STORAGE CAPACITY AS A LIMITATION ON YIELD

Storage capacity is more likely to limit yield in crops in which a more or less determinate inflorescence is harvested than in those where yield derives from vegetative organs. In rice, for example, the number of ears, panicle branches and spikelets, and even the upper limit of grain size, are all determined by the time of flowering, whereas sugar cane can continue to add internodes for sugar storage indefinitely, and the potato tuber and sugar beet root continue meristematic growth throughout the storage phase.

In many crops the *rate* of storage remains fairly constant over the greater part of the storage period, provided temperature does not vary greatly. The storage rate is sensitive to temperature, as in wheat (Figure 5.5) and cotton (Figure 10.3) but not potatoes (p. 241), although relatively insensitive to incident radiation, as noted above (p. 333). Storage rate per unit ground area in potato crops can be fairly constant in spite of marked fluctuations in the rate for individual tubers (p. 240), and in spite of changes in LAI or incident light.

The *duration* of storage is a powerful determinant of yield. Comparing wheat, barley, potatoes and sugar beet, Watson (1971) found differences in the rate of storage, but final yields were most clearly related to differences in duration of storage, as also in soybeans (p. 159). In a few crops, such as rice, the duration of storage may be limited by the physical capacity of the storage organs: so restrictive are the hulls of the rice grain that small pebbles or pieces of plastic enclosed within the glumes deform the grain (p. 79). Grains of the invested wheats may also be marked by glume veins as they are in rice, but those of the free-threshing wheats are free to expand. The potential size of storage organs is broadly related to the duration of their filling, and large grains may be an important yield attribute when storage is prolonged. On the other hand, in environments where storage is cut short by drought or heat, it may be more important to have many grains – and a fast rate of storage – than to have large grains. Another advantage of large storage organs is that the proportion of non-storage tissue, such as the seed coat, is smaller the larger the organ, but on the other hand diffusion of assimilates, oxygen and carbon dioxide to and from the centre of large organs without a vascular system may become limiting.

The duration of storage in several crops appears to be inversely related to temperature, as in rice (Sato, 1971), wheat (Figure 5.5) and cotton (Figure 10.3). Consequently, because temperature and incident radiation are often correlated, duration of storage may fall as the level of radiation rises (e.g. Figure 5.6), with the result that yield may increase with daily radiation only up to a point, beyond which it reaches a plateau, as for wheat in England (Welbank et al. 1968) and rice in Japan (Munakata et al. 1967). Just why high temperatures shorten the storage period so much is not clear. They may reduce the half-life of enzymes involved in storage or of leaf proteins, leading to earlier senescence. These processes may be influenced by cytokinins synthesized in the roots, and if they do play such a role it is possible that the cessation of root growth signals the end of storage activity in the shoot. Root growth in peas starts to decline once flower initation begins (p. 196), and in rice it may cease before heading but begin again towards the end of grain filling (Figure 4.1). Renewal of root growth in rice crops by mid-summer drainage may have very beneficial effects on grain yield (p. 87), of which one component could be renewed export to the shoot of root cytokinins (cf. Yoshida, R. et al. 1971). Exudation of root cytokinins in sunflowers reaches a peak just before flowering, and then falls rapidly as root growth presumably ceases (Sitton et al. 1967). We need to clarify to what extent continued storage of protein and carbohydrate in seeds depends on continued root growth, since past selection for high yield may often have been at its expense, and this consequence may need to be reversed if further yield increases are to be obtained.

CROP PHYSIOLOGY AND NEW IDEOTYPES

The crucial test for crop physiology is whether the insights it presents actually lead to new and productive approaches to plant breeding and agronomic practice. Much of the past increase in crop yields has been due to the progressive elimination by plant breeders of defects which are readily observable, such as susceptibility to various diseases, pests, frost, drought, high temperatures and lodging. Along with this defect elimination, there has been selection for the less readily observable characters influencing yield potential. Frankel (1947) and Donald (1968) are inclined to believe this latter process has been rather ineffective, but the methods of selecting for yield increase make some progress almost inevitable. Donald believes it could be far more rapid if plant breeders aimed to produce ideotypes, plants with specific combinations of characteristics favourable to photosynthesis, growth and grain production. Successful plant breeders have probably always had particular ideotypes in mind when making selections, but these have varied greatly from one breeder to another and from one environment to another. Farrer (1898) presented a clear prescription for Australian wheats, and Vavilov (1950) a much more comprehensive one for European conditions.

The problem, however, is that each physiological or morphological characteristic may affect yield in many ways, the net effect of which depends on other characteristics, on environmental conditions and on agronomic practice. Several counteracting effects of leaf inclination have already been noted (p. 336), as have the apparent complications of selecting for higher photosynthetic rate (p. 338), awned ears (p. 123), shorter stems (p. 109), or fewer tillers in wheat (p. 106). A weakness of the universal ideotype concept is that there may be many and subtle counter-productive features associated with the exaggeration of particular traits. Awns may be advantageous in dry climates but deleterious in wet conditions. Potentially large grains are wrinkled and unwanted when not filled. Extremely short stems may bring the leaves too close together for optimum distribution of light throughout the canopy. With increased physiological insight some of these undesirable associations may be broken, but for the present we need to know more about what limits yield before we can be confident of such specifications.

As agronomic practice changes, so must the ideotype. Because of the lag of more than a decade between inception and release of a new cultivar in most breeding programmes, crop production techniques may change substantially in the interval, requiring a change in breeding objectives (Jensen, 1967). Cheap nitrogen fertilizer, for example, generated a need for short strong stems long before suitable varieties were available, while herbicides likewise eliminated the need for large light-excluding leaves to suppress weed growth.

Currently there is a trend towards the replacement of locally adapted genotypes by those of wide adaptation, with adaptability specified as a valuable characteristic for many crops. This is undoubtedly so for varieties destined to be widely distributed in connection with the green revolution, but once this phase has passed there may well be a return to closer local adaptation. One reason for this would be to avoid the extreme genetic vulnerability to new races of disease organisms that exists when a few genotypes are grown very extensively. A second reason is that such adaptable varieties, while more reliable in their yield, may not take full advantage of the most favourable conditions.

Future crops may be selected not only for greater local adaptation, but also for genetic heterogeneity, to reduce the risk of epidemics and perhaps also to provide more physiological stability. Jensen (1967) has suggested that genetic and phenotypic diversity of crops should be accepted as a desirable goal in plant breeding programmes. Before this can be achieved, however, we will need to understand how different crop genotypes may complement one another physiologically. Weak competitors of the kind envisaged in Donald's monocultural ideotype are unlikely to be successful in such mixed stands.

Future crop physiology must also concern itself with the problems of multiple cropping, such as the need to minimize the initial lag period or to establish a crop before the preceding one is mature (cf. Bradfield, 1972). Increasing use of minimum tillage practices, to reduce energy inputs, may shift selection criteria, and the ability of crops to make more efficient use of phosphatic and other fertilizers will become more important in the future. Closed systems of crop production, especially in horticulture, will be more extensively used, as may desalinated water for irrigation, inevitably making efficiency of water use a more important criterion for selection. Greater response to applied regulants and resistance to pesticides may also be sought (cf. Pinthus *et al.* 1972). Crops may also be selected more intensively for the specialized production of carbohydrate, protein, oil, or fibre for industrial processing rather than as a direct source of balanced food and feed.

All these and other changes will lead to shifts not only in the selection criteria for crops, but also in agronomic practice. But whatever the extent of these shifts, the physiological basis of yield in the major crop plants will continue to be a major preoccupation of agronomists and plant breeders, and their understanding of it will continue to shape their approaches to the problems of feeding the world.

REFERENCES

Adams, M. W. (1967). Basis of yield component compensation in crop plants with special reference to the field bean, *Phaseolus vulgaris*. *Crop Sci.* 7, 505–510.

Alkire, W. H. (1965). Lamotrek Atoll and inter-island socio-economic ties. *Ill. Studies Anthropol.* No. 5. Univ. Ill., Urbana.

Allison, J. C. S. and Weinmann, H. (1970). Effect of absence of developing grain on carbohydrate content and senescence of maize leaves. *Pl. Physiol.* **46**, 435–436.

Angus, J. F., Jones, R. and Wilson, J. H. (1972). A comparison of barley cultivars with different leaf inclinations. *Aust. J. agric. Res.* **23**, 945–957.

Apel, P. and Lehmann, C. O. (1969). Variabilität und Sortenspezifität der Photosyntheserate bei Sommergerste. *Photosynthetica* **3**, 255–262.

Archbold, H. K. and Mukerjee, B. N. (1942). Physiological studies in plant nutrition. XII. *Ann. Bot. N.S.* **6**, 1–41.

Arias, P. J. and Butterworth, M. (1965). Crecimento del pasto elefante. *Proc. 9th Int. Grassl. Congr.* **1**, 407–412.

Beevers, H. (1970). Respiration in plants and its regulation. In *Prediction and Measurement of Photosynthetic Productivity*. Pudoc, Wageningen, pp. 209–214.

Begg, J. E. (1965). High photosynthetic efficiency in a low latitude environment. *Nature, Lond.* **205**, 1025–1026.

Begum, A. and Eden, W. G. (1965). Influence of defoliation on yield and quality of soybean. *J. Econ. Entomol.* **58**, 591–592.

Berdahl, J. D., Rasmusson, D. C. and Moss, D. N. (1972). Effect of leaf area on photosynthetic rate, light penetration and grain yield in barley. *Crop Sci.* **12**, 177–180.

Black, J. N. (1971). Energy relations in crop production – a preliminary survey. *Ann. appl. Biol.* **67**, 272–278.

Blackman, G. E. (1962). The limit of plant productivity. *Ann. Rep. East Malling Res. Sta. 1961*, pp. 39–50.

Borden, R. J. (1942). *Hawaiian Planters Record* **46**, 191–238.

Börger, H., Huhnke, W., Köhler, D., Schwanitz, F. and von Sengbusch, R. (1956). Untersuchungen über die Ursachen der Leistung von Kulturpflanzen. I. *Der Zuchter* **26**, 363–370.

Bowes, G., Ogren, W. and Hageman, R. H. (1972). Light saturation, photosynthesis rate, RuDP carboxylase activity, and specific leaf weight in soybeans grown under different light intensities. *Crop Sci.* **12**, 77–79.

Bradfield, R. (1972). Maximizing food production through multiple cropping systems centred on rice. In *Rice, Science and Man*. Intl. Rice Research Inst., Los Baños, pp. 143–163.

Cannell, R. Q., Brun, W. A. and Moss, D. N. (1969). A search for high net photosynthetic rate among soybean genotypes. *Crop Sci.* **9**, 840–842.

Canny, M. (1960). The rate of translocation. *Biol. Rev.* **35**, 507–532.

Curtis, P. E., Ogren, W. L. and Hageman, R. H. (1969). Varietal effects in soybean photosynthesis and photorespiration. *Crop Sci.* **9**, 323–327.

Das Gupta, D. K. (1972). Developmental physiology of sugar beet: IV. Effects of growth substances and differential root and shoot temperatures on subsequent growth. *J. exp. Bot.* **23**, 103–113.

Denmead, O. T. (1970). Transfer processes between vegetation and air: measurement, interpretation and modelling. In *Prediction and Measurement of Photosynthetic Productivity*. Pudoc, Wageningen, pp. 149–164.

Donald, C. M. (1968). The breeding of crop ideotypes. *Euphytica* **17**, 385–403.

Dornhoff, G. M. and Shibles, R. M. (1970). Varietal differences in net photosynthesis of soybean leaves. *Crop Sci.* **10**, 42–45.

Dreger, R. H., Brun, W. A. and Cooper, R. L. (1969). Effect of genotype on the photosynthetic rate of soybean. *Crop Sci.* **9**, 429–431.

Duncan, W. G. (1971). Leaf angles, leaf area, and canopy photosynthesis. *Crop Sci.* **11**, 482–485.

Duncan, W. G., Hatfield, A. L. and Ragland, J. L. (1965). The growth and yield of corn. II. Daily growth of corn kernels. *Agron. J.* **57**, 221–223.

Duncan, W. G. and Hesketh, J. D. (1968). Net photosynthetic rates, relative leaf growth rates, and leaf numbers of 22 races of maize grown at eight temperatures. *Crop Sci.* **8**, 670–674.

Dunstone, R. L. and Evans, L. T. (1973). The role of changes in cell size in the evolution of wheat. *Aust. J. Plant Physiol.* **1**, 157–165.

Dunstone, R. L., Gifford, R. M. and Evans, L. T. (1973). Photosynthetic characteristics of modern and primitive wheat species in relation to ontogeny and adaptation to light. *Aust. J. biol. Sci.* **26**, 295–307.

Dvorak, J. and Nátr, L. (1971). Carbon dioxide compensation points of *Triticum* and *Aegilops* species. *Photosynthetica* **5**, 1–5.

Dykyjova, D. (1971). Productivity and solar energy conversion in reed swamp stands in comparison with outdoor mass culture of algae in the temperate climate of Central Europe. *Photosynthetica* **5**, 329–340.

Eckern, P. C. (1964). *The evapotranspiration of pineapple in Hawaii. Pineapple Res. Inst. Rept.* No. 109, 233 pp.

Enyi, B. A. C. (1972). Effect of shoot number and time of planting on growth, development and yield of cassava (*Manihot esculenta* Crantz.). *J. hort. Sci.* **47**, 457–466.

Evans, L. T. (1971). Evolutionary, adaptive and environmental aspects of the photosynthetic pathway: assessment. In *Photosynthesis and Photorespiration*, eds M. D. Hatch, C. B. Osmond and R. O. Slatyer. Wiley-Interscience, New York, pp. 130–136.

Evans, L. T. and Dunstone, R. L. (1970). Some physiological aspects of evolution in wheat. *Aust. J. biol. Sci.* **23**, 725–741.

Evans, L. T., Dunstone, R. L., Rawson, H. M. and Williams, R. F. (1970). The phloem of the wheat stem in relation to requirements for assimilate by the ear. *Aust. J. biol. Sci.* **23**, 743–752.

Farrer, W. (1898). The making and improvement of wheats for Australian conditions. *Agric. Gazette NSW* **9**, 131–168, 241–250.

Finlay, K. W. (1970). Genetics of yield. *Second Int. Barley Genetics Symposium.* Pullman, Washington, pp. 338–345.

Fisher, D. B. (1970). Kinetics of ^{14}C translocation in soybean. I. Kinetics in the stem. *Plant Physiol.* **45**, 107–113.

Ford, M. A. and Thorne, G. N. (1967). Effect of CO_2 concentration on growth of sugar beet, barley, kale and maize. *Ann. Bot. N.S.* **31**, 629–644.

Frankel, O. H. (1947). The theory of plant breeding for yield. *Heredity* **1**, 109–120.

Geiger, D. R., Saunders, M. A. and Cataldo, D. A. (1969). Translocation and accumulation of translocate in the sugar beet petiole. *Plant Physiol.* **44**, 1657–1665.

Geiger, D. R. and Sovonick, S. A. (1970). Temporary inhibition of translocation velocity and mass transfer rate by petiole cooling. *Plant Physiol.* **46**, 847–849.

Gifford, R. M. (1974). A comparison of potential photosynthesis, productivity and yield of plant species with differing photosynthetic metabolism. *Aust. J. Plant Physiol.* **1**, 107–117.

Gifford, R. M., Bremner, P. M. and Jones, D. B. (1973). Assessing photosynthetic limitation to grain yield in a field crop. *Aust. J. agric. Res.* **24**, 297–307.

Gifford, R. M. and Millington, R. J. (1974). Energetics of food production with special emphasis on the Australian situation. In *Energy and how we live. Man and the Biosphere Symposium Proceedings*, UNESCO, Australia, in press.

Goeller, H. E. (1972). The ultimate mineral resource situation – an optimistic view. *Proc. Natl. Acad. Sci. USA* **69**, 2991–2992.

Habeshaw, D. (1969). The effect of light on the translocation from sugar beet leaves. *J. exp. Bot.* **20**, 64–71.

Hackett, C. (1973). An exploration of the carbon economy of the tobacco plant. I. Inferences from a simulation. *Aust. J. biol. Sci.* **26**, 1057–1071.

Hall, A. E. (1972). Photosynthesis in the genus *Beta*. *Crop Sci.* **12**, 701–702.

Hansen, P. (1970). ^{14}C studies on apple trees. VI. The influence of the fruit on the photosynthesis of the leaves, and the relative photosynthetic yields of fruits and leaves. *Physiol. Plantar.* **23**, 805–810.

Hansen, P. (1971). *Ibid.* VII. The early seasonal growth in leaves, flowers and shoots as dependent upon current photosynthates and existing reserves. *Physiol. Plantar.* **25**, 469–473.

Hanson, W. D. (1971). Selection for differential productivity among juvenile maize plants: associated net photosynthetic rate and leaf area changes. *Crop Sci.* **11**, 334–339.

Hanway, J. J. and Weber, C. R. (1971). Dry matter accumulation in eight soybean (*Glycine max* (L.) Merrill) varieties. *Agron. J.* **63**, 227–230.

Hardman, L. L. and Brun, W. A. (1971). Effect of atmospheric carbon dioxide enrichment at different developmental stages on growth and yield components of soybeans. *Crop Sci.* **11**, 886–888.

Harlan, J. (1967). A wild wheat harvest in Turkey. *Archaeology* **20**, 197–201.

Hartt, C. E. (1965). Light and translocation of ^{14}C in detached blades of sugarcane. *Plant Physiol.* **40**, 718–724.

Heichel, G. H. and Musgrave, R. B. (1969). Varietal differences in net photosynthesis of *Zea mays* L. *Crop Sci.* **9**, 483–486.

Hesketh, J. D., Baker, D. N. and Duncan, W. G. (1971). Simulation of growth and yield in cotton: respiration and the carbon balance. *Crop Sci.* **11**, 394–398.

Hicks, D. R. and Pendleton, J. W. (1969). Effect of floral bud removal on performance of soybeans. *Crop Sci.* **9**, 435–437.

Hiroi, T. and Monsi, M. (1966). Dry matter economy of *Helianthus annuus* communities grown at varying densities and light intensities. *J. Fac. Sci. Univ. Tokyo Sec. III Bot.* **9**, 241–285.

Hodges, C. N. and Hodge, C. O. (1971). An integrated system for providing power, water and food for desert coasts. *Hort. Sci.* **6**, 10–16.

Hofstra, G. and Nelson, C. D. (1969). A comparative study of translocation of assimilated ^{14}C from leaves of different species. *Planta, Berl.* **88**, 103–112.

Hozyo, Y. and Park, C. Y. (1971). Plant production in grafted plants between wild type and improved varieties of *Ipomoea*. *Bull. Natl. Inst. Agric. Sci.* **22**, 145–164.

Humphries, E. C. and French, S. A. W. (1969). Effect of seedling treatment on growth and yield of sugar beet in the field. *Ann. appl. Biol.* **64**, 385–393.

Inoue, E., Uchijima, Z., Udagawa, T., Horie, T. and Kobayashi, K. (1968). Studies of energy and gas exchange within crop canopies. (2) CO_2 flux within and above a corn plant canopy. *J. agric. Meteorol. Japan* **23**, 165–176.

Irvine, J. E. (1967). Photosynthesis in sugar cane varieties under field conditions. *Crop Sci.* **7**, 297–300.

Izhar, S. and Wallace, D. H. (1967). Studies on the physiological basis for yield differences. III. Genetic variation in photosynthetic efficiency of *Phaseolus vulgaris* L. *Crop Sci.* **7**, 457–460.

Jensen, N. F. (1967). Agrobiology: specialization or systems analysis. *Science* **157**, 1405–1409.

Johnson, D. R. and Tanner, J. W. (1972). Comparisons of corn (*Zea mays* L,) inbreds and hybrids grown at equal leaf area index, light penetration and population. *Crop Sci.* **12**, 482–485.

Jones, H., Martin, R. V. and Porter, H. K. (1959). Translocation of [14]carbon in tobacco following assimilation of [14]carbon dioxide by a single leaf. *Ann. Bot. N.S.* **23**, 493–508.

Kato, S. and Hozyo, Y. (1972). Translocation of ^{14}C-photosynthates in grafts between the wild type and improved variety in *Ipomoea*. *Proc. Crop Sci. Soc. Jap.* **41**, 496–501.

Kazaryan, V. O., Balagezyan, N. V. and Korapetyan, K. A. (1965). (Effect of apple fruit on the physiological activity of the leaves.) *Fiziol. Rast.* **12**, 313–319.

King, R. W., Wardlaw, I. F. and Evans, L. T. (1967). Effect of assimilate utilization on photosynthetic rate in wheat. *Planta, Berl.* **77**, 261–276.

Lemon, E. R. (1967). Aerodynamic studies of CO_2 exchange between the atmosphere and the plant. In *Harvesting the Sun*, eds A. San Pietro, F. A. Greer and T. J. Army. Acad. Press, New York, pp. 263–290.

Loomis, R. S. and Williams, W. A. (1963). Maximum crop productivity: an estimate. *Crop Sci.* **3**, 67–72.

Loomis, R. S., Williams, W. A. and Hall, A. E. (1971). Agricultural productivity. *Ann. Rev. Plant Physiol.* **22**, 431–468.

Lorenz, O. A. (1944). Studies on potato nutrition. II. Nutrient uptake at various stages of growth by Kern County potatoes. *Proc. Amer. Soc. Hort. Sci.* **44**, 389–394.

Ludwig, L. J., Saeki, T. and Evans, L. T. (1965). Photosynthesis in artificial communities of cotton plants in relation to leaf area. *Aust. J. biol. Sci.* **18**, 1103–1118.

Lush, W. M. and Evans, L. T. (1974). The translocation of photosynthetic assimilate from grass leaves, as influenced by environment and species. *Aust. J. Plant Physiol.* **1** (in press).

McAlister, D. F. and Krober, D. A. (1958). Response of soybeans to leaf and pod removal. *Agron. J.* **50**, 674–677.

McCree, K. J. (1970). An equation for the rate of respiration of white clover plants grown under controlled conditions. In *Prediction and Measurement of Photosynthetic Productivity*. Pudoc, Wageningen, pp. 221–229.

McCree, K. J. and Troughton, J. H. (1966). Non existence of an optimum leaf area index for the production rate of white clover grown under constant conditions. *Plant Physiol.* **41**, 1615–1622.

McNaughton, S. J. and Fullem, L. W. (1970). Photosynthesis and photorespiration in *Typha latifolia*. *Plant Physiol.* **45**, 703–707.

Mangelsdorf, P. C. (1966). Genetic potentials for increasing yields of food crops and animals. In *Prospects of the World Food Supply*. Natl. Acad. Sci., Washington, DC, pp. 66–71.

Meadows, D. H., Meadows, D. L., Randers, J. and Behrens, W. W. (1972). *The Limits to Growth*. Universe Books, New York, 205 pp.

Milthorpe, F. L. and Moorby, J. (1969). Vascular transport and its significance in plant growth. *Ann. Rev. Plant Physiol.* **20**, 117–138.

Moss, D. N. and Musgrave, R. B. (1971). Photosynthesis and crop production. *Adv. Agron.* **23**, 317–336.

Munakata, K., Kawasaki, T. and Kariya, K. (1967). Quantitative studies on the effects of the climate factors on the productivity of rice. *Bull. Chugoku Agr. Expt. Sta. Ser. A.* **14**, 59–96.

Muramoto, H., Hesketh, J. and El-Sharkawy, M. (1965). Relationships among rate of leaf area development, photosynthetic rate, and rate of dry matter production among American cultivated cottons and other species. *Crop Sci.* **5**, 163–166.

Murata, Y. (1965). Photosynthesis, respiration and nitrogen response. In *Mineral Nutrition of the Rice Plant*. Johns Hopkins, Baltimore, pp. 385–400.

Murayama, N., Yoshino, M., Oshima, M., Tsukahara, S. and Kawarazaki, K. (1955). Studies on the accumulation process of carbohydrates associated with growth of rice. *Bull. Nat. Inst. Agr. Sci. Japan Ser. B.* **4**, 123–166.

Neales, T. F., Patterson, A. A. and Hartney, V. J. (1968). Physiological adaptation to drought in the carbon assimilation and water loss of xerophytes. *Nature, Lond.* **219**, 469–472.

Oak Ridge Natl. Lab. (1968). *Nuclear energy centres. Industrial and Agro-Industrial Complexes. Summary Report.* ORNL–4291 UC80.

Odum, H. T. (1971). *Environment, Power and Society.* Wiley-Interscience, New York, 331 pp.

Ojima, M. and Kawashima, R. (1968). Studies on the seed production of soybean. 5. Varietal differences in photosynthetic rate of soybean. *Crop Sci. Soc. Jap. Proc.* **37**, 667–675.

Osada, A. (1966). Relationship between photosynthetic activity and dry matter production in rice varieties, especially as influenced by nitrogen supply. *Bull. Natl. Inst. Agr. Sci. Ser. D.* **14**, 117–188.

Passioura, J. B. and Ashford, A. E. (1974). Rapid translocation in the phloem of wheat roots. *Aust. J. Plant Physiol.* **1**, (in press).

Patefield, W. M. and Austin, R. B. (1971). A model for the simulation of the growth of *Beta vulgaris* L. *Ann. Bot.* **35**, 1227–1250.

Pearce, R. B., Carlson, G. E., Barnes, D. K., Hart, R. H. and Hanson, C. H. (1969). Specific leaf weight and photosynthesis in alfalfa. *Crop Sci.* **9**, 423–426.

Peel, A. J. and Ho, L. C. (1970). Colony size of *Tuberolachnus salignus* (Gmelin) in relation to mass transport of ^{14}C-labelled assimilates from the leaves in willow. *Physiol. Plantar.* **23**, 1033–1038.

Penning de Vries, F. W. T. (1972). Respiration and growth. In *Crop Processes in Controlled Environments*, eds A. R. Rees, K. E. Cockshull, D. W. Hand and R. G. Hurd. Acad. Press, New York, pp. 327–346.

Penning de Vries, F. W. T. (1974). Use of assimilates in higher plants. In *Photosynthesis and Productivity in Different Environments*, ed. J. P. Cooper. Cambridge University Press, Cambridge (in press).

Perelman, M. J. (1972). Farming with petroleum. *Environment* **14** (8), 8–13.

Pinthus, M. J., Eshel, Y. and Shchori, Y. (1972). Field and vegetable crop mutants with increased resistance to herbicides. *Science* **177**, 715–716.

Prokofyev, A. A., Zhdanova, L. P. and Sobolev, A. M. (1957). Certain regularities in the flow of substances from leaves into reproductive organs. *Fiziol. Rast.* **4**, 425–431.

Radley, R. W. (1963). The effect of season on growth and development of the potato. In *The Growth of the Potato*, eds J. D. Ivins and F. L. Milthorpe. Butterworth, London, pp. 211–220.

Rappaport, R. A. (1967). *Pigs for the Ancestors.* Yale Univ. Press, New Haven, 311 pp.

Rawson, H. M. and Hofstra, G. (1969). Translocation and remobilisation of ^{14}C assimilated at different stages by each leaf of the wheat plant. *Aust. J. biol. Sci.* **22**, 321–331.

Rhodes, I. (1972). Yield, leaf-area index and photosynthetic rate in some perennial ryegrass (*Lolium perenne* L.) selections. *J. agric. Sci., Camb.* **78**, 509–511.

Russell, W. A. (1972). Effect of leaf angle on hybrid performance in maize (*Zea mays* L.). *Crop Sci.* **12**, 90–92.

Sato, K. (1971). The development of rice grains under controlled environment. II. The effects of temperature combined with air humidity and light intensity during ripening on grain development. *Tohoku J. Agric. Res.* **22**, 69–79.

Shiroya, M., Lister, G. R., Nelson, C. D. and Krotkov, G. (1961). Translocation of C^{14} in tobacco at different stages of development following assimilation of $C^{14}O_2$ by a single leaf. *Canad. J. Bot.* **39**, 855–864.

Singh, O. S. and Sharma, V. K. (1972). Studies on the characterization of physiological sink in Anab-e-shahi grapes (*Vitis vinifera* L.). *Vitis* **11**, 131–134.

Sitton, D., Itai, C. and Kende, H. (1967). Decreased cytokinin production in the roots as a factor in shoot senescence. *Planta, Berl.* **73**, 296–300.

Slatyer, R. O. (1970). Comparative photosynthesis, growth and transpiration of two species of *Atriplex*. *Planta, Berl.* **93**, 175–189.

Sofield, I., Evans, L. T. and Wardlaw, I. F. (1974). The effects of temperature and light on grain filling in wheat. In *Mechanisms of Regulation of Plant Growth*. Roy. Soc. New Zealand, Wellington (in press).

Tanaka, A., Kawano, K. and Yamaguchi, J. (1966). Photosynthesis, respiration, and plant type of the tropical rice plant. *Int. Rice Res. Inst. Tech. Bull.* **7**, 1–46.

Tanaka, A., Navasero, S. A., Garcia, C. V., Parao, F. T. and Ramirez, E. (1964). Growth habit of the rice plant in the tropics and its effect on nitrogen response. *Int. Rice Res. Inst. Tech. Bull.* **3**, 1–80.

Tanner, J. W., Gardener, C. J., Stoskopf, N. C. and Reinbergo, E. (1966). Some observations on upright-leaf-type small grains. *Can. J. Plant Sci.* **46**, 690.

Thorne, G. N. and Evans, A. F. (1964). Influence of tops and roots on net assimilation rate of sugar beet and spinach beet and grafts between them. *Ann. Bot. N.S.* **28**, 499–508.

Thornley, J. H. M. and Hesketh, J. D. (1972). Growth and respiration in cotton bolls. *J. appl. Ecol.* **9**, 315–317.

Tsuno, Y. and Fujise, K. (1965). Studies on the dry matter production of the sweet potato. VIII. The internal factors influence on photosynthetic activity of sweet potato leaf. *Proc. Crop Sci. Soc. Japan* **33**, 230–235.

Vavilov, N. I. (1950). Scientific bases of wheat breeding. *Chron. Bot.* **13**, 169–349.

Wallace, D. H., Ozbun, J. L. and Munger, H. M. (1972). Physiological genetics of crop yield. *Adv. Agron.* **24**, 97–146.

Wardlaw, I. F. (1965). The velocity and pattern of assimilate translocation in wheat plants during grain development. *Aust. J. biol. Sci.* **18**, 269–281.

Wardlaw, I. F. (1968). The control and pattern of movement of carbohydrates in plants. *Bot. Rev.* **34**, 79–105.

Warren Wilson, J. (1972). Control of crop processes. In *Crop Processes in Controlled Environments*, eds A. R. Rees, K. E. Cockshull, D. W. Hand and R. G. Hurd. Acad. Press, New York, pp. 7–30.

Watson, D. J. (1958). The dependence of net assimilation rate on leaf area index. *Ann. Bot. N.S.* **22**, 37–54.

Watson, D. J. (1971). Size, structure, and activity of the productive system of crops. In *Potential Crop Production*, eds P. F. Wareing and J. P. Cooper. Heinemann, London, pp. 76–88.

Watson, D. J. and Witts, K. J. (1959). The net assimilation rates of wild and cultivated beets. *Ann. Bot. N.S.* **23**, 431–439.

Welbank, P. J., Witts, K. J. and Thorne, G. N. (1968). Effect of radiation and temperature on efficiency of cereal leaves during grain growth. *Ann. Bot.* **32**, 79–95.

Willey, R. W. and Holliday, R. (1971*a*). Plant population and shading studies in barley. *J. agric. Sci., Camb.* **77**, 445–452.

Willey, R. W. and Holliday, R. (1971*b*). Plant population, shading and thinning studies in wheat. *J. agric. Sci., Camb.* **77**, 453–461.

Williams, W. A., Loomis, R. S. and Lepley, C. R. (1965). Vegetative growth of corn as affected by population density. I. Productivity in relation to interception of solar radiation. *Crop Sci.* **5**, 211–215.

Wilson, D. and Cooper, J. P. (1969). Effect of temperature during growth on leaf anatomy and subsequent light-saturated photosynthesis among contrasting *Lolium* genotypes. *New Phytol.* **68**, 1115–1123.

Wittwer, S. H. (1970). Aspects of CO_2 enrichment for crop production. *Trans. Amer. Soc. Agric. Engg.* **13**, 249–252.

Yoshida, R., Oritani, T. and Nishi, A. (1971). Kinetin-like factors in the root exudate of rice plants. *Plant and Cell Physiol.* **12**, 89–94.

Yoshida, S. (1972). Physiological aspects of grain yield. *Ann. Rev. Plant Physiol.* **23**, 437–464.

Yoshida, S., Cock, J. H. and Parao, F. T. (1972). Physiological aspects of high yields. In *Rice Breeding.* IRRI, Los Baños, pp. 455–468.

Zelitch, I. (1971). *Photosynthesis, Photorespiration and Plant Productivity.* Acad. Press, New York, 347 pp.

Index

abscisic acid
  and dormancy of potato tuber, 227
  in shedding of cotton bolls, 313, 314
  in tuberization of potato, 239
in wheat, 134–5
abscission
  of cotton bolls, 306, 311, 313, 314
  of soybean buds, flowers, and pods, 164–5, 175
  of soybean leaves, 158
o-acetylhomoserine, in pea plants, 204
adaptability
  advantage of, in crops, 10, 347–8
  of hexaploid wheats, 101
Aegilops, phloem in, 131
Aegilops speltoides, goatgrass, 101
Aegilops squarrosa, in evolution of bread wheats, 101
agronomic practice
  ideotype depends on, 18, 347
  input energy for different systems of, compared with energy in crop, 329
  probable future changes in, 348
  for sugar beet: geared to processing capacity, 266; possible contribution of SUBGRO model to, 290
  and sugar cane cultivars, 67–8
albumins
  of pea seed, 210, 211
  of wheat grain, 125
alcohol dehydrogenase, in pea seed, 192
aleurone layer of wheat grain, basic amino acids in proteins of, 125
algae, pilot plants for culture of, 4
amino acids
  changing composition of pool of, in pea pods, 209
  in phloem sap of pea, 204
  in soybean seeds, 166

in wheat: in different proteins, 125; nitrogen fertilizer and amounts of different, in grain, 126; synthesized in roots, 105
ammonium salts, retard nodulation in soybean, and lessen nitrogenase activity in existing nodules, 155
anaerobiosis, in pea seed immediately after imbibition, 192
aphids: size of colony of, and amount of food obtained by individuals, 344
apical buds of potato tubers, dominance of, 227–8, 229, 233, 235, 344
apices of shoots
  dominance of: in pea, 195; in soybean, 178
  effects of removal of, in pea, 206–7
  life span of, in pea, 194
apple fruit, photosynthesis and growth of, 332, 335
asparagine
  exported from soybean root nodules, 155
  in pea plant, 202, 204, 205
  in xylem of sugar cane, 65
aspartic acid, in xylem of pea plant, 202, 205
assimilate, *see* photosynthate
auxins
  in cotton, 312
  in pea, 195, 208
  in sugar beet, 263
  in sugar cane, 64–5
  in wheat, 134
awns of wheat, 123, 124, 347

barley
  forms more tillers with extra carbon dioxide, 335